# ZERO-RANGE POTENTIALS and Their Applications in Atomic Physics

# PHYSICS OF ATOMS AND MOLECULES



*Recent volumes in the series:*

**ATOM–MOLECULE COLLISION THEORY: A Guide for the Experimentalist**
Edited by Richard B. Bernstein

**ATOMIC INNER-SHELL PHYSICS**
Edited by Bernd Crasemann

**ATOMS IN ASTROPHYSICS**
Edited by P. G. Burke, W. B. Eissner, D. G. Hummer, and I. C. Percival

**AUTOIONIZATION: Recent Developments and Applications**
Edited by Aaron Temkin

**COHERENCE AND CORRELATION IN ATOMIC COLLISIONS**
Edited by H. Kleinpoppen and J. F. Williams

**COLLISIONS OF ELECTRONS WITH ATOMS AND MOLECULES**
G. F. Drukarev

**DENSITY MATRIX THEORY AND APPLICATIONS**
Karl Blum

**ELECTRON AND PHOTON INTERACTIONS WITH ATOMS**
Edited by H. Kleinpoppen and M. R. C. McDowell

**ELECTRON-ATOM AND ELECTRON-MOLECULE COLLISIONS**
Edited by Juergen Hinze

**ELECTRON-MOLECULE COLLISIONS**
Edited by Isao Shimamura and Kazuo Takayanagi

**INNER-SHELL AND X-RAY PHYSICS OF ATOMS AND SOLIDS**
Edited by Derek J. Fabian, Hans Kleinpoppen, and Lewis M. Watson

**INTRODUCTION TO THE THEORY OF LASER-ATOM INTERACTIONS**
Marvin H. Mittleman

**ISOTOPE SHIFTS IN ATOMIC SPECTRA**
W. H. King

**PROGRESS IN ATOMIC SPECTROSCOPY, Parts A, B, C, and D**
Edited by W. Hanle, H. Kleinpoppen, and H. J. Beyer

**QUANTUM MECHANICS VERSUS LOCAL REALISM: The Einstein-Podolsky-Rosen Paradox**
Edited by Franco Selleri

**RECENT STUDIES IN ATOMIC AND MOLECULAR PROCESSES**
Edited by Arthur E. Kingston

**THEORY OF MULTIPHOTON PROCESSES**
Farhad H. M. Faisal

**VARIATIONAL METHODS IN ELECTRON-ATOM SCATTERING THEORY**
R. K. Nesbet

**ZERO-RANGE POTENTIALS AND THEIR APPLICATIONS IN ATOMIC PHYSICS**
Yu. N. Demkov and V. N. Ostrovskii

**A Continuation Order Plan is available for this series. A continuation order will bring delivery of each new volume immediately upon publication. Volumes are billed only upon actual shipment. For further information please contact the publisher.**

# ZERO-RANGE POTENTIALS and Their Applications in Atomic Physics

**Yu. N. Demkov and**
**V. N. Ostrovskii**
*Leningrad State University*
*Leningrad, USSR*

*Translated from Russian by*

**A. M. Ermolaev**
*University of Durham*
*Durham, England*

PLENUM PRESS • NEW YORK AND LONDON

Library of Congress Cataloging in Publication Data

Demkov, I͡U. N. (I͡Uriĭ Nikolaevich)
[Metod potent͡sialov nulevogo radiusa v atomonoĭ fizike. English]
Zero-range potentials and their applications in atomic physics / Yu. N. Demkov and V. N. Ostrovskii; translated from Russian by A. M. Ermolaev.
p. cm. – (Physics of atoms and molecules)
Translation of: Metod potent͡sialov nulevogo radiusa v atomnoĭ fizike.
Bibliography: p.
Includes index.

1. Scattering (Physics) 2. Approximation theory. 3. Quantum theory. I. Ostrovskiĭ, V. N. (Valentin Nikolaevich) II. Title. III. Series.
QC794.6.S3D4513 1988
539.7′54 – dc19 88-5790
CIP

ISBN-13:978-1-4684-5453-6 e-ISBN-13:978-1-4684-5451-2
DOI: 10.1007/978-1-4684-5451-2



Softcover reprint of the hardcover 1st edition 1988

A Division of Plenum Publishing Corporation
233 Spring Street, New York, N.Y. 10013

## CONTENTS

Chapter 4
SCATTERING BY A SYSTEM OF ZERO-RANGE POTENTIALS AND THE PARTIAL WAVE METHOD FOR A NONSPHERICAL SCATTERER



Chapter 5
ZERO-RANGE POTENTIALS IN MULTI-CHANNEL PROBLEMS



Chapter 6
MOTION OF A PARTICLE IN A PERIODIC FIELD OF ZERO-RANGE POTENTIALS



Chapter 7
WEAKLY BOUND SYSTEMS IN ELECTRIC AND MAGNETIC FIELDS

Chapter 8
ELECTRON DETACHMENT IN SLOW COLLISIONS BETWEEN A NEGATIVE ION AND AN ATOM



Chapter 9
TIME-DEPENDENT QUANTUM MECHANICAL PROBLEMS SOLVABLE BY CONTOUR INTEGRATION



Chapter 10
NONLINEAR APPROXIMATIONS IN THE THEORY OF ELECTRON DETACHMENT



Chapter 11
TIME-INDEPENDENT QUANTUM MECHANICAL PROBLEMS

# Chapter 1

# BASIC PRINCIPLES OF THE ZERO-RANGE POTENTIAL METHOD

## 1.1 INTRODUCTION

It is known that the Schrödinger equation admits solutions expressible in a closed analytic form in only a few cases. The stationary Schrödinger equation is solvable analytically for a harmonic oscillator, for a particle moving in a Coulomb field or in a rectangular potential well, and in some other cases. Even the one-particle Schrödinger equation cannot be solved exactly for the great majority of potentials. In the case of the many-particle and non-stationary problems the situation becomes even more intractable.

Various approximate methods have been developed in order to overcome this difficulty. Among the methods designed to solve quantum mechanical problems corresponding to certain real conditions, the variational method as well as quasi-classical and adiabatic approximations are most widely used. These methods are especially useful for one-dimensional problems where, alternatively, numerical integration can also be effectively carried out. For a three-dimensional equation, the situation is more complicated. Certain potentials possessing spatial symmetry allow separation of variables, so that the Schrödinger equation reduces to a set of one-dimensional equations. However, separation of variables can be achieved only in exceptional cases and, apart from that, the solution thus obtained can have some special properties which do not typify properties of the general equation where the variables cannot be separated. Among the approximate methods, the variational method is probably the most general and can be effectively applied to a nonseparable Schrödinger equation in order to find approximate solutions and the corresponding eigenenergies. The quasi-classical method, on the other hand, has not yet been developed sufficiently enough to deal with this general case.

The method of zero-range potentials (ZRPs) considered in this book belongs to a very limited class of approximate methods suitable for solving a wide range of problems. Similarly to the variational method, the ZRP method can be equally applied to one- and to three-dimensional problems. This method can treat both discrete and continuous spectra, and it can be used to obtain solutions of non-stationary problems.

In many difficult cases where the usual approximate methods fail (or where their applicability cannot be justified), analogous problems turn out to be solvable exactly when the ZRP method is applied. This book presents a collection of problems where such analytic solutions are possible. This is a relatively new direction of research, and the present monograph is based entirely on the original papers scattered in many, sometimes not readily accessible periodicals. Because the ZRP model does not require further approximation, the solutions obtained enable various general and sometimes subtle points of the theory to be analyzed.

Even simple ZRP models very often retain essential physical features of the real system (especially if the binding energy is small). Such models lead, on many occasions, to results which are correct not only qualitatively but also quantitatively and valid within a wide range of conditions.

Another positive feature of the ZRP method is its flexibility. The method allows accurate consideration of effects due to multiple scattering (including the case of an infinite number of scatterers), various details of the interaction between the discrete and continuous spectra, finite (not small) perturbations of physical systems, and some other effects. The ZRP method can be generalized to treat many-channel, nonstationary, and many-particle problems. This brief introductory review of some general characteristics of the ZRP method will be developed into a more detailed picture in the course of this monograph, which will also deal with some general questions of the quantum mechanical theory relevant to the exposition of the ZRP method.

The very first application of the ZRP method was made in atomic physics when Fermi solved in 1934 the problem of the shift of higher-order spectral lines in the presence of perturbing centers [1]. He replaced a large number of randomly distributed deep short-range potential wells by an effective potential independent of coordinates. (It is interesting to note that the concept of the scattering length was first introduced in the same paper.)

The beginning of intensive applications of ZRPs in nuclear physics dated from 1936, when Fermi considered neutron scattering in substances containing hydrogen [2]. The interaction between a neutron and a proton was described, in the Schrödinger equation, by a $\delta$-function, and the first Born approximation was used to obtain the scattering amplitude. Breit(3) gave a mathematically correct formulation of the problem by introducing the boundary conditions and showed that the Fermi formulation using a δ-function led, in comparison to the Born approximation, to a worse (incorrect) result. For neutron scattering, the Born approximation is usually accurate enough, because the internuclear separation in the media is much larger than the de Broglie wavelength associated with neutrons so that multiple scattering of a neutron on different nuclear centers may be neglected. However, there are certain cases where multiple scattering is important. Chew and Wick(4) and Brueckner (5) solved exactly the problem of scattering on a system of ZRPs and established the validity conditions for neglecting multiple scattering and for applying perturbation theory. Goldberger and Seitz(6) considered scattering on an infinite number of scatterers forming the periodic lattice (neutron diffraction on crystals). The Fermi method is presented in many monographs on the theory of nuclei(7-10) and is widely used in calculations.

In the three-body problem with a point interaction first studied by Skornyakov and Ter-Martirosyan(11), important results were obtained by Minlos and Faddeev(12).

In 1935 Bethe and Peierls(13) solved the problem of the photodisintegration of the deuteron* using the ZRP method when they replaced the potential of the neutron-proton interaction by a boundary condition. In a later work of Bethe and Longmire(15), the finite effective radius of the nucleus was also taken into account. A similar approach was applied later by Armstrong(16) to treat photo-detachment of an electron from the negative ion $H^-$ (the latter weakly bound system is an atomic analogue of the deuteron).

Wide applications of ZRPs in atomic physics were started in 1964 by Smirnov and Firsov(17) and by Demkov(18). These first works were soon followed by a considerable number of others, many of them carried out at the Department of Theoretical Physics, University of Leningrad. The ZRP method was applied to many problems in theoretical molecular physics, to electron and atomic collisions, and to some problems of solid state physics. The present monograph summarizes the main results obtained during a decade-long development of the method and its applications.

Finally we shall formulate two most important features which make applications of ZRPs in atomic physics different from those in nuclear physics. Firstly, in atomic physics there is a small parammete determined by the ratio of masses of the electron and nucleus. Consequently the adiabatic approximation assuming that the electron

---

* Disintegration of the deuteron caused by the Coulomb field was first considered by Landau and Lifshits in Ref.14.

velocity is much higher than that of the nucleus can be widely used. Secondly, in atomic physics it is nearly always necessary to take into account the multiple scattering of the electron by the potential centers, whereas the same effect is negligible for neutron scattering by nuclei.

## 1.2 FORMULATION OF THE METHOD

Baz' et al. (19) give a good description of the ZRP method and, in particular, discuss in detail the transition from a potential well of a finite width to a zero-range potential.

Let us assume that the particle is in an s-state and moves in a field of a central force acting in the domain $r_j < b$, where b is the radius of the well and $r_j = |\underline{r} - \underline{R}_j|$ is the distance between the position of the particle, $\underline{r}$, and that of the center of the well, $\underline{R}_j$. As is well known, for $r_j > b$, the wavefunction is*

$$\psi(\underline{r}) = c_j \frac{\exp(-\alpha_j r_j)}{r_j}, \quad r_j > b, \tag{1.2.1}$$

where $\alpha_j = \sqrt{2E_o}$, $E_o$ being the binding energy and $c_j$ being a constant. We will let the width of the well, b, tend to zero under the condition that this change in b is accompanied by a corresponding increase in the characteristic depth of the well, $U_o$, such that the position of the energy level $E_o$ inside the well is unchanged (this will be achieved if $U_o \sim 1/b^2$). Then the particle spends most of the time outside the well and it can be considered as being free provided that the potential well is replaced by a boundary condition at a corresponding point. This condition fixes the logarithmic derivative of the function $r_j\psi$ and has the form

$$\left. \frac{1}{r_j\psi} \frac{\partial(r_j\psi)}{\partial r_j} \right|_{r_j \to 0} = -\alpha_j. \tag{1.2.2}$$

Though the condition (1.2.2) has been obtained for a particle in a bound state with energy $E_o$ it can be extended to include particles whose energy E is near to $E_o$. Indeed, if b is small, then any change from $E_o$ to E which is small in comparison to the depth of the well will not affect the wavefunction for $r_j > b$. Therefore the boundary condition (1.2.2) will be unaltered. Similarly the addition of a slowly changing potential (an external field) to the narrow potential well will not affect the boundary condition. Hence the wavefunction $\psi$ will satisfy, at any point $\underline{r}$ except the point $\underline{R}_j$

*Atomic units ($\hbar = e = m = 1$) are used throughout, unless stated otherwise.

where the ZRP is situated, the Schrödinger equation

$$H\psi = E\psi\ , \tag{1.2.3}$$

$$H = T + V\ , \tag{1.2.4}$$

where the Hamiltonian H may include, apart from the kinetic energy operator $T = -\frac{1}{2}\nabla^2$, a potential $V(\underline{r})$ which is not singular at $\underline{R}_j$. The coefficient $\alpha_j$ in (1.2.2) may be of either sign. If $\alpha_j < 0$, the well is too shallow for the existence of a bound state.

The general solution of the Schrödinger equation can be expanded near $\underline{R}_j$ thus

$$\psi(\underline{r})\Big|_{r_j \to 0} = \frac{c_j}{r_j} - b_j + O(r_j)\ , \tag{1.2.5}$$

where $c_j$ and $b_j$ are constants (we consider a solution $\psi$ which is spherically symmetrical with respect to the point $\underline{R}_j$). Due to the boundary condition (1.2.2) there exists a constraint

$$b_j = \alpha_j c_j \tag{1.2.6}$$

so that

$$\psi(\underline{r})\Big|_{|\underline{r} - \underline{R}_j| \to 0} = c_j \left[\frac{1}{|\underline{r} - \underline{R}_j|} - \alpha_j\right] + O(|\underline{r} - \underline{R}_j|)\ . \tag{1.2.7}$$

An equivalent formulation can be obtained if one applies the Hamiltonian H to a function of the form $\psi$, allowing for (1.2.5). Then we have

$$(-\tfrac{1}{2}\nabla^2 + V(\underline{r}) - E)\psi = 2\pi c_j\, \delta(\underline{r} - \underline{R}_j) \tag{1.2.8}$$

or

$$(-\tfrac{1}{2}\nabla^2 + V(\underline{r}) - E)\psi = -\frac{2\pi}{\alpha_j}\, \delta(\underline{r}_j)\, \frac{\partial(r_j\psi)}{\partial r_j}\ . \tag{1.2.9}$$

For some further generalizations it is convenient to write the right-hand side of (1.2.9) in the form [10,20]

$$-\frac{2\pi}{\alpha_j}\, \delta(\underline{r}_j)\, \frac{\partial(r_j\psi)}{\partial r_j} = -\frac{2\pi}{\alpha_j}\, \delta(\underline{r}_j)\, (1 + \underline{r}_j \cdot \nabla_{\underline{r}_j})\psi\ . \tag{1.2.10}$$

In particular the latter is useful when the ZRP method is extended to partial waves with non-vanishing angular momentum [10,20]. However, this generalization has not yet been applied in atomic physics, and therefore we shall not discuss it here.

Equations (1.2.9) and (1.2.10) show that the quantity $2\pi/\alpha_j$ can be conveniently called the depth (or intensity) of the potential well. It also follows from (1.2.9) that a zero-radius well is not a potential well in the usual sense because it cannot be represented by a $\delta$-function alone.

For a shallow well, a perturbation treatment can be developed (see refs. 7, 8). However, in this monograph we consider a zero-range potential which is not a small perturbation.

According to (1.2.9), the wavefunction $\psi$ of an electron moving in a combined field of a ZRP and a potential $V(\underline{r})$ is identical, within a constant factor, to the Green's function for the Schrödinger equation with a potential $V(\underline{r})$. In a number of cases (such as motion of a free particle, motion of a particle in a uniform field, in a Coulomb field, or in a field of a harmonic oscillator) the Green's function can be obtained in a closed analytical form. This can be used in constructing solutions of various ZRP problems (see Chapter 7).

Any of the formulations derived as described above for a single ZRP can be generalized to a system of ZRPs. In the latter case, the boundary conditions (1.2.2) or (1.2.5) must be imposed at each of N points $\underline{R}_j$ $(j = 1,2,\ldots,N)$ and the equation (1.2.9) becomes

$$\left(-\tfrac{1}{2}\nabla^2 + V(\underline{r}) - E\right)\psi = -\sum_{j=1}^{N} \frac{2\pi}{\alpha_j}\,\delta(\underline{r}_j)\,\frac{\partial(r_j\psi)}{\partial r_j}, \qquad (1.2.11)$$

where that factors $2\pi/\alpha_j$, $j = 1,2,\ldots,N$, represent the depth of the wells.

We shall now prove that the Hamiltonian of the particle moving in the field of N ZRPs is Hermitian, that is, the condition

$$\int \phi_1^* H\phi_2\, d\underline{r} = \int (H\phi_1)^* \phi_2\, d\underline{r} \qquad (1.2.12)$$

is fulfilled for any pair of functions $\phi_1$ and $\phi_2$ which satisfy the boundary conditions (1.2.7) at each of the N points $\underline{R}_j$, are twice differentiable, and are square integrable. Taking into account (1.2.4), equation (1.2.12) can be replaced by the following:

$$\int \phi_1^* T\phi_2\, d\underline{r} = \int (T\phi_1)^* \phi_2\, d\underline{r}. \qquad (1.2.13)$$

Making use of Green's theorem, we shall transform the volume integral (1.2.13) into a surface integral thus:

$$\int_S \left[ \phi_1^* (\underline{n}\nabla)\phi_2 - \phi_2 (\underline{n}\nabla)\phi_1^* \right] dS = 0, \qquad (1.2.14)$$

where S is the total surface in question which consists of N small spheres $S_j$ around the force centers plus the infinite sphere $S_\infty$. The normal $\underline{n}$ is assumed to be directed outwards from the surface S (in the case of $S_j$ it is directed towards the force center). The surface integral over $S_\infty$ equals zero for a pair of any functions $\phi_1$ and $\phi_2$ which fall off fast enough as r increases. When small spheres $S_j$ contract to the force centers, the surface integrals $I_j$ over $I_j$ also vanish. Indeed,

$$I_j = \lim_{r_j \to 0} \int [ \phi_1^* (\underline{n}\,\nabla)\, \phi_2 - \phi_2 (\underline{n}\,\nabla) \phi_1^* ]\, dS$$

$$= \lim_{r_j \to 0} 4\pi c_j^{(1)} c_j^{(2)} \left[ \left( \frac{1}{r_j} - \alpha_j \right) \frac{1}{r_j^2} - \left( \frac{1}{r_j} - \alpha_j \right) \frac{1}{r_j^2} \right]$$

$$= 0 \qquad (1.2.15)$$

so that the hermiticity of H in (1.2.4) holds.

Above we have formulated a three-dimensional problem with ZRPs. There exist a considerable number of works studying one-dimensional systems with a point-interaction (including the examples of many-particle systems, see (21) and references in (19)). In the one-dimensional case, more general problems can be solved, though the model itself may be too inaccurate. In this monograph (except Sec. 3.6) we shall consider the more realistic three-dimensional approach, and therefore only a brief discussion will be given of the one- and two-dimensional ZRPs.

In the one-dimensional case, a $\delta$-type potential $V_j$ can be introduced into the Hamiltonian,

$$V_j(x) = -\alpha_j \delta(x_j), \quad x_j = x - X_j . \qquad (1.2.16)$$

A bound state exists if $\alpha_j > 0$. At the origin of the potential $X_j$ the first derivative of the solution $\psi$ is discontinuous, that is,

$$\left.\frac{d\psi}{dx}\right|_{x = X_j+0} - \left.\frac{d\psi}{dx}\right|_{x = X_j-0} = -2\alpha_j\delta(X_j). \tag{1.2.17}$$

In the case of a two-dimensional problem without any potential, the Green's function differs from the Hankel function $H_0^{(1)}(kr_j)$ only by a constant factor (here $r_j$ is a two-dimensional vector and k is the momentum). Expanding the function $H_0^{(1)}$ in small values of the argument, we find that instead of (1.2.7), the boundary condition for the wavefunction in the vicinity of the ZRP now takes the form

$$\psi\Big|_{r_j\to 0} = c_j\left[\ln r_j + \ln\frac{\alpha_j}{2} + C\right] + O(r_j^2 \ln r_j), \tag{1.2.18}$$

where C is the Euler constant.

## 1.3 THE ONE-CENTER PROBLEM AND ITS SIMPLE APPLICATIONS

Let us seek the solution of the problem for a particle scattering by a ZRP in the form

$$\psi(r) = \exp(i\underline{k}_0\cdot\underline{r}) + c\,G_0(\underline{r}, 0, E), \tag{1.3.1}$$

where the ZRP is assumed to be at the origin, $\underline{k}_0$ is the momentum of the projectile, and $G_0(\underline{r}, \underline{r}', E)$ is the Green's function for a free particle ($E = k^2/2$) satisfying

$$(-\tfrac{1}{2}\nabla^2 - E)\,G_0(\underline{r}, \underline{r}', E) = \delta(\underline{r} - \underline{r}'),$$

$$G_0(\underline{r}, \underline{r}', E) = \frac{\exp(ik|\underline{r} - \underline{r}'|)}{2\pi|\underline{r} - \underline{r}'|}. \tag{1.3.2}$$

The coefficient c in (1.3.1) is found either from the boundary condition (1.2.7) or from equation (1.2.9). Then the scattering amplitude $f = c/2\pi$ is

$$f = -(\alpha + ik)^{-1}. \tag{1.3.3}$$

The scattering is isotropic, and the total cross section is

$$\sigma = 4\pi/(\alpha^2 + k^2). \tag{1.3.4}$$

The only nonvanishing contribution to the cross section is that of s-wave scattering, the corresponding continuum eigenfunction $\psi$

being given by

$$\psi(r) = \sin(kr + \delta_o)/r \tag{1.3.5}$$

with the phase $\delta_o$ found from the boundary condition

$$k \cot \delta_o = -\alpha . \tag{1.3.6}$$

Let us compare this result with the well-known expansion of the phase for scattering by a short-range potential at low energy:

$$k \cot \delta_o = -\frac{1}{a} + \tfrac{1}{2}\rho k^2, \tag{1.3.7}$$

where a is the scattering length and $\rho$ is the effective range. For a ZRP, $\rho = 0$ and the scattering length is

$$a = 1/\alpha . \tag{1.3.8}$$

Note that it follows from (1.3.6) that $\delta_o \to \pm \pi/2$ as $k \to \infty$. This differs from the case of a finite-range potential where $\delta_j \to 0$ as k increases. The anomalous behavior of the phase is a result of the nonuniformity of the zero-range limit in k [22]. For a potential of a finite range, the limit $\lim \delta_o = 0$ is obtained if we require first that $k \to \infty$, leaving b finite, and then let $b \to 0$.

The scattering operator $\hat{t}_E$ in coordinate space is introduced in such a way that

$$\phi_{\underline{k}_o}(\underline{q}) = \delta(\underline{k}_o - \underline{q}) - \frac{t(\underline{q}, \underline{k}_o, E)}{q^2/2 - (E + i0)}, \tag{1.3.9}$$

if $\phi$ is a continuum wavefunction. Writing the wavefunction (1.3.1) in the momentum representation, we obtain immediately the following result (see also [19]):

$$t(\underline{q}, \underline{q}', E) = \frac{1}{4\pi^2 (\alpha + ik)} . \tag{1.3.10}$$

The operator $\hat{t}_E$ is local in coordinate space because it does not depend, in the momentum representation, upon $\underline{q}$ and $\underline{q}'$:

$$t_E(\underline{r}) = 2\pi (\alpha + ik)^{-1} \delta(\underline{r}) . \tag{1.3.11}$$

At $k = i\alpha$ both the scattering amplitude (1.3.3) and the scattering operator (1.3.10) have a pole. For $\alpha > 0$, this pole corresponds to a unique bound state of the one-center problem with the wavefunction

$$\psi_o(\underline{r}) = B \sqrt{\kappa/2\pi}\, \exp(-\kappa r)/r \tag{1.3.12}$$

and with the energy $E = -\kappa^2/2$ ($\kappa = \alpha > 0$). The usual normalization condition (the total probability equal to unity) requires $B = 1$. If $\alpha > 0$, the bound state does not exist and the pole corresponds to a virtual state (see Chapter 2).

The simplest atomic system which can be modelled by a particle moving in a field of ZRP is a negative ion of an atom with a weakly bound peripheral s-wave electron (in nuclear physics the deuteron represents a similar case).

The potential acting on the outer electron due to the neutral atom is the sum of a short-range potential and the polarization potential falling off as $1/r^4$. The binding energy of the outer electron is considerably smaller than that of other electrons of the atom. The outer electron spends much of the time beyond the potential well so that it may be treated as a free particle subject to boundary conditions imposed at the point where the nucleus is situated.

The asymptotic form of the wavefunction of the outer electron in the negative ion is given by equation (1.3.12). It is interesting to note that the correct asymptotic behavior happens to be of importance even in the variational calculation of energy. According to the usually adopted approach, the wavefunctions are expanded in terms of STOs $r^n \exp(-\kappa r)$, $n \geq 0$. When basis functions of the form (1.3.12), called Hulthen-type orbitals (HTOs), are included in the basis set, this improves noticeably the convergence of the variational calculations. The HTOs (in particular they will be used in Sec. 1.4) are given by

$$\psi(r) = B\sqrt{\frac{\kappa}{2\pi}}\,\frac{e^{-\kappa r} - e^{-\beta r}}{r}, \qquad (1.3.13)$$

$$B^2 = \frac{\beta(\beta+\kappa)}{(\beta-\kappa)^2},$$

where $\beta \gg \kappa$. For instance, a 45-parameter variational calculation of Rotenberg and Stein(23) gives the binding energy of the negative hydrogen ion $H^-$ which is better than that obtained with a 125-parameter wavefunction without the correct asymptotic behavior.*

The behavior of the wavefunction in the asymptotic region influences relatively insignificantly the energy functional. However, there exist some other physical quantities, such as dipole matrix elements, polarizabilities, etc., which depend chiefly on the wavefunctions in the asymptotic region. In the latter case, very good

---

* A list of other works using asymptotically correct wavefunctions can be found in Ref. 24.

results can be obtained even with the simplest variational function of the form (1.3.12) provided that the normalization constant B (depending on the affinity energy of the electron through $\kappa$) is determined correctly. Indeed (1.3.12), in comparison to the exact wavefunction for a finite potential well, tends to overestimate the normalization integral in the region of small r. By considering the departure of B from unity we actually take into account the finite effective size of the well, $\rho$, which is related to B, according to Bethe and Longmire (15), thus:

$$B^2 = (1 - \kappa\rho)^{-1} . \tag{1.3.14}$$

From a practical point of view, it is more convenient to determine B by comparing the asymptotic form (1.3.12) with the wavefunction obtained in the course of accurate variational calculations. In this way, it was found that for the negative hydrogen ion $H^-$, $B^2$= 2.65 (see, for instance, (25)). A detailed discussion of this parameter was also given by Smirnov (26), who recommended the value $B^2$ = 2.80. From the very accurate calculations of Pekeris (27), it follows that

$$\kappa = 0.23559 . \tag{1.3.15}$$

When the inner electron in $H^-$ is neglected, the wavefunction (1.3.12) leads to the following expression for the matrix elements of $\langle r_1^n + r_2^n \rangle$:

$$\langle r_1^n + r_2^n \rangle = 2 \int \psi_o(\underline{r})\, r^n\, \psi_o(\underline{r})\, d\underline{r} = \frac{2B^2 n!}{(2\kappa)^n} . \tag{1.3.16}$$

With the numerical values of $\kappa$ and B taken from (25,27), this matrix element differs, for $n \geq 2$, from the most accurate variational calculation of Adelman (24) by not more than by 0.5%.

For the cross section for the photodetachment* of an electron from a negative ion the calculation is reduced to the computation of the oscillator strengths $df/d\omega$ for the transitions between the bound state $\psi_o$ and a continuum state $\psi_E$, that is,

$$\sigma_{phot}(\omega) = \frac{2\pi^2}{c} \frac{df}{d\omega} ,$$

$$\frac{df}{d\omega} = 2\omega \left| \int \psi_o(\underline{r})\; Z\psi_E(\underline{r})\, d\underline{r} \right|^2 , \tag{1.3.17}$$

---

* The reverse process of the radiative capture of an electron can be obtained from $\sigma_{phot}$ using the principle of detailed balancing.

where $\omega$ is the frequency of the incident light and c is the velocity of light. For a negative ion, the phase shift $\delta_1$ for p-wave scattering is small for energies in the region of interest ($\delta_1 \sim k^3$) so that the departure of $\psi_E$ from the free motion function can be neglected.* Armstrong used a ZRP to obtain the following expression for the photodetachment cross section (16):

$$\sigma_{phot}(\omega) = \frac{32\pi\sqrt{E_{av}}}{3c} B^2 \frac{(\omega - E_{av})^{3/2}}{\omega^3} ,$$

$$E_{av} = \kappa^2/2 . \tag{1.3.18}$$

In the region of the cross section maximum (i.e., for $\omega_{max} = 2E_{av}$) this simple formula produces a result which agrees with the experimental data for $H^-$ no worse than does the calculation of Geltman (28) with a 70-parameter bound state wavefunction and a 6-parameter continuum wavefunction (see (26) for a comparative discussion).

The cross section of photodetachment $\sigma_{phot}$ is the imaginary part of the dynamic polarizability $\alpha(\omega)$ and corresponds to absorption of the incident light:

$$\sigma_{phot}(\omega) = \frac{4\pi\omega}{c} \operatorname{Im} \alpha(\omega) , \qquad \omega > E_{av} . \tag{1.3.19}$$

The analytical function $\alpha(\omega)$ can be reconstructed if its imaginary part is known. For a case where there are no excited bound states,

$$\alpha(\omega) = \int \frac{df(u)}{du} \frac{du}{u^2 - \omega^2} . \tag{1.3.20}$$

In the ZRP approximation (24),

$$\alpha(\omega) = \frac{2}{3}\kappa^2\alpha_{st} \left\{ \left( \frac{1}{\kappa + (\kappa^2 - 2\omega)^{1/2}} \right)^2 \frac{2\kappa\left[\kappa + 2(\kappa^2 - 2\omega)^{1/2}\right] - 2\omega}{\kappa^2 + \kappa(\kappa^2 - 2\omega)^{1/2} - \omega} + \left( \frac{1}{\kappa + (\kappa^2 + 2\omega)^{1/2}} \right)^2 \frac{2\kappa\left[\kappa + 2(\kappa^2 + 2\omega)^{1/2}\right] + 2\omega}{\kappa^2 + \kappa(\kappa^2 + 2\omega)^{1/2} + \omega} \right\} . \tag{1.3.21}$$

Here $\alpha_{st} = \alpha(0) = B^2/4\kappa^4$ is the static polarizability of the negative ion first obtained in the ZRP approximation by Demkov and

---

*Note that the relation $\delta_1 \sim k^3$ is only true if we neglect the polarization potential seen by the ejected electron. Otherwise $\delta_1 \sim k^2$ .

Drukarev [29]. The accuracy of this simple formula (1.3.21) competes successfully with that of the most complicated variational calculations (for a discussion see [24]). For imaginary frequencies $\omega$, $\alpha(\omega)$ enables the van der Waals coefficients describing the interaction between atoms at large separations to be found.

A system similar to the negative ion exists in solid state physics and is known as a color F'-center. It is formed by an electron weakly bound to a neutral F-center. The theory of photodetachment developed above is applicable to the F'-center and can describe the absorption bands of the spectrum.* The experimental determination of the position of the maximum of the band ($\omega_{max} = 2E_{av}$) enables the binding energies to be found [31,32] (in fact, a similar method of determining $E_{av}$ is also used for negative ions). These results have found application in the theory of the Auger effect on defects in crystals [31], and in the theory of color M'- and R'-centers [32] (see Sec. 3.2 of the the present monograph).

We shall now consider the application of ZRPs to the scattering of electrons by atoms. For a given energy of the incident electron, there exists a lower limit if the long-range polarization interaction between the electron and the atom is neglected. The polarization effects are significant only in the case of low-energy incident electrons and for small scattering angles, and they can be taken into account, within the framework of the ZRP method, using, for instance, the theory of Smirnov [33]. However, our main concern will be scattering of electrons at energies higher than the potential of polarization forces so that the latter can be completely neglected.

The validity condition for replacing the atomic field by a single ZRP is $\lambda \gg r_{at}$, where $\lambda$ is the de Broglie wavelength of the incident electron and $r_{at}$ is the effective radius of the atomic potential. For $r_{at} \sim 1$, the upper limit set on the energy of the incident electron is $|E| < 0.5$ a.u.

The parameter $\alpha$ in (1.2.7) can be chosen from the scattering length data, making use of (1.3.8). This is a natural way of determining $\alpha$ for a system which has no bound state. On the other hand, if the negative ion does exist, then there is an alternative way of determining $\alpha$, that is, from the affinity energy of the electron. The two methods may lead to somewhat different results. For instance, in the singlet state of the system $e + H$, $a = 5.7$ (see [34]) and therefore

$$\alpha = 0.175, \tag{1.3.22}$$

* This theory gives also the correct position of the maximum and the shape of the absorption curve for the photoionization of impurities in semiconductors, provided that the binding energy of the electron is not too low (Ref. 30).

which is smaller than the value of 0.236, formula (1.3.15), derived from the affinity of the electron. The difference between the two values of $\alpha$ can be explained, for an atom of low polarizability, by the neglect of the second term in the expansion (1.3.7) for k cot $\delta$. Including this term leads to a very good description of electron scattering from H at low energies (see Sec. 1.4).

We have shown above that, for certain physical quantities, the finite radius of interaction can be taken into account simply by choosing the corresponding normalization constant B. However, this method cannot be used in scattering problems.

Smirnov [26,33] suggested a generalization of the ZRP model to take into account the finite size of the atom. According to this, the boundary condition is imposed on the wavefunction on the surface of the sphere of radius $\rho_o$. However, the exact solution of the corresponding two-center problem is not known so that an approximate wavefunction has to be used and therefore the boundary conditions are not satisfied exactly. Apart from this difficulty, the radius $\rho_o$ is taken the same for both singlet and triplet states of the system e + A, which is difficult to justify for an atom A with spin ½.

Within the ZRP method the finite size of the atom can be taken into account by introducing several potentials, as is shown in Sec. 3.6. However, such a model potential is not spherically symmetrical, unlike the exact atomic potential.

Another way to account for the finite size of the atom is to assume that the parameter $\alpha$ in the boundary condition depends upon the energy of the particle. This, however, leads to difficulties with respect to the ortho-normalization of eigenfunctions (see [10]). Below we shall consider the method of separable potentials, which accounts for the finite size of the atom and, at the same time, retains such advantages of the ZRP method as simplicity in applications and the possibility to solve exactly the many-center problem.

## 1.4 SEPARABLE POTENTIALS AND SCATTERING OF SLOW ELECTRONS BY ATOMS

A separable, or factorizable, potential is defined [35] as the operator

$$\hat{V} = |\phi\rangle\, v \,\langle\phi| \, , \tag{1.4.1}$$

where v is a constant. We assume that the vector $|\phi\rangle$ in the Hilbert space is normalized to unity, that is, $\langle\phi|\phi\rangle = 1$. The operator $\hat{V}$ differs only by a constant factor from the projection operator $\hat{P} = |\phi\rangle\langle\phi|$ whose property is $\hat{P}^2 = \hat{P}$.

The Schrödinger equation with a superposition of any finite number of such potentials,

$$\hat{V} = \sum_j |\phi_j> v_j <\phi_j|, \tag{1.4.2}$$

can be reduced to a system of linear algebraic equations. On the other hand, even a one-term potential (1.4.1) enables an arbitrary number of parameters to be introduced.

The Schrödinger equation with a separable potential can be solved without difficulties in both momentum and coordinate representations. We shall use a coordinate representation below. Assume that the Hamiltonian is

$$H = -\tfrac{1}{2}\nabla^2 + |\phi_j> v_j <\phi_j| \tag{1.4.3}$$

and introduce a function

$$|\chi_j(E)> = (-\tfrac{1}{2}\nabla^2 - E)^{-1} |\phi_j>, \tag{1.4.4}$$

where the Green's function appears on the right-hand side of (1.4.4).

Eigenfunctions of the continuum for energy $E = k^2/2$ which correspond to the scattering problem are sought in the following form:

$$\psi(\underline{r}) = \exp(i\underline{k}_o \cdot \underline{r}) + b_j(k)\ \chi_j(E,\underline{r}). \tag{1.4.5}$$

On substituting (1.4.5) into the Schrödinger equation it is found that

$$b_j = -\frac{v_j <\phi_j| \exp(i\underline{k}_o \cdot \underline{r})>}{1 + v_j <\phi_j|\chi_j>}. \tag{1.4.6}$$

For large r, we shall write

$$\chi_j(E,\underline{r}) = \eta_j(k)\ \frac{\underline{r}}{r}\ \exp(ikr)/r + O(r^{-2}). \tag{1.4.7}$$

This gives the following scattering amplitude, for a finite momentum $\underline{k}$:

$$f(\underline{k}_o,\underline{k}) = b_j \eta_j(\underline{k}) = -\eta_j(\underline{k})\ \frac{v_j <\phi_j|\exp(i\underline{k}_o \cdot \underline{r})>}{1 + v_j <\phi_j|\chi_j>}. \tag{1.4.8}$$

For a spherically symmetric function $\phi_j$, $\eta_j$ does not depend on the angle, and the scattering is isotropic. The bound state energy is obtained from the equation

$$1 + v_j \langle\phi_j|\chi_j(E)\rangle = 0, \tag{1.4.9}$$

and this energy is a pole of the scattering amplitude.

Separable potentials have some properties which distinguish them from the local potentials. However, we note that, strictly speaking, the effective potential of an electron in the atom is also nonlocal.

The separable potentials were first used in nuclear theory by Yamaguchi (36) to describe the deuteron and then by many authors in nuclear physics and general scattering theory (37,38).

We now consider the scattering of an electron by an atom in the separable potential approximation (39). We shall adapt the simplest operator of the form (1.4.1) for the atomic potential and discuss some possible choices of the function $\phi$ and parameter v.

Let us suppose that the wavefunction of the bound state $\psi_o(\underline{r})$ and the corresponding energy $E_o = -\kappa^2/2$ (the latter being determined by the asymptotic behavior of $\psi_o(r)$ at large r) are known. Then it follows from the Schrödinger equation with the separable potential (1.4.1) that

$$|\phi\rangle = \frac{1}{2v\langle\phi|\psi_o\rangle}(\nabla^2 - \kappa^2)\,|\psi_o\rangle, \tag{1.4.10}$$

$$v = \frac{\langle\phi|\nabla^2 - \kappa^2\,|\psi_o\rangle}{2\,\langle\phi|\psi_o\rangle}, \tag{1.4.11}$$

that is, the one-term separable potential is uniquely determined by the wavefunction.

If the wavefunction $\psi_o(\underline{r})$ is chosen to have the form (1.3.13), then it follows from (1.4.10) that

$$\phi(r) = \sqrt{\beta/2\pi}\,\exp(-\beta r)/r\ . \tag{1.4.12}$$

For a singlet state of the system electron + hydrogen atom we set $\kappa$ in (1.3.13) according to (1.3.15), put $B^2 = 2.65$ as explained in Sec. 1.3, obtain $\beta$ from $\kappa$ and $B^2$, and, finally, find v from (1.4.11). Thus, for a singlet state

$$\beta_s = 0.785 \quad \text{and} \quad v_s = -0.52\ . \tag{1.4.13}$$

Some quantities related to low-energy scattering on this potential are presented in Table 1.1 and in Fig. 1.1 together with data obtained from accurate calculations (34,40,41) of the electron scattering

Table 1.1 Summary of the exact and model calculations of some parameters related to the e + H system *

| Method and Reference | Singlet state | | Triplet state | | $\kappa$ | $B^2$ |
|---|---|---|---|---|---|---|
| | $a_s$ | $\rho_{os}$ | $a_t$ | $\rho_{ot}$ | | |
| Calculations of e + H scattering | | | | | | |
| Polarized orbitals, Ref. 34 | 5.7 | - | 1.76 | - | - | - |
| Strong coupling, Ref. 40 | 6.74 | 3.28 | 1.89 | 1.23 | - | - |
| Variational method, Ref. 41 | 5.966 | 2.86 | 1.769 | 1.20 | - | - |
| Variational calculation of $H^-$, Ref. 27 | - | - | - | - | 0.23556 | - |
| Calculations of $B^2$ | | | | | | |
| From comparison of the asymptotic eq. (1.3.12) with the variational wavefunction, Ref. 25 | - | - | - | - | - | 2.65 |
| Same, Ref. 26 | - | - | - | - | - | 2.80 |
| Hulthen potential (1.4.14) with $\kappa$ = 0.236 and $\beta$ = 1.76 | 5.06 | - | - | - | 0.236 | 1.51 |
| Separable potential (1.4.1) with: | | | | | | |
| $\phi$ (1.4.12) and parameters (1.4.13) | 5.22 | 1.42 | - | - | 0.236 | 2.30 |
| same $\phi$ and parameters (1.4.16) | - | - | 2.66 | 1.50 | - | - |
| same $\phi$ and parameters (1.4.19) | 6.22 | 2.7 | 2.08 | 0.80 | 0.231 | 2.48 |
| $\phi$ (1.4.20) and parameters (1.4.21) | 6.22 | 2.7 | 2.08 | 1.02 | 0.231 | 2.41 |

* In atomic units.

by hydrogen. It is seen that the separable potential provides a good approximation to the atomic potential.

It is interesting to compare these results with those using the local potential whose bound state is the same as that described by the wavefunction $\psi(r)$, Eq. (1.3.13),

$$V(\underline{r}) = \frac{(\nabla^2-\kappa^2)\psi_0(\underline{r})}{2\psi_0(\underline{r})} = -\frac{\beta^2-\kappa^2}{2\{\exp[(\beta-\kappa)r]-1\}} \qquad (1.4.14)$$

This potential is known as the Hulthen potential and, for s-states, it admits solutions of the Schrödinger equation expressible in terms of hypergeometrical functions [22]. The potential (1.4.14) falls off exponentially as r increases and has a Coulomb singularity, $V(r) \sim -(\beta+\kappa)/2r$ as r goes to zero. If we interpret this singularity as the attraction of the electron by a nucleus of charge Z, we then have to put $\beta + \kappa = 2Z$. For $H^-$, this leads to $\beta = 1.764$, which differs considerably from the numerical value $\beta_s = 0.785$ in (1.4.13). Scattering parameters obtained using this Hulthen potential are also given in Table 1.1 and Fig. 1.1. We note that the scattering length is reproduced better (a = 5.7 a.u.) if the numerical value of $\beta$ is selected according to (1.4.13).

We shall now consider the triplet state of the e + H system, which has no bound state (no negative ion $H^-(^3S)$ exists). Let us write an approximate wavefunction of the system in the following antisymmetized form:

$$\psi(\underline{r}_1,\underline{r}_2) = \phi_a(\underline{r}_1)\, F(\underline{r}_2) - \phi_a(\underline{r}_2)\, F(\underline{r}_1), \qquad (1.4.15)$$

where $\phi_a(r) = \exp(-r)/\sqrt{\pi}$ is the atomic wavefunction. The wavefunction $F(\underline{r})$ of the incident electron satisfies, in the static exchange

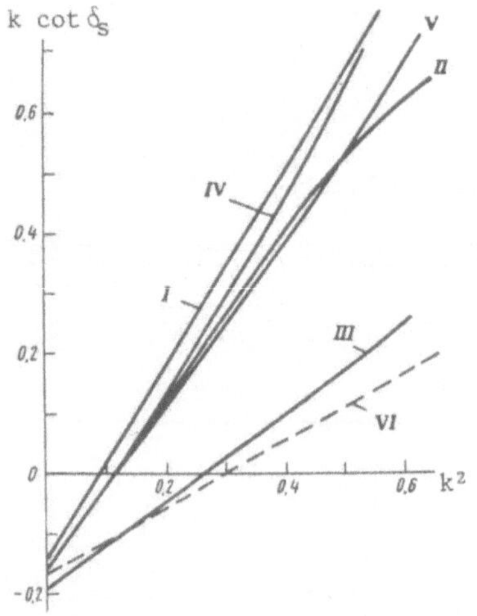


Fig. 1.1. Dependence of k cot $\delta_s$ on $k^2$. ($\delta_s$ is the phase of the singlet s-wave scattering of an electron by a hydrogen atom). Curve I: the strong coupling approximation [40]. Curve II: variational method [41]. Curves III–V: calculations with the separable potential (1.4.1). III: wavefunction (1.4.12) and parameters (1.4.13). IV: as in III with parameters (1.4.19). V: wavefunction $\phi$ (1.4.20) and parameters (1.4.21). Curve VI: calculations with the Hulthen potential (1.4.14) with $\kappa$ = 0.236 and $\beta$ = 1.76.

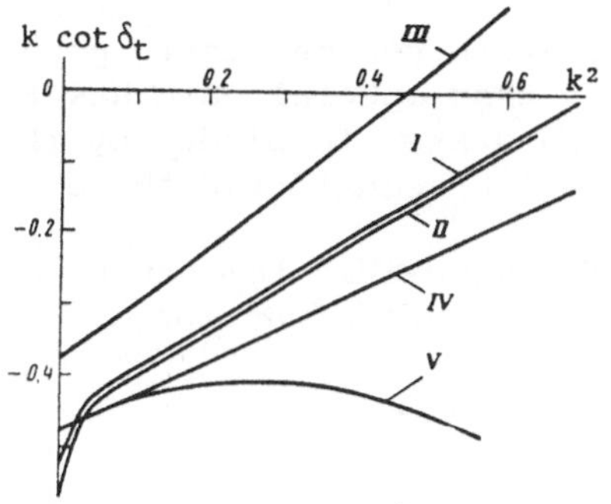


Fig. 1.2. Dependence of $k \cot \delta_t$ on $k^2$ ($\delta_t$ is the phase of the triplet s-wave scattering of an electron by a hydrogen atom). Curve I: the strong coupling approximation (40). Curve II: variational method (41). Curves III-V: calculations with separable potential (1.4.1) (39). III: wavefunction (1.4.12), parameters (1.4.16). IV: as in III with parameters (1.4.19). V: wavefunction (1.4.20), parameters (1.4.21).

approximation (35,42), an integrodifferential equation whose explicit form can be easily derived. For the purpose of the present discussion, it is essential that the function $F(\underline{r}) = \phi_a(\underline{r})$ be a solution of this equation. The total wavefunction corresponding to this solution vanishes identically due to the antisymmetry condition (43,44). Though this indicates that the state is forbidden by the Pauli exclusion principle,* it does mathematically exist and can be used to reconstruct a model separable potential with the help of (1.4.10) and (1.4.11). The function $\phi$ has the form (1.4.12), with the parameters being

$$\beta_t = \kappa = 1 \quad \text{and} \quad v_t = -2. \tag{1.4.16}$$

The scattering data obtained with the help of this potential are presented in Fig. 1.2 and Table 1.1. Accurate calculations by Burke and Schey (40) and by Schwartz (41) show that, for very low energy the dependence of $k \cot \delta$ on $k^2$ departs from linear due to the polarization interaction (see Fig. 1.2). On extrapolating the linear dependence established by Schwartz (41) for the range $0.05 < k^2 < 0.6$ down to $k^2 = 0$, we find that $a_t = 2.17$. This is the numerical value of $a_t$ that is to be meaningfully compared with the model potential value. The way the model potential has been chosen influences the scattering length, which is closer to the static potential value with exchange ($a_t = 2.347$ from Ref. 42) than to that found in more accurate calculations. Similar conclusions will be reached below in the case of the helium atom.

The triplet scattering length for e + H is positive. Therefore in the corresponding ZRP model, there formally exists a deep bound $^3S$ state of $H^-$. The present discussion suggests that the ZRP can

* The existence of such states requires a modification of Levinson's theorem (38) for this case of scattering.

be treated as an approximation to the static potential with exchange. We shall see in Chapter 5 that special care has to be taken to eliminate effects of this bound state, forbidden by the Pauli principle, on the model scattering e + $H_2$ considered there.

The ground state of the helium atom can be described accurately enough by a simple variational wavefunction

$$\phi(\underline{r}_1,\underline{r}_2) = \phi_o(\underline{r}_1)\ \phi_o(\underline{r}_2),$$
$$\phi_o(\underline{r}) = \frac{1}{\sqrt{\pi}} e^{-\kappa r}, \quad \kappa = \frac{27}{16}. \tag{1.4.17}$$

For scattering e + He, the static potential equation has a solution of the form $\phi_o(\underline{r})$ [38,43,45] which corresponds to a state forbidden by the Pauli principle. For $\phi$, we again come to (1.4.12) where the numerical values of the parameters are

$$\beta = \kappa = \frac{27}{16} \quad \text{and} \quad c = -5.696. \tag{1.4.18}$$

In the case of the model potential, $a = 1.58$ and $\rho = 0.89$, whereas the static exchange approximation [46] gives $a = 1.483$ and $\rho = 0.69$. Taking into account correlation leads to $a = 1.282$ [46].

The suggested method [39] of determining the potential with the help of the wavefunction and energy of a bound state (even if the state is forbidden according to physical principles) is very simple and gives good results for low-energy scattering. This is true even when the energy level is so deep (as in the last two examples considered above) that the concept of the effective range cannot be introduced.

Generally speaking, the problem of reconstructing a nonlocal potential from the wavefunction is not unique. In the cases considered above, uniqueness of the potentials was due to additional assumptions such as separability of the potential and its one-term form (1.4.1). This approach can be extended to determine a many-term separable potential of the form (1.4.2), provided that several bound states of the system are known.

It will be shown later that, making use of the results of accurate calculations of the low-energy scattering e + H, it is possible to improve upon the model results presented above by a suitable choice of parameters in the model potentials. It is sufficient to use the atomic wave functions, instead of wavefunctions of the system "electron + atom," as in the cases of the triplet scattering e + H and scattering e + He considered above. For a complex atom, this method has a great advantage because it uses one-electron wavefunctions. If necessary, further improvement can be achieved by treating perturbationally the difference between the exact and model potentials.

In the examples considered in this chapter, the functions $\phi$ of the singlet and triplet states of e + H had the same analytic form (1.4.12) but numerical values of the parameter $\beta$ were different. In Chapter 5 where the system e + $H_2$ is considered, it will be convenient to use the same functions $\phi$ for both singlet and triplet states of e + H.* It was suggested in [47] to use the following two sets of parameters in $\phi$ given by (1.4.12):

$$\beta_s = 0.8214, \qquad c_s = -0.554;$$
$$\beta_t = 0.8214, \qquad c_t = 1.977. \tag{1.4.19}$$

For $\phi$ taken in the form

$$\phi(r) = \sqrt{\frac{\beta^3}{\pi}}\, e^{-\beta r}, \tag{1.4.20}$$

these two sets are

$$\beta_s = 1.235, \qquad c_s = -0.261;$$
$$\beta_t = 1.235, \qquad c_t = 0.62. \tag{1.4.21}$$

Some low-energy scattering results obtained with these functions are presented in Table 1.1 and in Figs. 1.1 and 1.2. Curve V of Fig. 1.2 differs substantially from a straight line because of a large numerical value of the coefficient of $k^4$ in the expansion of $k \cot\delta$, with this single exception the model potentials give better results than those obtained with potentials used earlier, as is clearly seen from Figs. 1.1 and 1.2.

* This condition simplifies calculations and is not of any essential significance.

Chapter 2

# TRAJECTORIES OF THE POLES OF THE S-MATRIX AND RESONANCE SCATTERING

## 2.1 PRELIMINARY REMARKS

An important advantage of the ZRP method is that it offers the possibility of obtaining in a uniform way the exact solution of both bound state and scattering problems. Transitions between states of the discrete spectrum and the continuum caused by various perturbations as well as resonance scattering will be discussed and their relation to the properties of the poles of the S-matrix in the complex k plane (or complex E). The relation between the poles of the S-matrix and resonance scattering is well known and described in many monographs [19,22,48]. Here we shall only give a brief discussion of some important results of the general theory which are relevant to the problems considered in this monograph. Among these are the relation between various formulae for resonance scattering, the existence of second-order poles in the S-matrix, and the movement of these poles in the complex k plane caused by variation in the depth of the potential well.

Let us consider scattering of a particle of energy E by a central potential $V(r)$ of a finite range $b$, i.e., $V(r) = 0$ if $r \geq b$. Assuming that the motion of the particle is slow, $kb \ll 1$, we shall take into account only the s-wave ($\ell = 0$).

A distinction is usually made between two types of resonance scattering [9,48]:

(i) scattering by a potential well with either a bound state level lying near to the continuum or with a virtual level,

and

(ii) scattering by a potential well with a quasi-stationary level, for instance, in the case of a well surrounded by a potential barrier.

In the case (i), the S-matrix $S = \exp(2i\delta)$, where $\delta$ is the scattering phase, is written as follows (9,48):

$$S = \exp(2i\phi_p(k)) \frac{1 + k/k_1}{1 - k/k_1}, \tag{2.1.1}$$

$k_1 = i\kappa_1$ being purely imaginary. In (2.1.1) $\kappa_1 > 0$ if there exists a bound state, and $\phi_p(k)$ is the potential (non-resonance) scattering phase. For low energies,

$$\phi_p(k) = kr_{op}. \tag{2.1.2}$$

For $|\kappa_1| r_{op} << 1$, (2.1.1) and (2.1.2) lead to the well-known expansion of the phase (1.3.7), where the scattering length a and the effective range $\rho$ are expressed in terms of $\kappa_1$ and $r_{op}$ thus:

$$a = \kappa_1^{-1} \quad \text{and} \quad \rho = 2r_{op}. \tag{2.1.3}$$

In the case (ii), the S-matrix is written as

$$S = \exp(2i\phi_p(k)) \frac{(1 + k/k_1)(1 - k/k_1^*)}{(1 - k/k_1)(1 + k/k_1^*)}, \tag{2.1.4}$$

where the complex parameters $k_1$ and $k_1^*$ lie in the lower half k plane symmetrically with respect to the immaginary axis.

The position of the S-matrix poles depends on the potential V(r). One might have expected that, changing the depth of the potential well, it would be possible to observe the transition of resonance scattering given by (i) into the resonance scattering (ii). This is not possible if one uses formula (2.1.1). Indeed, the S-matrix can be expressed as a ratio

$$S = f(-k)/f(k), \tag{2.1.5}$$

where f(k) is the Jost function defined in Sec. 2.2 which is analytic in the entire k plane if $V(r) = 0$ for $r \geq b$. Therefore the S-matrix poles are zeros of the Jost function. Formula (2.1.4) is written in the approximation where the Jost function has two zeros. Because the number of zeros of an analytic function cannot change by a continuous change of this function, it is clear that it is not possible to obtain (2.1.1) from (2.1.4) by any movement of the zeros of f(k) in the k plane.

In order to be able to observe the transition from one type of resonance scattering to another, it is necessary to have in both cases formulae with <u>two</u> Jost zeros. These Jost zeros must lie either symmetrically with respect to the imaginary axis or be on the imaginary axis itself. Then one may expect that in the course of the transition from one type of scattering to another a two-fold zero of $f(k)$ will appear on the imaginary axis in the complex k plane. Therefore we conclude that the corresponding S-matrix must have a second-order pole. The existence of second (and higher) order poles is important from the point of view of the general structure of the S-matrix [49] as well as for the theory of the decay of unstable states [22].

If the energy E is used instead of k, then the upper half k plane will correspond to the physical sheet and the lower half k plane will correspond to the so-called nonphysical sheet of the Riemann surface of the complex E plane.

In Secs. 2.2 and 2.3 we follow a work by Demkov and Drukarev [50]. We shall investigate movement of the Jost zeros in the complex k plane, caused by changes in the potential $V(r)$, and we shall establish the conditions under which two zeros come together. The case of a two-fold zero near the origin $k = 0$ will also be considered. The S-matrix derived in Sec. 2.3 is a two-pole approximation which includes, in particular, a generalization of the single-pole expression (2.1.1). It allows the transition between the two types of resonance scattering to be studied.

In the two-pole approximation for the S-matrix, the expansion of $k \cot \delta$ assumes the same form as in (1.3.7). However, we shall see that, for a two-fold zero of $f(k)$ near $k = 0$, the effective range $\rho$ becomes large and negative.

## 2.2 TRAJECTORIES OF THE ZEROS OF THE JOST FUNCTION FOR $\ell = 0$

The Jost function $f(k)$ is defined as $f(k,r)$ at $r = 0$, where $f(k,r)$ is a solution of the equation

$$\frac{d^2 f(k,r)}{dr^2} + (k^2 - 2V)\, f(k,r) = 0 \qquad (2.2.1)$$

with asymptotic form

$$f(k,r) = e^{ikr} \quad \text{as} \quad r \to \infty . \qquad (2.2.2)$$

The properties of the Jost function are discussed, for instance, in Ref. (19). For the potential V considered here, (2.2.2) is satisfied identically for $r \geq b$. Then the function $f(k)$ is analytic in the entire k plane. On the real axis,

$$f(k) = |f(k)|\, e^{-i\delta(k)} . \qquad (2.2.3)$$

Apart from that, the function f(k) satisfies the symmetry relation

$$f(k^*) = f^*(-k).$$

In the upper half k plane, f(k) has zeros only on the imaginary axis, and these zeros correspond to bound states of the particle moving in the field V.

In the lower half k plane, f(k) may have zeros on the imaginary axis as well as away from it. It follows from the symmetry relation that in the latter case, the zeros appear in pairs, being positioned symmetrically with respect to the imaginary axis. The zeros lying on the imaginary axis close to the origin k = 0 correspond to virtual (anti-bound) states of the particle, whereas the pairs of zeros lying in the lower half k plane close to the real axis correspond to quasi-stationary states.

We now consider the zeros of f(k) on the imaginary axis. Let us put $k = i\kappa$ and denote

$$f(i\kappa) = g(\kappa), \tag{2.2.4}$$

where $g(\kappa) = g(\kappa,r)$ at $r = 0$, and $g(\kappa,r)$ satisfies the equation

$$\frac{d^2g(\kappa,r)}{dr^2} - (\kappa^2 + 2V)\, g(\kappa,r) = 0 \tag{2.2.5}$$

and has the form

$$g(\kappa,r) = e^{-\kappa r}, \quad r \geq b. \tag{2.2.6}$$

For real $\kappa$ and r, the functions $g(\kappa,r)$ and $g(\kappa)$ are real. Let us write an equation for $g(\kappa',r)$:

$$\frac{d^2g(\kappa',r)}{dr^2} - (\kappa'^2 + 2V)\, g(\kappa',r) = 0. \tag{2.2.7}$$

Multiplying (2.2.5) by $g(\kappa',r)$, (2.2.7) by $g(\kappa,r)$, taking the difference between the two products, and integrating the difference from r = 0 to r = b' gives

$$\left[ g(\kappa',r)\,\frac{\partial g(\kappa,r)}{\partial r} - g(\kappa,r)\,\frac{\partial g(\kappa',r)}{\partial r} \right]_{r=0}^{r=b'} =$$

$$(\kappa^2 - \kappa'^2) \int_0^{b'} g(\kappa,r)\, g(\kappa',r)\, dr \,.$$

Assuming that $b' \geq b$ and making use of the asymptotic form (2.2.6), we obtain

$$(\kappa' - \kappa)\, e^{-(\kappa+\kappa')b'} - g(\kappa') \left.\frac{\partial g(\kappa,r)}{\partial r}\right|_{r=0} + g(\kappa) \left.\frac{\partial g(\kappa',r)}{\partial r}\right|_{r=0} = (\kappa^2 - \kappa'^2) \int_0^{b'} g(\kappa,r)\, g(\kappa',r)\, dr \,.$$

Differentiating this formula with respect to $\kappa'$ and then demanding that $\kappa' \to \kappa$, we obtain

$$e^{-2\kappa b'} - \frac{dg(\kappa)}{d\kappa} \left[\frac{\partial g(\kappa,r)}{\partial r}\right]_{r=0} + g(\kappa) \left[\frac{\partial^2 g(\kappa,r)}{\partial\kappa\partial r}\right]_{r=0} = - 2\kappa \int_0^{b'} g^2(\kappa,r)\, dr.$$

Let $\kappa_n$ be a zero of the Jost function so that $g(\kappa_n) = 0$. We then obtain the equation

$$e^{-2\kappa_n b'} - \frac{dg(\kappa_n)}{d\kappa} \left[\frac{\partial g(\kappa_n,r)}{\partial r}\right]_{r=0} = -2\kappa_n \int_0^{b'} g(\kappa_n,r)^2\, dr \,. \qquad (2.2.8)$$

The derivative of $g(\kappa,r)$ with respect to $r$ at $r = 0$ can be expressed in terms of $g(-\kappa)$. In order to obtain this relation we shall use the Wronskian of two independent solutions of the equation (2.2.5) that is of $g(\kappa,r)$ and $g(-\kappa,r)$. The Wronskian is independent of $r$, and calculating its value at $r = 0$ and at $r > b$, we find that

$$W = \left[g(\kappa,r) \frac{\partial g(-\kappa,r)}{\partial r} - g(-\kappa,r) \frac{\partial g(\kappa,r)}{\partial r}\right]_{r>b} = 2\kappa \,,$$

and

$$W = -g(-\kappa_n)\left[\frac{\partial g(\kappa_n,r)}{\partial r}\right]_{r=0} = 2\kappa_n .$$

In this way we come to the identity

$$-\frac{1}{2\kappa_n} e^{-2\kappa_n b'} - \frac{1}{g(-\kappa_n)} \frac{\partial g(\kappa_n)}{\partial \kappa} = \int_0^{b'} g^2(\kappa_n,r)\,dr, \tag{2.2.9}$$

first derived by Drukarev [51].

For $\kappa_n > 0$, i.e., in the case where the zero corresponds to a bound state, formula (2.2.9) can be simplified by requiring $b' \to \infty$. Thus

$$g'(\kappa_n)/g(-\kappa_n) = -\int_0^{\infty} g^2(\kappa_n,r)\,dr < 0 . \tag{2.2.10}$$

Now we assume that $\kappa_n > \kappa_{n+1} > 0$ are two successive zeros and therefore $g'(\kappa_n)$ and $g'(\kappa_{n+1})$ have opposite signs. It follows then from (2.2.10) that $g(-\kappa_n)$ and $g(-\kappa_{n+1})$ also have opposite signs; hence there exists at least one zero of the function $g(\kappa)$ in the interval $(-\kappa,-\kappa_{n+1})$. We conclude, in particular, that the Jost function $f(k)$ does not have multiple zeros in the upper half-plane. Indeed it would then follow that $f(-k)$ also vanishes, which contradicts the uniqueness theorem.

It can be shown for a potential of a sufficiently general form (a well, a barrier, or a well surrounded by a barrier) that the function

$$\xi(\kappa) = e^{-2\kappa b} + 2\kappa\int_0^{b} g^2(\kappa,r)\,dr \tag{2.2.11}$$

vanishes only at a unique value of $\kappa = \bar{\kappa} < 0$, $\xi(\kappa)$ being positive for $\kappa > \bar{\kappa}$, and $\xi(\kappa)$ being negative for $\kappa < \bar{\kappa}$. Then we can make use of formula (2.2.9) to show that all real zeros of $g(\kappa)$ (i.e., all imaginary zeros of the Jost function $f(k)$) can be classified into two groups:

(a) $\kappa_1 > \kappa_2 > \ldots \kappa_N > \bar{\kappa}$,

where the last zero, $\kappa_N$, may be of either sign, whereas all others must be positive;

$$\text{(b)} \qquad \kappa_2' < \kappa_3' < \ldots \kappa_N' < \bar{\kappa} \,,$$

being in the intervals

$$-\kappa_1 < \kappa_2' < -\kappa_2 \,,$$
$$-\kappa_2 < \kappa_3' < -\kappa_3 \,,$$
$$\ldots\ldots\ldots\ldots\ldots$$
$$-\kappa_{N-1} < \kappa_N' < \bar{\kappa} \,.$$

The function $g(\kappa)$ goes to unity as $\kappa \to \infty$ so that $g'(\kappa_1) > 0$ and $g(-\kappa_1) < 0$. Therefore the existence of a zero $\kappa_1'$ in the interval $-\infty < \kappa_1' < \kappa_1$ depends on the sign of $g(\kappa)$ as $\kappa \to -\infty$. Using the equation

$$g(\kappa) = 1 + \frac{2}{\kappa} \int_0^b \sinh \kappa r \, V(r) \, g(\kappa, r) \, dr \qquad (2.2.12)$$

one can obtain that if $V(r) > 0$ as $r \to b$ (the case of a well surrounded by a barrier), then $g(\kappa) > 0$ as $\kappa \to -\infty$; hence there exists a zero $\kappa_1'$ in the interval $-\infty < \kappa_1' < \kappa_1$ and the total number of zeros is even. In the opposite case of $V(r) < 0$ as $r \to b$, $g(\kappa) < 0$ as $\kappa \to -\infty$; hence no zero $\kappa_1'$ exists in the interval and the total number of zeros is odd. The result shows particularly clearly that an infinitesimal change in the potential $V(r)$ at the end of the force range may affect the number of zeros the function $g(\kappa)$ has at large negative $\kappa$. Therefore these zeros cannot have any particular physical significance.

For more complicated potentials such as an oscillating potential, the function $\xi(\kappa)$ may have several zeros. In this case the resulting picture of the distribution of the Jost zeros will not be as simple as that described above.

Now we shall investigate what change in the position of the zeros $\kappa_n$, $\kappa_n'$ in the complex k plane will occur as a result of variation on the potential $V(r)$. Making use of equation (2.2.5), where we put $\kappa = \kappa_n$, together with equations (2.2.6) and (2.2.9) it is easy to show that a small variation $\delta V$ of the potential leads, in the first approximation, to the following shift $\delta\kappa_n$:

$$\delta\kappa_n = -\frac{2}{\xi(\kappa_n)} \int_0^b g^2(\kappa_n, r)\, \delta V\, dr, \tag{2.2.13}$$

where $\xi(\kappa)$ is given by (2.2.11)*. We shall consider variations of V such that

$$\int_0^b g^2(\kappa_n, r)\, \delta V\, dr > 0 \tag{2.2.14}$$

for real $\kappa_n$. In particular, a decrease in the depth of the potential well satisfies the condition (2.2.14). Making use of the properties of $\xi(\kappa)$ established above, we find that

$$\begin{aligned} &\text{(a)} \quad \delta\kappa_n < 0, \quad \kappa_n > \bar{\kappa}, \\ &\text{(b)} \quad \delta\kappa_n' > 0, \quad \kappa_n' < \bar{\kappa}. \end{aligned} \tag{2.2.15}$$

In other words, the variation (2.2.14) of V(r) shifts the zeros of the two groups, (a) and (b), towards each other (the position of $\bar{\kappa}$ will also be changed when V changes).

Formula (2.2.13) cannot be applied at $\kappa = \bar{\kappa}$ since $\xi(\kappa)$ vanishes at that point. We shall modify (2.2.13) by expanding $\xi(\kappa_n)$ near $\kappa_n = \bar{\kappa}$:

$$\xi(\kappa_n) = \delta\kappa_n\, (d\xi/d\kappa)_{\bar{\kappa}}. \tag{2.2.16}$$

Then we obtain, instead of (2.2.13), the modified expression

$$(\delta\kappa_n)^2 = -\frac{2}{(d\xi/d\kappa)_{\bar{\kappa}}} \int_0^b g^2\, \delta V\, dr. \tag{2.2.17}$$

It follows from the properties of $\xi(\bar{\kappa})$ that $(d\xi/d\kappa) > 0$ at $\kappa = \bar{\kappa}$; hence $(\delta\kappa_n)^2$ is negative and $\delta\kappa_n$ is purely imaginary, for variations of V satisfying equation (2.2.14).

---

* Formula (2.2.13) is analogous to Zel'dovich's formula [19] for a change in the energy of a quasi-stationary state caused by a small variation of the potential V.

Returning to the variable $k = i\kappa$, we find that a real variation $\delta\kappa_n$, which causes the zeros to leave the imaginary axis, corresponds to an imaginary variation $\delta\kappa_n$. Therefore we can conclude that due to the variation (2.2.14) of the potential V(r) zeros with $\kappa_n > \bar{\kappa}$ move towards zeros with $\kappa_n < \bar{\kappa}$. Forming pairs (one zero from each group) they move towards the point $i\bar{\kappa}$ where they first coincide and then move away from each other and from the imaginary axis.

If an odd zero exists, it moves down along the imaginary axis and always remains above $i\bar{\kappa}$. This general picture of the Jost zeros under variation of V has been confirmed in a numerical calculation of Nussenzveig (52) for a particular choice of the rectangular potential well. An odd zero of the Jost function existing in the case of this potential is marked with a square in Fig. 2.1, where trajectories (52) of the Jost zeros in the complex k plane are shown. Ferreira and Teixeira (53) found that trajectories of the Jost zeros in the case of a cutoff Coulomb potential display features similar to those in Fig. 2.1.

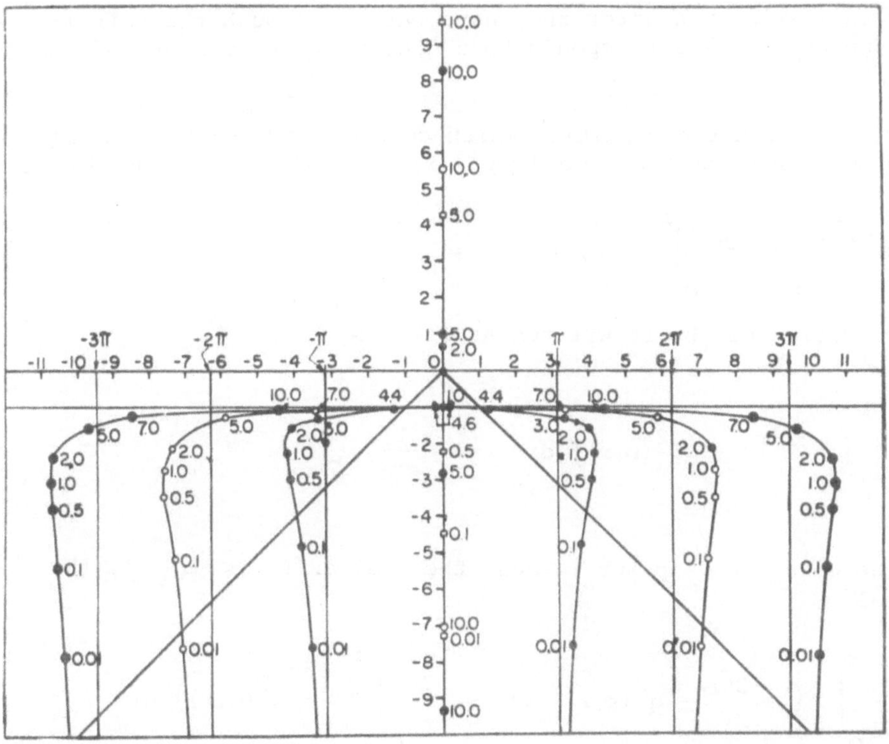


Fig. 2.1. The s-wave scattering by a square well potential of radius b and depth $V_o$. Trajectories of the Jost zeros in the complex plane kb (52). The numbers along the trajectories are values of the parameter $b(V_o/2)^{\frac{1}{2}}$.

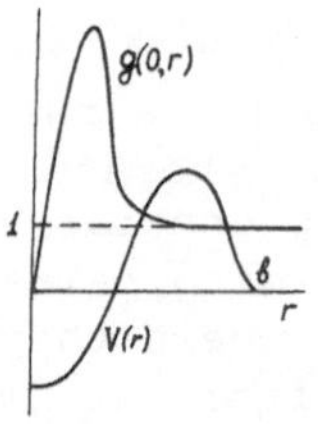


Fig. 2.2. A barrier potential V(r) and a wavefunction g(0,r). The case corresponds to a bound state with zero energy.

For a rectangular well where all pairs of zeros come together at the same point $i\bar{\kappa}$, this point lies at a comparatively large distance $b^{-1}$ from the real axis. This distance is of the same order as the spacing between zeros so that there is no particular reason here to prefer one pair of zeros to another. However, the situation changes for a potential well surrounded by a barrier. We will see that in the latter case the point where zeros come together is very close to the real axis and zeros continue moving along this axis even after they have passed through the coincidence point. This corresponds to a quasistationary state of the particle.

Let us find the condition which determines the coincidence point for a pair of Jost zeros. According to equation (2.2.9), we shall require

$$e^{-2\bar{\kappa}b} + 2\bar{\kappa}\int_0^b g^2(\bar{\kappa},r)\,dr = 0.$$

This condition can be re-written as follows:

$$\int_0^b \left( e^{-2\bar{\kappa}r} - g^2(\bar{\kappa},r)\right) dr = \frac{1}{2\bar{\kappa}}.$$

When the coincidence point is near the real axis $|\bar{\kappa}|\,b \ll 1$, the integral

$$\frac{1}{2}\int_0^b \left( e^{-2\bar{\kappa}r} - g^2(\bar{\kappa},r)\right) dr \approx \frac{1}{2}\int_0^b \left(1 - g^2(0,r)\right) dr$$

is practically identical with the effective range $\rho$ (see, for instance, reference (54)), and $\rho = 1/\bar{\kappa}$. It then follows that $\rho$ is negative and $|\rho| \gg b$.

This case arises if $V(r)$ is a potential well surrounded by a barrier. Figure 2.2 gives an example of such a potential and shows the general behavior of the function $g(0, r)$. It is evident from Fig. 2.2 that $\rho < 0$ and $|\rho| >> b$. The correct order of magnitude of $\bar{\kappa}$ is given by the transmission coefficient for the barrier, i.e., it is exponentially small. Therefore, $\rho$ is exponentially large. Direct calculations for a rectangular barrier confirm Fig. 2.2.

## 2.3 THE S-MATRIX IN A TWO-POLE APPROXIMATION

Let $k_1, k_2, \ldots$ be zeros of the Jost function. If the distribution of zeros is such that the series $\Sigma_n \; 1/k_n$ converges, then the Jost function can be written as follows:

$$f = e^{-i\varphi_p(k)} \prod_{n=1}^{\infty} (1 - \frac{k}{k_n})$$

$$= e^{-i\varphi_p(k)} \{1 - k \sum_n \frac{1}{k_n} + k^2 \sum_{n,m} \frac{1}{k_n k_m} + \ldots\}.$$

We shall consider now the case where two zeros $k_1$ and $k_2$ are near the coincidence point, the latter being close to $k = 0$.

Under these assumptions, only $k_1$ and $k_2$ will be important for low-energy scattering. We shall neglect all other terms in the sums above, replacing the first sum by $1/k_1 + 1/k_2$ and the second sum by $1/(k_1k_2)$. Assuming the approximate expression (2.1.2) for the phase, we obtain the following two-zero formula for the Jost function:

$$f(k) = e^{-ikr_{op}}\{1 - k(1/k_1 + 1/k_2) + k^2/(k_1k_2)\}.$$

Then the S-matrix in the two-pole approximation is

$$S(k) = e^{2ikr_{op}} \frac{(1 + k/k_1)(1 + k/k_2)}{(1 - k/k_1)(1 - k/k_2)} . \qquad (2.3.1)$$

For $k_2 = -k_1^*$, formula (2.3.1) reduces to (2.1.4). When both $k_1$ and $k_2$ are purely imaginary, (2.3.1) is a two-zero generalization of formula (2.1.1). Let us write $k_1$ and $k_2$ thus:

$$k_1 = \sqrt{\Delta} - i\alpha, \quad k_2 = - \sqrt{\Delta} - i\alpha . \qquad (2.3.2)$$

The quantity $\alpha = (k_1 + k_2)/2$ is the "center of gravity" of the two zeros on the imaginary axis. The signs in (2.3.2) have been chosen in such a manner that $\alpha > 0$ if the center of gravity is in the lower half-plane and $2\sqrt{\Delta}$ is the spacing between the zeros. The case (i) of Sec. 2.1 corresponds to $\Delta < 0$, the case (ii) to $\Delta > 0$, and the transition between those two takes place through the point $\Delta = 0$.

It follows from Sec. 2.2 that $\alpha \neq 0$. In a two-zero approximation, $\alpha$ is always positive.

Making use of the definition $S = e^{2i\delta}$, as well as of formula (2.3.1) and relations (2.3.2), we obtain the following expression for the scattering length:

$$a = -\lim_{k\to 0} \{k \cot \delta\}^{-1} = -\{2\alpha/(\alpha^2 + \Delta) + r_{op}\}. \tag{2.3.3}$$

For $r_{op} << |a|$, equation (2.3.1) leads to

$$k \cot \delta = -1/a + (2r_{op} - 1/\alpha)\, k^2/2, \tag{2.3.4}$$

which is similar to (1.3.7) if the effective range $\rho$ is replaced by the new expression

$$\rho = 2r_{op} - 1/\alpha. \tag{2.3.5}$$

For small $\alpha$, the effective range $\rho$ defined by (2.3.5) becomes large and negative. The necessary condition to realize this is $r_{op} < 1/\alpha$. It should be satisfied simultaneously with the condition $r_{op} < \alpha/(\alpha^2 + \Delta)$, which ensures the validity of the equation (2.3.4) itself. If $\rho$ given by (2.3.5) is exponentially large, the scattering length a is also large, $|a| >> b$. No conventional interpretation of $\rho$ as being an effective range of the potential well is possible in this case. A new meaning of $\rho$ can be established by observing that $\rho$ defines, through the relation $k = -i/\rho$, the midpoint between the two Jost zeros nearest to $k = 0$.

We see that for a potential well surrounded by a barrier, it is possible to have for $k \cot \delta$ a two-term expression formally coinciding with the formula (1.3.7) for a potential well without a barrier. In the latter case, the second term in (1.3.7) depends on many zeros of the Jost function, and the expression is valid only if the second term in the expansion is small in comparison with the first one. For a potential well surrounded by a barrier, the two-term form of (2.3.4) reflects the two-zero approximation to the Jost function rather than truncation of the power series in $k^2$. Consequently, there is no condition imposed on the magnitude of the second term and, as a matter of fact, in the example considered above the second term entirely dominates the first one. Similarly, it is possible to write an expression for k cot δ, which accounts for three and four Jost zeros, in the form

$$k \cot \delta = \frac{a + bk^2}{c + dk^2},$$

and

$$k \cot \delta = \frac{a + bk^2 + ck^4}{d + ek^2}.$$

The effective cross section for scattering is then found by using the expansion for k cot δ given above. If $r_{op}$ is neglected in (2.3.3) and (2.3.5), the corresponding expression for the effective cross section becomes

$$\sigma(k) = \frac{16\pi\alpha^2}{(k^2 - \alpha^2 - \Delta)^2 + 4k^2\alpha^2}. \tag{2.3.6}$$

For $\alpha^2 << \Delta$, (2.3.6) gives the usual Breit-Wigner formula for resonance elastic s-wave scattering. For $|\Delta| << \alpha^2$ (two Jost zeros coincide or are close to each other) the cross section depends on a single parameter α:

$$\sigma(k) = 16\pi\alpha^2/(k^2 + \alpha^2)^2 .$$

Instead of the usual $E^{-1}$ dependence of the cross section upon energy E, the above formula rapidly falls off as $E^{-2}$. The metastable states corresponding to these positions of the Jost zeros will have long lifetimes and will decay according to $w = t^2 \exp(-\Gamma t)$ rather than exponentially (22). However, when such states exist, the allowed limits for the variation of parameters entering the potential become more restricted, the longer the lifetime of the state. This makes it difficult to observe such states experimentally, for instance, in nuclei.

## 2.4 THE CASE OF $\ell \neq 0$ AND PERTURBATION THEORY FOR A BOUND STATE CLOSE TO THE CONTINUUM

We shall now show how the theory developed for the Jost zeros in the case of the s-wave scattering can be generalized to include the case of $\ell \neq 0$ scattering. The starting point is the differential equation for the radial function $\psi_\ell(r)$ for an arbitrary angular momentum $\ell$. Due to the presence of the centrifugal potential in such an equation, that is, the term $\ell(\ell + 1)/r^2$, the potential well V(r) representing an attractive nuclear field in the atom is always surrounded by a repulsive barrier. We shall write the equation thus:

$$\frac{d^2\psi_\ell}{dr^2} - \frac{\ell(\ell+1)}{r^2}\psi_\ell + (k^2 - 2V)\psi_\ell = 0 . \qquad (2.4.1)$$

For $k \to 0$, the transmission coefficient through the barrier is infinitely small. Indeed, an estimate of this coefficient can be obtained using the integral

$$-2\int_{b_1}^{(\ell+\frac{1}{2})/k} \left(\frac{(\ell+\frac{1}{2})^2}{r^2} - k^2\right)^{\frac{1}{2}} dr \approx$$

$$2(\ell+\tfrac{1}{2})\ln(kb_1) + \text{Const} \qquad (2.4.2)$$

to compute the barrier factor. In deriving the final result in (2.4.2) we have neglected $V(r)$ everywhere except at the lower limit of integration, which was assumed independent of k. Formula (2.4.2) suggests that a pair of the Jost zeros corresponding to a bound state of vanishing energy and of non-zero angular momentum $\ell$ come together at the origin $k = 0$ itself.

More rigorous results can be obtained following Ostrovskii and Solov'ev [55]. Let us assume that there exists a bound state of zero energy, and find out what happens to the corresponding Jost zero if a small perturbation is applied to the system.

In the case of a bound state close to the continuum, the validity criterion for ordinary perturbation theory is violated. This situation requires re-formulation of perturbation theory in a way which would ensure an effective account of the infinite number of the interacting continuum states. A simple way to construct such a theory is to consider the Jost function $f_\ell(k,\lambda)$ for a perturbed system whose potential depends upon a small parameter $\lambda$.

Let us assume that, for $\lambda = 0$, $k = 0$ is the Jost zero, that is,

$$f_\ell(0,0) = 0. \qquad (2.4.3)$$

Consider now the equation

$$f_\ell(k,\lambda) = 0, \qquad (2.4.4)$$

which determines the position of the Jost zeros $k(\lambda)$. In order to obtain an expansion of $k(\lambda)$ in terms of the perturbation parameter, first we expand (2.4.4) in terms of $\lambda$ and k and demand the forms containing the same powers of $\lambda$ to be zero. The derivatives $f_{mn}$,

$$f_{mn} = \left. \frac{\partial^{m+n} f_\ell(k,\lambda)}{\partial k^m \, \partial \lambda^n} \right|_{\lambda = k = 0} ,$$

can be generally expressed in terms of the wavefunction of the unperturbed system [55]. Here it is sufficient to use the property [42] that $f_\ell(k)$ can be written as follows:

$$f_\ell(k) = A_\ell(k) + i\, k^{2\ell+1} B_\ell(k),$$

where $A_\ell$ and $B_\ell$ are even functions of k. Then it follows readily that the coefficients $f_{mn}$ in the expansion of (2.4.4) vanish if m is odd and $m < 2\ell$, and the expansion of $k(\lambda)$, for $\ell > 0$, may be written in the form

$$k(\lambda) = \sum_{j=1}^{\ell} a_j \lambda^{j-\frac{1}{2}} + \sum_{j=2\ell} b_j \lambda^{j/2}, \tag{2.4.5}$$

where the coefficients $a_j$ and $b_j$ can be determined successively.

In the expansion for $E(\lambda) = k^2(\lambda)/2$, terms up to $\ell$th order are the same as those obtainable from ordinary perturbation theory. However, the next term in $E(\lambda)$ is of order $\lambda^{\ell+\frac{1}{2}}$ and it becomes imaginary for negative $\lambda$. The corresponding state is quasi-stationary, with the width being

$$\Gamma(\lambda) = 2 \operatorname{Im} E(\lambda) \sim \lambda^{\ell+\frac{1}{2}} .$$

It follows from (2.4.5) that after the Jost zeros have passed through the coincidence point $k = 0$, they continue to depart from the real axis according to the law [22,50,55]

$$\operatorname{Im} k(\lambda) = \{\operatorname{Re} k(\lambda)\}^{2\ell}, \tag{2.4.6}$$

that is, they are moving along a parabola of order $2\ell$. This general result embraces the behavior of the Jost zeros in a particular case $\ell = 1$ established for a rectangular barrier [52] and for a cut-off Coulomb potential [53].

For an s-state, modified perturbation theory gives $k(\lambda) \sim \lambda$, that is, $E(\lambda) \sim \lambda^2$. Comparing this result with (2.4.6), we conclude that the term $E(\lambda)$ is a tangent to the continuum boundary $E(0)$ if $\ell = 0$, and it crosses the spectrum boundary if $\ell > 0$.

In this analysis of problems with cutoff potentials, such properties of the Jost function as analyticity in the whole complex plane, etc. have not been used while considering the behavior of the Jost zeros close to the imaginary axis. Therefore the present conclusions may be extended to a wider class of exponentially decaying potentials such as, for instance, $\exp(-\beta r)$, $\beta < \kappa < \infty$. For potentials decreasing as an inverse power of r, particularly when the power is low, a similar study may lead to substantially different results (56).

Up to now it has been assumed that the potential well is spherically symmetrical. This property of the potential becomes less and less important as the energy level approaches the continuum boundary (see Chapter 4). In the limit of zero energy, the wavefunction beyond the potential well is the same as that for a spherically symmetrical potential. A definite value of $\ell$ can be assigned to the wavefunction, and therefore the results obtained above can be extended to the case of a non-spherical potential well.

Another aspect of the problem discussed by Perelomov and Popov (56) and by Migdal et al. (57) is the influence of the Jost zeros situated close to the origin k = 0, on wavefunctions of the continuum. It was first shown by Galitskii and Chel'tsov (58) that in the region $r < b$ (i.e., inside the potential well) the dependence of the continuum wavefunctions on energy (momentum) is factorable. Let $\phi(k,r)$ be such a function normalized to the delta-function of momentum,

$$\phi(k,r) = \sqrt{2/\pi}\ \sin(kr + \delta), \quad r > b,$$

and let $\phi_o(r)$ be a solution of the Schrödinger equation for zero energy, subject to the normalization condition $\phi_o(b) = 1$. For a sufficiently deep potential well and small k, the wavefunction inside the well does not depend upon k, except perhaps through the normalization factor. Therefore inside the well, $\phi(k,r)$ is proportional to $\phi_o(r)$. The proportionality coefficient can be obtained directly by comparing these two solutions:

$$\phi(k,r) = \sqrt{2/\pi}\,\sin\delta\ \phi_o(r), \tag{2.4.7}$$

$$\sin\delta = k\sqrt{\sigma(k)/4\pi}\ , \tag{2.4.8}$$

where $\sigma(k)$ is the scattering cross section (for simplicity, we are discussing here the s-wave). Thus factorization of the k-dependent part of the solution has been achieved. The resonance dependence on k is realized if there exist Jost zeros near k = 0. In the latter case, formula (2.3.6), which accounts for poles near a coincidence point, has to be used for the cross section $\sigma(k)$. A factorized form of the wavefunction is very convenient for computation of matrix elements.

For instance,

$$|\langle\phi_k|U|\phi_o\rangle|^2 = k^2\ \sigma(k)\,|U_{oo}|^2/2\pi^2.$$

Due to factorization, the integral equation for the spectrum of two interacting particles moving inside a potential well reduces to an algebraic equation. The latter can be used, particularly, to show that an additional bound state may arise, under certain conditions, which cannot be predicted from perturbation theory (59).

## 2.5 TRAJECTORIES OF THE POLES OF THE S-MATRIX IN THE CASE OF ZRP AND SEPARABLE POTENTIALS

As we have seen, low-energy resonance scattering requires, as a rule, a two-pole rather than a single-pole approximation to the S-matrix. For $\ell \neq 0$, the two-pole approximation has to be always used.

The scattering by a potential well is an exception since the two-pole approximation for the S-matrix fails in this case. This happens because the separation between the coincidence point in the lower half-plane, i.e., the point where two poles come together, and the origin $k = 0$ is of the same order of magnitude as the separation between $k = 0$ and the next pole lying in the upper half-plane. In the case of a potential well surrounded by a barrier with a low transmission coefficient, the two-pole approximation is applicable for $E < \Delta E$, where $\Delta E$ is the energy level spacing as $E \to 0$. At the same time, the one-pole approximation is either inapplicable or applicable in a much smaller range of energies, $E < \varepsilon\Delta E$, provided that the separation of one of the poles of the S-matrix from the origin is essentially smaller than $\varepsilon\Delta E$.

Any stabilization of an s-state in the continuum means that there exists a coincidence point close to the origin. This requires a two-pole approximation to be used for the S-matrix. The stabilization itself can be a result of either the existence of a potential barrier or (in more complicated problems) such factors as weakness of dynamic interaction with the scatterers, special features of many-particle systems involved in the process of scattering, etc.

The properties of the poles of the S-matrix on the nonphysical sheet of E considered here are of importance also in the theory of slow collisions between negative ions and atoms. As colliding particles come close to each other, the bound state of the outer electron in the negative ion will be "repelled" into the continuum. Qualitatively different results can be obtained for the probability of electron detachment in the course of such collisions, depending on how close is the coincidence point to the origin. Approximations for ZRP and separable potentials are used in Chapters 8-11 to solve dynamical problems of collisions between negative ions and atoms. As an introduction to these methods, in the last section of this chapter we consider trajectories of the poles of the S-matrix in the case of a ZRP or separable potential.

For a one-center ZRP, the S-matrix has only one pole $k = i\alpha$, where $\alpha$ is a parameter of the boundary condition. When $\alpha$ changes in the domain $-\infty < \alpha < \infty$, the pole moves in the complex k plane along the imaginary axis. This corresponds to the case of an odd pole considered in Sec. 2.2.

For a two-center problem with ZRPs, the total number of poles is infinite. It is possible to obtain in this case resonance scattering as well as coincidence of the poles similar to that found for a one-center problem with a finite range potential. These questions will be considered in Chapter 3.

As already mentioned, trajectories of poles lying far away from the real axis have no obvious physical meaning. For a finite range potential, the total number of poles is infinite. A finite number of them lie on the imaginary axis; all others are in the lower half k plane. The number of poles in the horizontal strip lying below the real axis is always finite as well as the number of the poles below any ray which starts at $k = 0$ and lies in the lower half-plane [22]. Indeed, making some simple assumptions concerning the potential, it is easy to derive the following equation for the distant poles:

$$2\left(\frac{e^{ikb}}{2k}\right)^2 V(b-0) = (-1)^{\ell} \tag{2.5.1}$$

(for $\ell = 0$, this equation follows directly from (2.2.12) if $|k| \to \infty$). It can be concluded from (2.5.1), for instance, that the poles move to infinity in the lower half kb plane as the strength of the potential decreases. Their trajectories approach asymptotically lines $\mathrm{Im}(k_N b) = \pm N\pi$, where N and $l$ have the same parity if the potential is attractive and opposite parity if the potential is repulsive. For any specified potential, the trajectories of distant poles ($N \to \infty$) are such that the equation

$$\arg(k_N b) = -\frac{2}{\pi N}(\ln N + O(1))$$

holds [22, 52, 60]. These features of the pole trajectories can be observed in the particular case of a two-center system with ZRPs (see Chapter 3). For a separable potential, there may exist either a finite or an infinite number of S-matrix poles. This is true even for the simplest one-term separable potential of the form (1.4.1). Trajectories of the poles may have various forms, repelling of a bound state into the continuum being accompanied by "fusing" of the poles. For instance, if the function $\phi$ in the separable potential (1.4.1) has been taken in the form (1.4.12), the trajectories along which the poles move as the parameter v varies are cross-like [61, 62]:

$$k = \pm \sqrt{2v} - i\beta , \tag{2.5.2}$$

i.e., the poles move along a line parallel to the real axis after they have fused at $k = -i\beta$ and $v = 0$. Different choices of $\phi$ in the separable potential can lead to trajectories which either approach or go away from the real axis [61, 62]. Special choices of the potential can lead to even more complicated behavior of trajectories near the coincidence point. For instance, if $\phi$ is chosen to have the form

$$\phi = \frac{(2\beta)^{n + 3/2}}{\sqrt{4\pi(n + 2)!}} r^n e^{-\beta r} ,$$

where n is an integer, then at the point $k = - i\beta$, $v = 0$, $(2n + 4)$ poles fuse together [62].

Beregi [61] gives numerous examples of pole trajectories for various separable potentials. In some cases there exist trajectories tangent to the real axis k. At the point of contact there appears a bound state with a positive energy, on the continuum background. This state is not stable and, under the influence of small perturbation, it becomes a resonance (quasi-stationary state) as shown by Beregi et al. [38]. As $v \to \pm\infty$, the poles can move asymptotically to fixed points in the complex k plane. Pairs of the poles moving in the lower half-plane may describe semi-circular trajectories symmetric with respect to the imaginary axis (similar trajectories were obtained by Ferreira and Teixeira [53] for $\ell = 2$ states in the cutoff Coulomb field).

## Chapter 3

# ZERO-RANGE POTENTIALS FOR MOLECULAR SYSTEMS. BOUND STATES

## 3.1 MANY-CENTER PROBLEMS WITHOUT EXTERNAL FIELDS

The determination of the bound-state energy levels of an electron moving in the potential field of a complicated molecule is a typical problem in quantum chemistry. The resulting field is a superposition of several spherically symmetrical potentials centered on the nuclei of the molecule. The solution of such problems with real potentials encounters great numerical difficulties. Therefore it is important to see whether simpler models can be developed to give a qualitatively correct description of the molecules. In this chapter, we shall consider a superposition of several ZRPs forming a many-center potential field. The corresponding Schrödinger equation for an arbitrary number of force centers admits an analytical solution. This model can be applied to study quantum-mechanically the motion of outer electrons in negative molecular ions. With a suitable choice of parameters, it can also be applied to the inner electrons of these molecular systems.

In this chapter, we shall consider simple static problems for a particle moving in a combined field of several ZRPs. Let us suppose that there exist N (finite number) ZRPs of strength $2\pi/\alpha_j$ each, which are positioned at $\underline{R}_j$ $(j = 1,2,3,\ldots,N)$. The wavefunction of the particle, $\psi(\underline{r})$, satisfies the Schrödinger equation for a free particle everywhere except at the points $\underline{R}_j$, and falls off as $r \to \infty$. It can be written as the linear combination

$$\psi(\underline{\vec{r}}) = \sum_{j=1}^{N} c_j \phi_j , \qquad (3.1.1)$$

where the functions $\phi_{j\kappa}$ can be expressed in terms of the Green's function for a free particle of negative energy (see formula (1.3.2) in Sec. 1.3).

$$\phi_{j\kappa} = 2\pi\, G_o(\underline{r}, \vec{\underline{R}}_j, -\kappa^2/2) = \frac{e^{-\kappa r_j}}{r_j}, \tag{3.1.2}$$

$$\underline{r}_j = \underline{r} - \underline{R}_j .$$

Imposing the boundary conditions (1.2.7) on the wavefunction (3.1.1) at the points $\underline{r} = \underline{R}_j$, $j = 1,2,\ldots N$, we obtain a homogeneous system of N algebraic equations

$$(\alpha_i - \kappa)c_i + \sum_{\substack{j=1\\ j\neq i}}^{N}{}' c_j \frac{e^{-\kappa R_{ij}}}{R_{ij}} = 0, \tag{3.1.3}$$

$$\underline{R}_{ij} = \underline{R}_i - \underline{R}_j, \quad i = 1,2,\ldots,N.$$

This system* has a nonvanishing solution $\{c_j\}$ if the determinant of the system is zero:

$$\det \| W_{ij} \| = 0 , \tag{3.1.4}$$

where

$$W_{ij} = \begin{cases} \alpha_j - \kappa, \text{ if } i = j \\ \dfrac{e^{-\kappa R_{ij}}}{R_{ij}}, \text{ if } i \neq j. \end{cases}$$

This gives a transcendental equation for $\kappa$, i.e., for the energy of the bound state $E = -\kappa^2/2$.

The equation (3.1.4) has several (N') solutions $\kappa > 0$ corresponding to the bound states of the system. The number of solutions does not exceed the number of the potential wells, $N' \leq N$, as shown in Sec. 4.4. Then N - N' solutions belong to the continuum. Equation (3.1.4) also determines virtual ($\kappa < 0$) and quasi-stationary (complex $\kappa$) states.

---

* This system of equations (3.1.3) becomes identical with the system of equations for linear coefficients in the MO LCAO method, provided that the matrix elements $\alpha_j$ and $\beta_{ij} = \exp(-\kappa R_{ij})/R_{ij}$ in (3.1.3) are interpreted as Coulomb integrals and resonance integrals, respectively.

Let us consider now a more general problem where the total potential is a superposition of ZRPs and a long-range potential V(r) whose Green's function $G(\underline{r},\underline{r}',E)$, which satisfies the equation

$$\left(-\tfrac{1}{2}\nabla^2 + V(\underline{r}) - E\right) G(\underline{r},\underline{r}',E) = \delta(\underline{r}-\underline{r}') , \qquad (3.1.5)$$

is known. We shall seek the Green's function $L(\underline{r},\underline{r}',E)$ for a particle moving in the combined field of all potentials, in the form

$$L(\underline{r},\underline{r}',E) = G(\underline{r},\underline{r}',E) + \sum_{i,j=1}^{N} c_{ij}\, G(\underline{r},\underline{R}_i,E)\, G(\underline{r}',\underline{R}_j,E), \qquad (3.1.6)$$

which is symmetrical with respect to interchanging $\underline{r}$ and $\underline{r}'$. The function L satisfies the corresponding Schrödinger equation, and from the boundary conditions we obtain a system of inhomogeneous equations for the coefficients $c_{ij}$ in (3.1.6):

$$\left(\alpha_i/(2\pi) + G_r(\underline{R}_i,\underline{R}_i,E)\right) c_{ij} + \sum_{\substack{k=1\\k\neq i}}^{N}{}' c_{kj}\, G(\underline{R}_i,\underline{R}_k,E) = -\delta_{ij}, \qquad (3.1.7)$$

where $G_r$ is the regularized Green's function:

$$G_r(\underline{r},\underline{r}',E) = G(\underline{r},\underline{r}',E) - \frac{1}{2\pi|\underline{r}-\underline{r}'|} . \qquad (3.1.8)$$

For a single ZRP, we obtain

$$c_{11} = c = -2\pi\left(\alpha + 2\pi\, G_r(\underline{R},\underline{R},E)\right)^{-1} . \qquad (3.1.9)$$

The poles of the function L (i.e., of the coefficient c) determine bound states of the electron moving in the combined field of ZRPs and the potential $V(\underline{r})$. Examples of such fields will be considered in Chapter 7.

The two-center problem with ZRPs was first considered by Smirnov and Firsov [17]. General equations (3.1.1) - (3.1.4) were derived by Adamov et al. [63] and by Dalidchik and Ivanov [64]. The Green's function was obtained in various forms by many authors [19, 65-67].

Let us consider now the bound-state problem for a particle moving in a field of a finite number N of separable potentials (1.4.2). Making use of the Green's function, it is possible to write the wavefunction in a form similar to that of the one-center problem considered in Sec. 1.4, that is,

$$|\psi> \quad = \quad \sum_{j=1}^{N} c_j |\chi_j(E)> \; , \qquad (3.1.10)$$

where

$$|\chi_j(E)> \quad = \quad (H_o - E)^{-1} \; |\phi_j> \; .$$

In order to determine coefficients $c_j$, we shall substitute (3.1.10) into the Schrödinger equation and use the linear independence of the vectors $|\phi_j>$. Thus we obtain a linear system of homogeneous algebraic equations:

$$\sum_{j=1}^{N} \left( c_j \delta_{ij} + c_j v_j <\phi_i|(H_o - E)^{-1}|\phi_j> \right) \quad = \quad 0, \qquad (3.1.11)$$

$$i = 1,2,...,N,$$

which determines a non-zero set $\{c_j\}$, $j = 1,2,...,N$, provided that

$$\det \| \; \delta_{ij} + v_j <\phi_i|(H_o - E)^{-1}|\phi_j> \; \| \quad = \quad 0 \; . \qquad (3.1.12)$$

Equation (3.1.12) determines a number of bound and quasi-stationary states of the molecular system. The number of bound states cannot exceed the number of negative coefficients $v_j$ in the separable potential (1.4.2). However, the total number of poles of the S-matrix has no upper limit [61]. A case of a combined field of separable potentials and a long-range potential $V(\underline{r})$ can be included in the theory by assuming $H_o$ in equations (3.1.10 - 3.1.12) to be of the form $H_o = -\frac{1}{2}\nabla^2 + V(\underline{r})$. As in the case of ZRPs, the Green's function for this modified problem can be easily found.

For molecules, the correct choice of parameters in ZRPs becomes a task which is even more difficult than that in the one-center case. The $R_j$ are usually interpreted as being the distance between the atomic nuclei in the molecule (even this assumption is not necessary as we shall see in Sec. 3.2, where the theory is applied to color centers). The determination of $\alpha_j$ by relating them to the depth of the potential wells obtained from atomic data may be too crude because, generally speaking, the effective depth of the well for an isolated atom and that for the atom in a molecule are not the same, but depend upon the valence state of the atom in the molecule. Another difficulty appears because of the non-physical singularities* arising within the ZRP model as $R_{ij} \to 0$, that is, when two of the centers come together. Existence of these singularities may become noticeable even at finite separations $R_{ij}$.* One of the possible ways

---

* There are known methods, within the ZRP approach, to avoid this complication (see Sec. 3.2 for discussion). For separable potentials the difficulty does not exist.

to choose parameters of the problem is to use some known properties of the molecular system. Then the ZRP model can be employed to obtain some other properties of the system. The inner shell electrons present considerable difficulties and may require special modifications of the ZRP method (see Sec. 3.5). General semi-empirical rules for the selection of the parameters must be based on a wide and systematic experience in ZRP calculations for various molecular systems. Such experience is gradually being gained at the moment.

## 3.2 POTENTIAL CURVES FOR A TWO-CENTER SYSTEM AND SOME APPLICATIONS

The simplest many-center problem is two ZRPs separated by a distance R. Equation (3.1.4) takes the form (17)

$$(\kappa - \alpha_1)(\kappa - \alpha_2) - e^{-2\kappa R}/R^2 = 0, \qquad (3.2.1)$$

and we obtain from (3.1.3) for the wavefunction of a bound state:

$$\frac{c_1}{c_2} = \frac{e^{-\kappa R}/R}{\alpha_1 - \kappa} = \frac{\alpha_2 - \kappa}{e^{-\kappa R}/R} . \qquad (3.2.2)$$

This is a model of a negative molecular ion $AB^-$ in the Born-Oppenheimer approximation when the positions of both nuclei are assumed to be fixed. If R is treated as the internuclear separation, then the function $E(R) = -\kappa^2(R)/2$ is a family of potential curves or terms of the negative molecular ion, the energy E(R) being measured with respect to the potential curve of the corresponding neutral molecule.

A system of a weakly bound electron moving in the field of two identical potential centers is also known to exist in alkali-halide crystals. It is called a color M'-center and consists of an electron and a neutral M-center. The latter is a system of two F-centers so that an M'-center is analogous to a negative ion $A_2^-$. A system of an electron moving in the field of three F-centers is also known to exist. It is called an R'-center and analogous to $A_3^-$.

Let us consider how the potential curves may depend upon the internuclear separation R in the case of a two-center system which has been modelled by two ZRPs. For $\alpha_1 < 0$ and $\alpha_2 < 0$ there is no bound state of an electron centered on either nucleus. A bound state of zero energy first appears at $R_0 = \sqrt{\alpha_1\alpha_2}$, the binding energy being increased as R decreases. In the limit, $\kappa \to \infty$ as $R \to 0$. We conclude (17) that the negative ion $He_2^-$ could exist only if the internuclear separation were less than 1.3-1.6 a.u. (The numerical value of $\alpha$ used in this estimate has been obtained from the scattering length (46)). This distance, however, is outside the domain where the theory is valid ($R > 2r_{at}$).

For $\alpha_1 > 0$ and $\alpha_2 < 0$, there exists only one bound state (with $\kappa \to \alpha_1$ as $R \to \infty$, and $\kappa \to \infty$ as $R \to 0$).

For $\alpha_1 > 0$ and $\alpha_2 > 0$, there exists a bound state of an electron centered on either nucleus. The behavior of the potential energy curves as a function of R is shown in Fig. 3.1 for a particular choice $\alpha_1 = 1$ and $\alpha_2 = 2$ made in Reference (63). The binding energy of the lower-lying state becomes even greater as R decreases. At the same time, the binding energy of the higher-lying state decreases and, at $R_c = \sqrt{\alpha_1\alpha_2}$, it vanishes, giving rise to either the virtual or quasi-stationary state. The disappearance of bound states and the subsequent behavior of the poles of the S-matrix play an important role in the theory of the electron detachment discussed in Chapters 8-11. In an enlargement inserted in Fig. 3.1, it can be seen that the curve corresponding to the disappearing state is tangent to the continuum boundary (E = 0) at the point of contact $R_c$, in accord with Sec. 2.4. This discussion of the behavior of the curve will be continued below in Sec. 3.3.

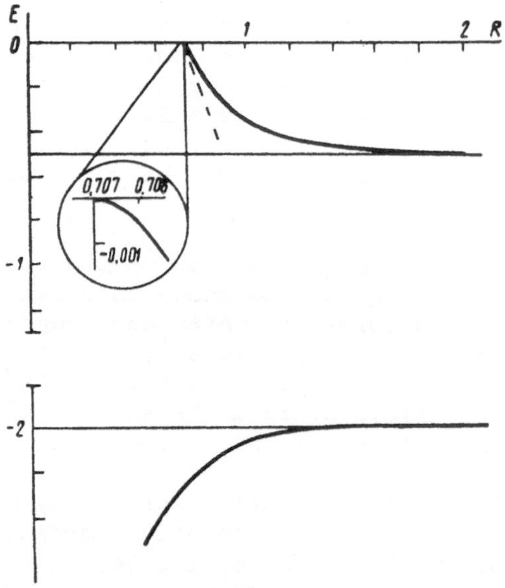


Fig. 3.1. Molecular potential energy curves for a system of two zero-range potentials; $\alpha_1 = 1$ and $\alpha_2 = 2$ (63). In the inset above: an enlarged part of the graph showing that the curve is tangent, at the point of entering the continuum, to the boundary E = 0.

As a simple example of the theory developed above we shall consider now resonance charge exchange in slow collisions between an atom A and its negative ion $A^-$:

$$A^- + A \rightarrow A + A^- . \tag{3.2.3}$$

In this particular case, the colliding particles are modelled by two identical ZRPs. Due to the symmetry of the system with respect to the center of mass, the state of a particle at fixed R has to be either symmetrical ($c_1 = c_2$) or antisymmetrical ($c_1 = -c_2$). Correspondingly, the equation (3.2.1) splits into two:

$$\alpha - \kappa \pm e^{-\kappa R}/R = 0, \tag{3.2.4}$$

where "+" relates to the symmetrical terms (g-terms) and "-" relates to the antisymmetrical terms (u-terms) of the system.

For large R, the system is degenerate because the last term in (3.2.4) disappears, as $R \rightarrow \infty$, and both u- and g-terms come together. More precisely, the energy splitting

$$E_u(R) - E_g(R) = 2\alpha e^{-\alpha R}/R \tag{3.2.5}$$

is exponentially small if $\alpha R \gg 1$. For small R, the splitting $E_u - E_g$ becomes more significant and, as $R \rightarrow 0$, it has a non-physical singularity. As is well known, the cross sections of low-energy resonance charge exchange can be expressed in terms of the splitting $E_u - E_g$ [26].

If the numerical value of $\alpha$ is selected using data for free atoms, the best result is obtained when the determined value of $\alpha$ is used for an asymptotic estimate of the potential energy curves for large separations R, which, in turn, are used to calculate the resonance charge exchange cross sections. When the colliding atoms come closer to each other, the original charge distribution in their electron shells become mutually distorted. This effect can be taken into account assuming that $\alpha = \alpha(R)$. For $R \rightarrow 0$, the energy terms exhibit a non-physical singularity pointed out above. Due to this singularity it is impossible to perform continuous transition from a molecular system to the united atom limit. However, this difficulty can be circumvented if we assume, following Rebane et al. [68], that

$$\alpha(R) = R^{-1} + \tilde{\alpha}(R), \tag{3.2.6}$$

where $\tilde{\alpha}(R)$ is a smooth function of R such that the united atom limit $E_{ua}$ of the potential energy curve in question is $E_{ua} = -\tilde{\alpha}(0)^2/2$. In the intermediate region of finite R, the function $\tilde{\alpha}(R)$ can be determined from exact calculations [68].

Another application of the calculated splitting between g- and u-terms can be found in the semi-empirical theory of color centers considered by Berezin and Kiri [32]. Transitions between g- and u-states in color M'-centers may give rise to absorption bands observable in the spectrum. Making use of the value of $\alpha$ obtained from the one-center theory (for F'-centers considered in Sec. 1.3) and of the experimental term splitting $E_u(R) - E_g(R)$ in M'-centers, it is possible to determine the effective separation $R_{M'}$ between the M'-centers. $R_{M'}$ obtained in this way turns out to be smaller than the spacing between two neighboring anion vacancies in the undistorted lattice.

A color R'-center can be modelled by three identical ZRPs centered on the vertices of an equilateral triangle of side $R_{M'}$. In this case equation (3.1.4) takes the form

$$(\kappa - \alpha)^3 - 3(\kappa - \alpha)\frac{e^{-2\kappa R}}{R^2} - 2\frac{e^{-3\kappa R}}{R^3} = 0\,. \tag{3.2.7}$$

This equation has one simple root $\kappa \approx \alpha + 2e^{-\alpha R}/R$ corresponding to the ground (symmetrical) state, and one double root $\kappa \approx \alpha - e^{-\alpha R}/R$ corresponding to a doubly degenerate excited state. The transition energy for a R'-center obtained with the help of this model is between the maxima of the $R_1'$- and $R_2'$-absorption bands arising due to the Teller splitting of the excited state of the R'-center. For crystals of KCl, the theoretical value of the transition energy is found to be 0.81 eV, whereas the experimentally determined maxima $R_1'$ and $R_2'$ of the bands are 0.83 eV and 0.77 eV. For crystals of KBr, the corresponding values are 0.73 eV, 0.77 eV, and 0.72 eV. In these calculations, the parameters $\alpha$ and $R_{M'}$ have been obtained from experimental data on F'- and M'-centers and they are independent of the data for R'-centers. Hence the agreement between the theoretical and experimental results obtained for R'-centers confirms that ZRPs are a good approximation for the description of the negative electron color centers in ionic crystals [32].

## 3.3 ANALYTIC PROPERTIES OF THE POTENTIAL CURVES AND TRAJECTORIES OF THE POLES OF THE S-MATRIX

It is obvious from equations (3.1.4) and (3.2.1) that the function $\kappa(R)$ (or the energy $E(R)$) they determine is an analytic function of the nuclear separation R. The analyticity of the energy E(R) is expected to hold also in the general case of a many-electron molecule [62], though the rigorous proof of that has not yet been obtained. The analytic properties of E(R) are important because they play an essential role in the description of dynamic processes occuring during the collision when the parameter R has to be treated as a function of time. The type and position of the branch points $R_b$ of the energy E(R) are especially important. This is because all potential energy curves of a molecular system are different branches of the same analytic function E(R). While moving along a path encircling the branch point, we go from one sheet to another sheet of the complex E plane, i.e., we go from one potential energy curve to another potential energy curve of the system. Some of these may be non-physical (corresponding for real R to complex values of E). The curves belonging to different sheets of E come together (intersect) at the branch points. The intersection point of two discrete curves determines, as is well known, the probability of non-adiabatic transitions between these potential energy curves in the non-stationary problem with slow changing R(t). The branch points where a discrete curve intersects with a virtual or quasi-stationary curve determine the dynamics of the transition to the continuum.

If a quantum-mechanical system possesses a certain type of symmetry (for all values of parameter R in the Hamiltonian), the terms corresponding to different symmetries are not connected with each other, and transitions between such terms are impossible. Then there exists a set of analytic functions $E_s(R)$, each of them being related to certain symmetry properties of the wavefunctions.

For fixed real R, the set of function values E(R) on all sheets of the complex E plane determines the set of all poles of the S-matrix (or the Green's function), that is, the poles which are related to bound states as well as those related to virtual and quasi-stationary states. In Chapter 2 we have discussed the general importance of trajectories described by the poles in the complex E (or momentum) plane due to changes in parameters of the problem. The practical applications of this theory will be considered below as well as in Chapter 5.

We have seen earlier that a pair of S-matrix poles lying on the imaginary k axis may come together, at a certain depth of the well, at a point on this axis. If the change in the depth persists, these two points move away from the coincidence point in opposite directions, remaining on a trajectory normal to the imaginary axis (the case of quasi-stationary states). The point where the two poles "fuse" together is, obviously, a branch point of function E(R). The particular physical significance of this point is that it determines non-adiabatic transitions of the system to the continuum.

At present the analytical properties of energy E(R) have not yet been studied sufficiently fully in the general case of multi-center (Coulomb) potentials. The ZRPs offer an exceptional opportunity to introduce a solvable and, at the same time, realistic model where energy (or $\kappa$) is a simple analytical function of R. This may be used in studying some interesting and important properties of E(R).

We start our discussion of a two-center problem by pointing out the invariance of Eq. (3.2.1) with respect to the simultaneous transformation $\kappa \to -\kappa$, $R \to -R$, and $\alpha_i \to -\alpha_i$ $(i = 1, 2)$. Generally speaking, one could have expected that the case of negative R is non-physical. We see now that $R < 0$ corresponds, within the ZRP method, to some other Hermitian problem where the sign of $\kappa$ and the sign of the depth of both potential wells must be changed.

In order to determine the branch points $R_b^{(n)}$ of $\kappa(R)$, we consider

$$\{\kappa(R) - \kappa_b^{(n)}\} \sim \{R - R_b^{(n)}\}^{1/2}, \tag{3.3.1}$$

where

$$\kappa_b^{(n)} \equiv \kappa\{R_b^{(n)}\}.$$

We differentiate (3.3.1) with respect to $\kappa$ and take into account that

$$\left.\frac{dR}{d\kappa}\right|_{\kappa = \kappa_b^{(n)}} = 0. \tag{3.3.2}$$

The resulting equation to determine $\kappa_b^{(n)}$ and $R_b^{(n)}$ is

$$\kappa - (\alpha_1 + \alpha_2)/2 + e^{-2\kappa R}/R = 0. \tag{3.3.3}$$

First we shall consider two identical wells where two systems of terms (g- and u-terms) exist. Instead of (3.3.3) we obtain, in this case, from (3.2.4), the following equations:

$$e^{-\kappa R} = \pm 1, \tag{3.3.4}$$

the branch points being determined explicitly from (3.2.4) and (3.3.4) [62] as follows:

$$\kappa_b^{(n)} = -\alpha \frac{(2\pi n + \frac{\pi}{2} \pm \frac{\pi}{2})}{1 - (2\pi n + \frac{\pi}{2} \pm \frac{\pi}{2})}, \tag{3.3.5}$$

$$R_b^{(n)} = \left(-\alpha + \kappa_b^{(n)}\right)^{-1}, \tag{3.3.6}$$

$$n = 0, \pm 1, \pm 2, \ldots .$$

The branch point of the u-term for $n = 0$, that is, $\kappa_b^{(0)} = 0$, $R_b^{(0)} = \alpha^{-1}$, is of special physical interest because it corresponds to the point described above where two S-matrix poles fuse together and the curve crosses into the continuum. The wavefunction is antisymmetric (for low energies, it is close to a p-function) so that the poles come together exactly at E = 0. In the neighborhood of this point, we shall use expansions in terms of a small dimensionless parameter $q^{1/2}$, where $q = 1 - R/R_c$. Thus we obtain (69)

$$\kappa(R) = (2q)^{1/2}/R_c + q/(3R_c) + O(q^{3/2}), \quad R_c = R_b^0 . \tag{3.3.7}$$

This is in agreement with the results of the general theory for an arbitrary potential (see Sec. 2.4). The g-term (symmetric state) crosses into the continuum if $\alpha < 0$ (as R increases). However, in the latter case, the poles do not fuse, and the pole remains to be on the real $\kappa$ axis.

If the two potential wells are not identical, the solution of equations (3.2.1) and (3.3.3) cannot be obtained in a closed form. An approximate position of the branch points can be found, making use of the expansions in terms of a parameter $\delta$ thus:

$$\delta = \frac{(\alpha_1^{1/2} - \alpha_2^{1/2})^2}{2(\alpha_1\alpha_2)^{1/2}} . \tag{3.3.8}$$

The parameter $\delta$ is small if the two potential wells are not very different from each other, and it was first used by Demkov et al. (69). Let us compare this case with the case of two identical wells considered above. There are no new branch points, but each old branch point of either g- or u-symmetry undergoes a shift which depends on $\delta$. For a new position of the most interesting branch point $\kappa_b^{(0)} = 0$ when $\alpha_1 = \alpha_2 > 0$, we obtain

$$\kappa_b^{(0)} = -(\alpha_1\alpha_2)^{1/2}\delta ,$$
$$R_b^{(0)} - R_c = (\alpha_1\alpha_2)^{-1/2}\delta^2/2 , \tag{3.3.9}$$

where $R = (\alpha_1\alpha_2)^{-1/2}$ is the contact point of the curve at the continuum boundary (due to $\kappa_b^{(0)} \neq 0$, the curve is tangent to the boundary and does not cross it). Near the point $R_b^{(0)}$ the curve $R(\kappa)$ is a parabola, which explains why, in terms of $\delta$, the order of the difference $R_b^{(0)} - R_c$ is higher than that of $\kappa_b^{(0)}$. For numerical values of $\alpha_1$ and $\alpha_2$ for typical atomic problems, $\kappa_b^{(0)}$ turns out to be very small.

It may be concluded that in a two-center ZRP model of a negative molecular ion, there exists an infinite number of branch points and corresponding sheets of the function $E(R)$. The point $R = \infty$ is an essential singular point of $\kappa(R)$ (branch points converge to it). The point $R = 0$ is a pole of $\kappa(R)$ (62), thus:

$$\kappa(R) = \frac{c}{R} + \frac{\alpha}{1 - c} + O(R), \tag{3.3.10}$$

where c is a root of the equation*

$$e^{-c} = \pm c \, . \qquad (3.3.11)$$

This pole is a special feature of the solution, and it is due to ZRPs used in the model. Unlike the branch points, generally it is difficult to ascribe any physical meaning to these poles. It is probable that these poles do not exist in the case of molecular curves.**

$\alpha = 0$ is a special case because all branch points fuse together and the potential energy curve is exactly $E(R) = -c^2/(2R)^2$, i.e., it is inversely proportional to $R^2$, where c is defined by (3.3.11); for a physical curve, $c = 0.567$.

Maxima of the function $\kappa(R)$ are branch points of the inverse function $R(\kappa)$ and can be found in a way analogous to that used for $\kappa(R)$. It turns out that there exist only two maxima:

$$\kappa_{max} = \frac{(\alpha_1 + \alpha_2) \pm \{(\alpha_1 + \alpha_2)^2 + 4\alpha_1\alpha_2 e^2\}^{\frac{1}{2}}}{2(1 - e^2)} \, ,$$

$$R_{max} = -\kappa_{max}^{-1} \, . \qquad (3.3.12)$$

For $\alpha_1 \to \alpha_2$, there is one maximum of the g-term and one maximum of the u-term. We note that $\kappa_{max}$ is real if $\alpha_1$ and $\alpha_2$ are of the same sign and the ratio $\alpha_1/\alpha_2$ lies within the interval

$$\frac{1}{27.5} < \alpha_1/\alpha_2 < 27.5 \, .$$

The maxima of $\kappa(R)$ are related to the behavior of the S-matrix poles. For two ZRPs of opposite signs, the S-matrix pole corresponding to a bound state first moves away from the origin ($|\kappa|$ increases, $\kappa < 0$) as R increases. However, with further increase of R, $R > R_{max}$, the pole moves back towards the origin and reaches it as $R \to \infty$,

$$\kappa(R) = -\frac{\ln(\alpha R)}{R}\{1 + O(R^{-1})\} \, . \qquad (3.3.13)$$

These properties of $\kappa(R)$ for real R are illustrated in Figs. 3.2 and 3.3, where a typical behavior of $\kappa(R)$ is shown for $\alpha_1$ and $\alpha_2$ having

---

* This equation has an infinite number of (complex) solutions. This property corresponds to the existence of poles of the function $\kappa(R)$ on all sheets of the Riemann surface at $R = 0$.

** Provided that the Coulomb interaction between the nuclei has not been included into the electron part of the total Hamiltonian.

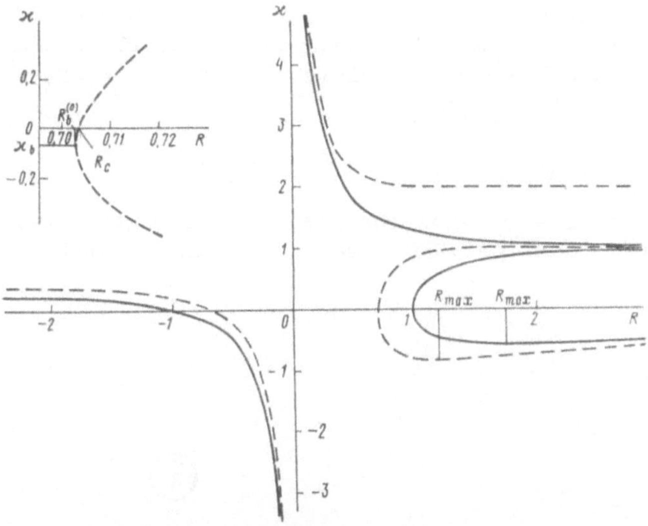


Fig. 3.2. Function $\kappa(R)$ for a system of two zero-range potentials ($\alpha_1$ and $\alpha_2$ being of the identical sign). Continuous line: $\alpha_1 = \alpha_2 = 1$. Broken line: $\alpha_1 = 1$, $\alpha_2 = 2$.

the same sign (Fig. 3.2) and opposite signs (Fig. 3.3). Taking into account the properties of $\kappa(R)$ for $R < 0$ (see the invariance of the problem under the transformation $\alpha_i \to -\alpha_i$, $R \to -R$, and $\kappa \to -\kappa$ mentioned above), we conclude that the graphs presented in Fig. 3.2 and Fig. 3.3 exhaust all possible qualitatively different cases for a system of two zero-range potentials.

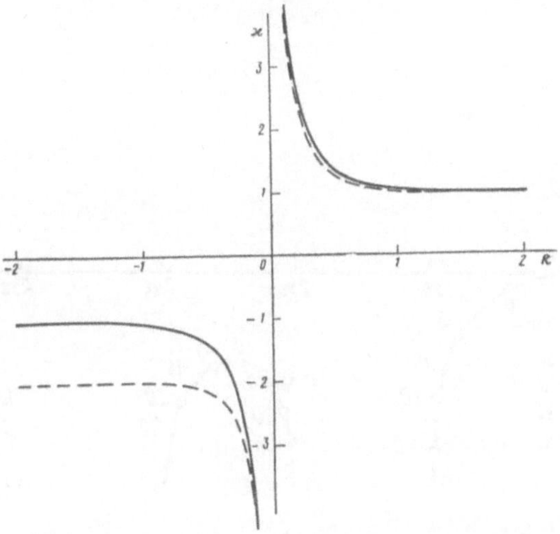


Fig. 3.3. Function $\kappa(R)$ for a system of two zero-range potentials ($\alpha_1$ and $\alpha_2$ being of opposite signs). Continuous line: $\alpha_1 = -\alpha_2 = 1$. Broken line: $\alpha_1 = 1$, $\alpha_2 = -2$.

Now we shall turn to the trajectories described by the poles of the S-matrix in the complex k plane when $\alpha$ is changing. It is convenient to introduce new dimensionless variables $\tilde{\kappa} = \kappa R$, and $\tilde{\alpha}_i = \alpha_i R$. Equation (3.2.1) becomes

$$(\tilde{\kappa} - \tilde{\alpha}_1)\,(\tilde{\kappa} - \tilde{\alpha}_2) = e^{-2\tilde{\kappa}}\,. \tag{3.3.14}$$

Figures 3.4(a) and 3.4(b) show the trajectories in the complex $\tilde{k}$ plane ($\tilde{k} = i\tilde{\kappa}$), for two identical ZRPs, separately for symmetric and anti-

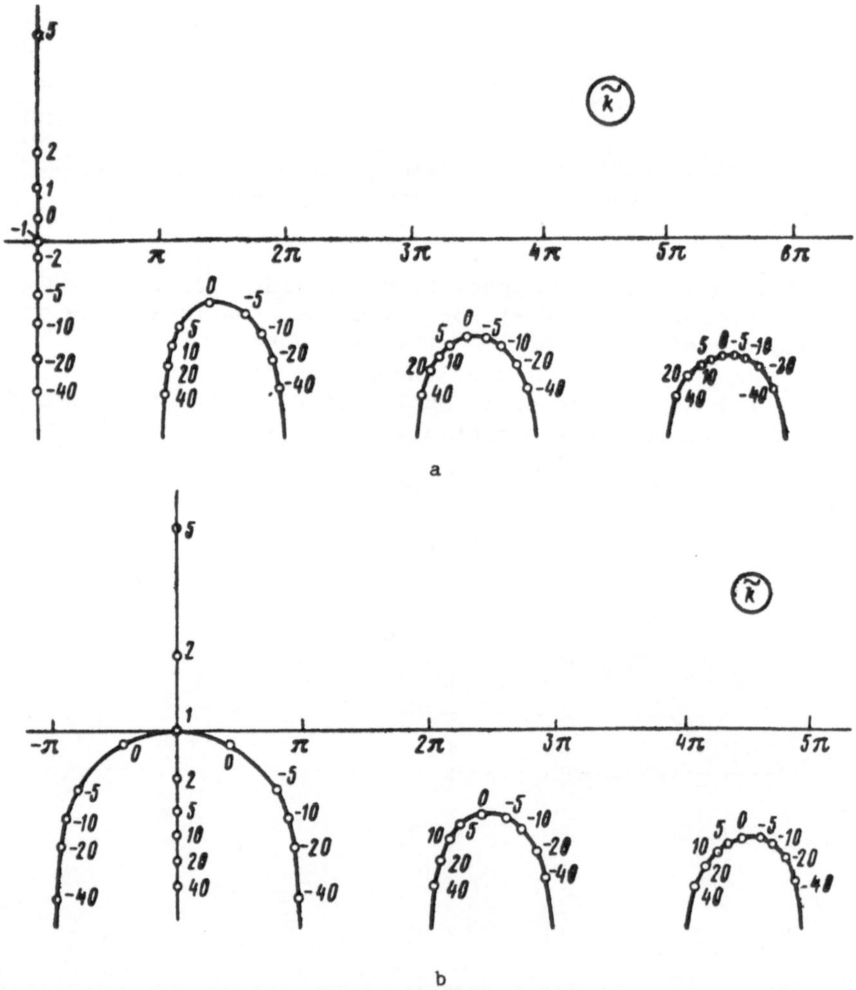


Fig. 3.4. A system of two identical zero-range potentials. Dependence of the poles of the S-matrix in the complex $\tilde{k} = kR$ plane upon the parameter $\tilde{\alpha} = \alpha R$. (a) g-terms, (b) u-terms. Numbers on the graph are values of $\tilde{\alpha}$.

symmetric states. For large $|\tilde{\alpha}|$, all poles move to infinity. This result is different from the previous one, where $\alpha$ was fixed and R was changing, since then the trajectories could tend to finite point ($\kappa = \alpha_i$, $\kappa = 0$) as $R \to \infty$. The trajectories of the poles are in agreement with the general picture obtained earlier in Chapter 2.

Even for a relatively simple problem of scattering on a system of two ZRPs the S-matrix is of a complicated structure. We shall emphasize here that the existence of this structure is entirely due to multiple scattering of the projectile on the ZRPs (for multiple scattering see Sec. 4.5).

Analytic properties of the wavefunction $\psi(\underline{r},R)$ were considered by Solov'ev (70). The wavefunction (3.1.1) is uniquely determined by $\kappa(R)$ (the wavefunction $\psi$ also depends upon the ratio $c_1/c_2$ but this ratio is a unique function of $\kappa$, see equation (3.2.2)). Therefore the branch points of function $\kappa(R)$ are also branch points of the wavefunction $\psi(\underline{r},R)$ if the latter is considered as a function of R at fixed $\underline{r}$. For this molecular system, there exist a unique function $\psi(\underline{r},R)$ and a unique energy E(R) whose values on the different sheets of the complex R plane give the wavefunctions and energies of different curves.

At a branch point, functions $\kappa(R)$ for different curves coincide. So do the wavefunctions. This reduction in the number of the eigenfunctions at $R_b$ is related to the properties of the Hamiltonian which is not Hermitian for complex R. For degenerate eigenvalues, it cannot be diagonalized and is reduced to a Jordan form which always has one eigenvector. This result is important for transitions between adiabatic curves (70). Both curves are physical, and it is necessary to use two different potential wells within a two-center model in order to obtain pseudocrossing of the curves (the branch point in question corresponds to this pseudocrossing). It is not difficult to obtain

$$\int \psi^2(\underline{r},R)\,d\underline{r} = \frac{4\pi c_1^2\,\{-\kappa + (\alpha_1 + \alpha_2)/2 - R(\alpha_1 - \kappa)(\alpha_2 - \kappa)\}}{\kappa R(\alpha_2 - \kappa)}, \quad (3.3.15)$$

where equation (3.2.2) has been used. Evidently, the integral (3.3.15) can be analytically continued to complex R, and the condition that (3.3.15) vanishes coincides with the condition of pseudocrossing. This is a sufficiently general result (70).

## 3.4 PERTURBATION THEORY IN THE PRESENCE OF AN EXTERNAL ELECTRIC FIELD

We shall apply the ZRP model to describe an electron moving in a combined field of N centers and a weak homogeneous external electric field of strength $\mathcal{E}$. In this section, we shall assume the field strength to be small and apply perturbation treatment to the problem (the case of a strong field will be considered for a one-center system in Chapter 7 without applying perturbation theory).

If the wavefunction of the system is known up to terms linear in $\mathcal{E}$, that is,

$$\psi(\underline{r}) = \psi^{(0)}(\underline{r}) + \mathcal{E}\psi^{(1)}(\underline{r}) + O(\mathcal{E}^2), \tag{3.4.1}$$

then the energy of the system is given accurately, up to terms cubic in $\mathcal{E}$, by

$$\bar{E}_\psi = \frac{\langle\psi|H|\psi\rangle}{\langle\psi|\psi\rangle} = E(\mathcal{E}) + O(\mathcal{E}^4), \tag{3.4.2}$$

where the Hamiltonian of the system H includes the external field potential $V = \mathcal{E}z$ (it is assumed that the external electric field is directed along the z-axis). Considering quadratic terms in (3.4.2), we obtain the polarizability of the system:

$$\alpha_{zz} = -2\int \psi^{(0)} z\, \psi^{(1)} d\underline{r}\,. \tag{3.4.3}$$

The first-order correction to the wavefunction, $\psi^{(1)}$, satisfies the standard equation of perturbation theory:

$$\{\nabla^2 - \kappa^2\}\, \psi^{(1)}(\underline{r}) = 2z \sum_{j=1}^{N} c_j \psi_{j\kappa}(r_j), \tag{3.4.4}$$

where the origin is placed at such a point that $\bar{z} \equiv \langle\psi^{(0)}|z|\psi^{(0)}\rangle = 0$ and the functions $\psi_{j\kappa}(r_j)$ are defined by (3.1.3). We shall seek the solution of (3.4.4), i.e., $\psi^{(1)}$, in the form of a sum of one-center functions thus:

$$\psi^{(1)} = \sum_{j=1}^{N} \{-\frac{c_j}{2\kappa}(z_j + 2Z_j)\, r_j \phi_{j\kappa} + b_j \phi_{j\kappa}\}\,. \tag{3.4.5}$$

In (3.4.5), $Z_j$ is the projection of $\underline{R}_j$, formula (3.1.2), along the direction of the external electric field and the terms $b_j\phi_{j\kappa}$ are solutions of the homogeneous equation (3.4.4). The coefficients $b_j$ in (3.4.5) are to be found from the boundary condition (1.2.7). The latter must be satisfied by $\psi(\underline{r})$ in all orders of perturbation theory. Thus we obtain a system of algebraic equations for $b_j$:

$$(\alpha_i - \kappa)\, b_i + \sum_{\substack{j=1 \\ j\neq i}}^{N}{}' \, b_j \, e^{-\kappa R_{ij}}/R_{ij}$$

$$= \frac{1}{2\kappa} \sum_{j=1}^{N} c_j (Z_j + Z_i)\, e^{-\kappa R_{ij}} \, . \tag{3.4.6}$$

The condition $\bar{z} = 0$ above automatically ensures that the right-hand side of equation (3.4.4) is orthogonal to $\psi^{(0)}$ and therefore the system (3.4.6) has a non-vanishing solution. It is evident, from the present discussion, that solutions accurate up to higher orders of perturbation theory can be constructed in a similar way.

Now we can obtain the polarizability of the molecule. For the $\alpha_{zz}$ component we find (63)

$$\alpha_{zz} = \frac{1}{\kappa} \sum_{i,j}^{N} c_i c_j (K_{ij} + 3Z_i L_{ij} + 2Z_i^2 M_{ij})$$

$$- \sum_{i,j}^{N} b_i c_j (Z_i + Z_j)\, S_{ij} \, . \tag{3.4.7}$$

In (3.4.7) we have used the following notations:

$$K_{ij} = \int r_i z^2 \phi_{i\kappa} \phi_{j\kappa} d\underline{r} \, ,$$

$$L_{ij} = \int r_i z \, \phi_{i\kappa} \phi_{j\kappa} d\underline{r} \, ,$$

$$M_{ij} = \int r_i \phi_{i\kappa} \phi_{j\kappa} d\underline{r} \, , \tag{3.4.8}$$

$$S_{ij} = \int \phi_{i\kappa} \phi_{j\kappa} d\underline{r} \, .$$

All these integrals can be solved in terms of elementary functions.

Formula (3.4.7) for the polarizability of a molecule has been obtained for arbitrary positions of the centers. We shall point out that in earlier calculations of the polarizability, which used a $\delta$-function model for the atomic potentials, it was essential to assume one-dimensional motion of the electron in the molecule (for references, see (63)). The difficulty in satisfying continuity conditions at the nodes, recorded in earlier work, was in fact due to the one-dimensional model of the electron motion rather than due to the ZRP model itself.

Below we shall consider several examples of using (3.4.7) for calculations of the polarizability of molecules.

1. <u>Molecule of type $A_2^-$</u>. For a symmetric state,

$$c_1 = c_2 \equiv c = \frac{1}{\sqrt{2(1+S)}},$$

$$b_1 = -b_2 \equiv b = cR^2 e^{\tilde{\kappa}}/2\kappa, \quad \tilde{\kappa} = \kappa R, \tag{3.4.9}$$

$$Z_1 = -Z_2 \equiv -R/2 .$$

Taking into account that

$$K_{11} = K_{22} \equiv K,$$

$$L_{11} = L_{22} = 0, \tag{3.4.10}$$

$$M_{11} = M_{22} \equiv M,$$

and introducing the further notations

$$K_{12} = \nu, \quad L_{12} = \lambda, \quad M_{12} = \mu, \tag{3.4.11}$$

we obtain

$$\alpha_{zz}^{g} = \frac{1}{\kappa(1+S)}\{K + \frac{R^2}{2}M + \nu - \frac{3}{2}R\lambda + \frac{R^2}{2}\mu + \frac{R^3}{4}e^{\tilde{\kappa}}\}, \tag{3.4.12}$$

where the z-axis has been directed along the axis of the molecule.

For an antisymmetric state,

$$c_1 = -c_2 \equiv c = \frac{1}{\sqrt{2(1-S)}} \tag{3.4.13}$$

(we shall remind the reader that this bound state does not exist if $R < R_c = 1/\alpha$). Also we obtain

$$b_1 = b_2 \equiv b = cR^2 e^{\tilde{\kappa}}/2\kappa \tag{3.4.14}$$

and

$$\alpha_{zz}^{u} = \frac{1}{\kappa(1-S)}\{K - \frac{R^2}{2}M - \nu + \frac{3}{2}R\lambda - \frac{R^2}{2}\mu - \frac{R^3}{4}e^{\tilde{\kappa}}\}. \tag{3.4.15}$$

Let us mention some interesting features of these results. For a symmetric state, the polarizability $\alpha_{zz}^{g} \to \infty$ as $R \to \infty$. This is an obvious consequence of a localization of the electron on the center of the lower potential as the internuclear separation increases. For an antisymmetric state, $\alpha_{zz}^{u} \to -\infty$ as $R \to \infty$. This happens because the electron localizes, in this case, on the center of the higher potential. The polarizability of $H_2^+$ shows a similar type of behavior (for fixed nuclei) when $R \to \infty$.

It is easy to prove for $\alpha_{xx}$ that $b_1 = b_2 = 0$. Introducing the notation

$$K_{ij}^{x} = \int r_i x^2 \phi_{i\kappa}\phi_{j\kappa} d\underline{r},$$

we obtain

$$\alpha_{xx}^{g} = \frac{1}{\kappa(1 + S)} (K_{11}^{x} + K_{12}^{x}), \quad (3.4.16)$$

and

$$\alpha_{xx}^{u} = \frac{1}{\kappa(1 + S)} (K_{11}^{x} - K_{12}^{x}). \quad (3.4.17)$$

2. Molecule of type $AB^-$. Formulae (3.4.9) - (3.4.17) derived above for an $A_2^-$ molecule do not contain $\alpha$ in them. For an $AB^-$ molecule, which we are considering now, we have to assume that $\alpha_1$ and $\alpha_2$ are known. For $c_1$ and $c_2$ we obtain

$$c_1 = c2,$$
$$c_2 = c\delta, \quad (3.4.18)$$
$$c = (1 + \delta^2 + 2\delta S)^{-1/2},$$

where

$$\delta = \frac{R}{2} (\alpha_2 - \alpha_1) e^{-\tilde{\kappa}} \pm \{R^2 (\alpha_2 - \alpha_1)^2 e^{-2\tilde{\kappa}}/4 + 1\}^{\frac{1}{2}}. \quad (3.4.19)$$

For $b_1$ and $b_2$, we have

$$b_1 = b_2 = c\delta R^2 e^{\tilde{\kappa}}/2\kappa \quad (3.4.20)$$

and

$$\alpha_{zz} = \frac{1}{\kappa} \frac{1}{(1 + \delta^2 + 2\delta S)} \Big( (1 + \delta)^2 + \{K + \delta( \frac{R^2}{2} M + \nu - \frac{3}{2} R\lambda + \frac{R^2}{2} \mu + \frac{R^3}{4} e^{\tilde{\kappa}} )\}\Big) . \quad (3.4.21)$$

For $\alpha_1 = \alpha_2$, $\delta = \pm 1$ and formulae (3.4.18) - (3.4.21) reduce to those for an $A_2^-$ molecule. We restrict ourselves to considering only these two examples because in other cases the scheme of calculation remains the same as above.

## 3.5 PERTURBATION THEORY IN THE PRESENCE OF AN EXTERNAL MAGNETIC FIELD

We shall follow Adamov et al. (63) and Rebane and Sharibdzhanov (71) and show that the boundary conditions imposed on the wavefunction at each center must be modified in the presence of an external magnetic field.

If there is an external magnetic field with vector-potential $\underline{A}$, the Hamiltonian H which is, in the present case, identical to the kinetic energy operator T takes the form

$$H = T = \frac{1}{2}(i\nabla + \frac{1}{c}\underline{A})^2, \quad c = 137 \text{ a.u.} \tag{3.5.1}$$

The Hermiticity condition (1.2.12) with H given by (3.5.1) becomes

$$\int \operatorname{div}\{\phi_1^* \nabla\phi_2 - \phi_2\nabla\phi_1^* - \frac{2i}{c}\phi_1^*\phi_2\underline{A}\}\, d\underline{r} = 0. \tag{3.5.2}$$

Transforming (3.5.2) into a surface integral, we obtain

$$\int_S \{\phi_1^*(\underline{n}\nabla)\phi_2 - \phi_2(\underline{n}\nabla)\phi_1^* - \frac{2i}{c}\phi_1^*\phi_2\,\underline{A}\,\underline{n}\}\, dS = 0, \tag{3.5.3}$$

where the total surface S was defined in Sec. 1.2.

It is not difficult to check that the condition (3.5.3) is satisfied by a wavefunction whose expansion in the neighborhood of center j has the form

$$\phi_m = c_j^{(m)}\left(\frac{1}{r_j} - \alpha_j + \frac{i}{c}(\underline{A}_j \frac{\underline{r}_j}{r_j})\right) + O(r_j), \; m = 1,2, \tag{3.5.4}$$

where $\underline{A}_j \equiv \underline{A}(\underline{R}_j)$ is the value of the vector-potential at the jth center. In order to prove this result we shall calculate the surface integral over the $S_j$-th sphere (see Sec. 1.2) as its radius $r_j \to 0$:

$$I_j = \lim_{r_j \to 0} \int_{S_j} \{-\phi_1^* \frac{\partial}{\partial r_j}\phi_2 + \phi_2 \frac{\partial}{\partial r_j}\phi_1^*\}\, dS$$

$$= \lim_{r_j \to 0} 4\pi c_j^{(1)*} c_j^{(2)}\, r_j^2 \left[\left(\frac{1}{r_j} - \alpha_j - \frac{i}{c}\frac{(\underline{A}_j\cdot\underline{r}_j)}{r_j}\right)\frac{1}{r_j^2} - \right.$$

$$- \left( \frac{1}{r_j} - \alpha_j + \frac{i}{c} \frac{(\underline{A}_j \cdot \underline{r}_j)}{r_j} \right) \frac{1}{r_j^2} - \frac{2i}{c} (\underline{A}_j \cdot \underline{r}_j) \frac{1}{r_j^2} \Bigg\}$$

$$= - \frac{8\pi i}{c} c_j^{(1)*} c_j^{(2)} \lim_{r_j \to 0} \left\{ (\underline{A}_j \cdot \underline{n}) + \frac{(\underline{A}_j \cdot \underline{r}_j)}{r_j} \right\} = 0.$$

If there is no external magnetic field ($\underline{A} = 0$), condition (3.5.4) reduces to the usual boundary condition.

Under the gauge transformation $\underline{A}' = \underline{A} + \nabla f$ the wavefunction corresponding to the new vector-potential $\underline{A}'$ takes the form $\psi_{\underline{A}'} = \psi_{\underline{A}} \exp(if/c)$. In the vicinity of the jth center the new wavefunction has an expansion in the old form (3.5.4) provided that $\underline{A}$ has been replaced by $\underline{A}'$. Therefore the boundary condition (3.5.4) ensures the gauge invariance of the theory for an arbitrary magnetic field.

The continuity of the current at any point in space, except the force centers, follows from the Schrödinger equation. It is easy to check that the boundary condition (3.5.4) ensures that the current is also continuous at the singular points of the force centers. If there is an external magnetic field, the current density $\underline{j}$ is given by [54]

$$\underline{j} = \frac{i}{2} \{\psi\nabla\psi^* - \psi^*\nabla\psi + \frac{2i}{c} \psi^*\psi \underline{A}\}. \qquad (3.5.5)$$

Hence the total current flowing into the jth center through surface $S_j$ is

$$\int_{S_j} (\underline{j}\,\underline{n})\, dS = \frac{i}{2} \int_{S_j} \{\psi(\underline{n}\nabla)\psi^* - \psi^*(\underline{n}\nabla)\psi + \frac{2i}{c} \psi^*\psi(\underline{A}\,\underline{n})\} dS. \qquad (3.5.6)$$

The integral on the right-hand side of (3.5.6) is a particular case of the general type (3.5.3), for $\phi_1 = \phi_2 = \psi$. Making use of the previous result concerning (3.5.3) we conclude that (3.5.6) vanishes as the sphere $S_j$ contracts to the jth center. Therefore the current distribution has no singularity at the force centers. For the time-independent wavefunction $\psi$, the current satisfies the continuity condition div $\underline{j} = 0$ over all space.

We note that the gauge invariance of the theory and Hermiticity of the energy operator as well as continuity of the current in the presence of external magnetic fields hold only if the boundary conditions have been modified according to (3.5.4). If the old boundary conditions (1.2.7) had been applied in the case of $\underline{A} \neq 0$, all these properties would have been violated and the theory would not have physical meaning. In this respect, the situation in the ZRP model resembles that in the one-dimensional branching current theory (the metal model). In this model, developed for $\pi$-electrons in aromatic substances, it is also required to modify the matching conditions for wavefunctions and their derivatives at the junctions in order to ensure the gauge invariance of the theory, Hermiticity of the Hamiltonian, and continuity of the current in the presence of magnetic fields (for references see [63, 71]).

Now we shall turn to calculations of the corrections to the energy which are quadratic in the magnetic field strength $H$ [71]. It is sufficient to find an approximate wavefunction which would (i) satisfy the Schrödinger equation with the Hamiltonian (3.5.1) up to terms linear in $H$ and (ii) satisfy exactly the boundary conditions (3.5.4) at each center. Then the magnetic moment $\mu_z$ and magnetic susceptibility $\chi_{zz}$ are given by

$$\mu_z = -\left.\frac{dE}{dH}\right|_{H=0} = -\left.\frac{d\bar{E}_{\tilde{\psi}}}{dH}\right|_{H=0},$$

and

$$\chi_{zz} = -\left.\frac{d^2E}{dH^2}\right|_{H=0} = -\left.\frac{d^2\bar{E}_{\tilde{\psi}}}{dH^2}\right|_{H=0}, \tag{3.5.7}$$

where the energy $\bar{E}_{\tilde{\psi}}$ has been obtained with the help of (3.4.2).

Let us now determine the wavefunction $\tilde{\psi}$ in the case when the vector-potential $\underline{A}$ has been taken in the form

$$\underline{A}(\underline{r}) = \frac{1}{2} H(\underline{e}_z \times \underline{r}). \tag{3.5.8}$$

Then

$$T_{\underline{A}} = \frac{1}{2}\{-\nabla^2 + \frac{i}{c}H\frac{\partial}{\partial\Phi} + \frac{H^2}{4c^2}\rho^2\}, \tag{3.5.9}$$

where $\rho = (x^2 + y^2)^{1/2}$, and the angle $\Phi$ describes the rotation around the z-axis. Apart from the vector-potential $\underline{A}$ given by (3.5.8) we shall consider a family of vector-potentials given as follows:

$$\underline{A}_j(\underline{r}) = \underline{A}(\underline{r} - \underline{R}_j) = \underline{A}(\underline{r}) - \nabla(\underline{A}(\underline{R}_j)\underline{r}). \tag{3.5.10}$$

The energy operator $T_{\underline{A}_j}$ which corresponds to the vector-potential $\underline{A}_j$ has the origin of the coordinate system at the jth center:

$$T_{\underline{A}_j} = \frac{1}{2}\{-\nabla^2 + \frac{i}{c}H\frac{\partial}{\partial\Phi_j} + \frac{H^2}{4c^2}\rho_j^2\}. \tag{3.5.11}$$

In (3.5.11), $\rho_j = (x_j^2 + y_j^2)^{1/2}$ and the angle $\phi_j$ describes the rotation around the $z_j$-axis which passes through the jth center and is parallel to the z-axis.

It is easy to check that the function $\phi_{j\kappa}$, formula (3.1.3), satisfies the equation

$$T_{\underline{A}_j}\, \phi_{j\kappa} = \left(-\frac{\kappa^2}{2} + \frac{H^2}{8c^2}\rho_j^2\right)\phi_{j\kappa}; \tag{3.5.12}$$

in other words, $\phi_{j\kappa}$ is, up to terms linear in $H$, an eigenfunction of the operator $T_{\underline{A}_j}$. Taking into account that the vector-potentials $\underline{A}$ and $\underline{A}_j$ are related to each other via a gauge transformation, formula (3.5.10), it is easy to show that the function

$$g_j(\underline{r}) = \exp\{\frac{i}{c}\,\underline{A}(\underline{R}_j)\,\underline{r}\}\ \phi_{j\kappa} \tag{3.5.13}$$

satisfies the equation

$$T_{\underline{A}}\, g_j = \left(-\frac{\kappa^2}{2} + \frac{H^2}{8c^2}\rho_j^2\right) g_j, \tag{3.5.14}$$

or, up to terms linear in $H$,

$$T_{\underline{A}}\, g_j = -\frac{\kappa^2}{2}\, g_j\ . \tag{3.5.15}$$

Therefore the wavefunction $\tilde{\psi}$ can be sought in the form of a linear combination of functions $g_j$:

$$\tilde{\psi} = \sum_{j=1}^{N} d_j g_j(\underline{r}) = \sum_{j=1}^{N} (d_j/r_j)\,\exp\{\frac{i}{c}(\underline{A}(\underline{R}_j)\,\underline{r}) - \kappa r_j\}. \tag{3.5.16}$$

The coefficients $d_j$ in (3.5.16) can be obtained if we demand that function $\tilde{\psi}$ satisfies exactly the boundary conditions (3.5.4) at each of N centers. This leads to the following system of homogeneous algebraic equations:

$$(\alpha_j - \kappa)d_j + \sum_{k\neq j}^{N} (d_k/R_{jk})\,\exp\{\frac{i}{c}(\underline{A}(\underline{R}_k)\underline{R}_j) - \kappa R_{jk}\} = 0, \tag{3.5.17}$$

which has a non-zero solution provided that

$$\det\left\|(\alpha_j - \kappa)\delta_{jk} + \frac{1-\delta_{jk}}{R_{jk}}\exp\{\frac{i}{c}(\underline{A}(\underline{R}_k)\underline{R}_j) - \kappa R_{jk}\}\right\| = 0. \tag{3.5.18}$$

The phase factors in (3.5.17) and (3.5.10) have a simple physical meaning. Indeed, if we take into account the explicit expression (3.5.8) for the vector-potential $\underline{A}(\underline{r})$, then

$$(\underline{A}(\underline{R}_k)\underline{R}_j) = \frac{1}{2}H((\underline{e}_z \times \underline{R}_k)\,\underline{R}_j) =$$

$$= \frac{1}{2}H((\underline{R}_k \times \underline{R}_j)\,\underline{e}_z) = \Phi_{kj}\,, \tag{3.5.19}$$

where $\Phi_{kj}$ is the magnetic flux through the area bounded by the perimmeter of a triangle whose vertices are at the origin and at centers k and j. The flux is positive if vectors $\underline{R}_k$, $\underline{R}_j$, and $\underline{e}_z$ form a right-handed coordinate system.

Let us denote the real positive roots of equation (3.5.18) as $\kappa_n$, functions $\phi_j$ (3.1.3) corresponding to $\kappa_n$, as $\phi_{jn}$, and sets of coefficients $d_j$ corresponding to $\kappa_n$ and $\phi_{jn}$, as $d_{jn}$. Taking into consideration (3.5.14), we obtain

$$\overline{E}_{\tilde{\psi}} = -\frac{\kappa_n^2}{2} + \frac{H^2}{8c^2}\,\frac{\sum\limits_{j,k}^{N} d^*_{jn}\, d_{kn}\, R^{(n)}_{jk}(\underline{A})}{\sum\limits_{j,k}^{N} d^*_{jn}\, d_{kn}\, N^{(n)}_{jk}(\underline{A})}\,, \tag{3.5.20}$$

where $\overline{E}_{\tilde{\psi}}$ is the average energy of the state $\tilde{\psi}$, and the matrix elements $R_{jk}$ and $N_{jk}$ are expressed thus:

$$R^{(n)}_{jk}(\underline{A}) = \langle f_{jk}(\underline{A})\phi_{jn}|\rho_k^2|\,\phi_{kn}\rangle\,,$$

$$N^{(n)}_{jk}(\underline{A}) = \langle f_{jk}(\underline{A})\phi_{jn}|\,\phi_{kn}\rangle\,,$$

with

$$f_{jk}(\underline{A}) = \exp\{\frac{i}{c}(\underline{A}(\underline{R}_j) - \underline{A}(\underline{R}_k))\cdot\underline{r}\}\,.$$

If we are interested in the terms of order up to $H^2$, the above formula for the average energy can be simplified. The phase factors $f_{jk}$ can be omitted by using $f_{jk}(0)$ and the coefficients $d_j$ can be replaced by coefficients $d_j$ which satisfy equations (3.5.17) without a magnetic field. In this way we obtain

$$\overline{E}_{\tilde{\psi}} = -\frac{\kappa_n^2}{2} + \frac{H^2}{8c^2}\,\frac{\sum\limits_{j,k}^{N} c^*_{jn}\, c_{kn}\, R^{(n)}_{jk}(0)}{\sum\limits_{j,k}^{N} c^*_{jn}\, c_{kn}\, N^{(n)}_{jk}(0)}\,. \tag{3.5.21}$$

Finally, making use of (3.5.7) we find that

$$\mu_{nz} = \left( \kappa_n \frac{d\kappa_n}{dH} \right)_{H=0} \tag{3.5.22}$$

and

$$\chi_{nzz} = \left( \kappa_n \frac{d^2\kappa_n}{dH^2} + \left(\frac{d\kappa_n}{dH}\right)^2 \right)_{H=0} - \frac{1}{4c^2} \frac{\sum_{j,k}^{N} c^*_{jn} c_{kn} R^{(n)}_{jk}(0)}{\sum_{j,k}^{N} c^*_{jn} c_{kn} N^{(n)}_{jk}(0)} . \tag{3.5.23}$$

We remember that $\rho_j^2 = x_j^2 + y_j^2$, where $\underline{r_j}$ determines the position with respect to the jth force center. Derivatives of the energy parameter $\kappa$ with respect to the magnetic field strength $H$ can be obtained with the help of matrix perturbation theory. Sometimes it is possible to find the explicit dependence $\kappa(H)$ from (3.5.18), which, of course, simplifies the calculations of these terms.

Considerable simplifications in the expressions obtained above are possible for linear molecules. The vector-potential can be specified in such a way that it vanishes on the molecular axis itself. Then all quantities $\Phi_{kj}$ in the secular equation (3.5.18) vanish and the roots of the equation $\kappa_n$ become independent of the magnetic field so that the differential terms in (3.5.22) and (3.5.23) vanish altogether.

Rebane and Sharibdzhanov (72) used the exact expression for the magnetic susceptibility obtained in the ZRP model, in order to study a general problem of paramagnetism in molecular systems with closed electron shells. The question whether such a system can be paramagnetic, i.e., possess a positive magnetic susceptibility, had been discussed for a long time without obtaining a conclusive answer. These authors used a three-center ZRP model* of a molecule forming an isosceles triangle, with non-interacting electrons. It was found that a four-electron ** molecule of such structure is paramagnetic,

* Apart from ZRPs, the authors considered a model using an anisotropic oscillator potential, which is also solvable in an analytic form.

** A rigorous theorem states that the ground state of a system of two fermions with spin ½ is always diamagnetic. Therefore the least number of electrons in a closed shell system which could display paramagnetic properties is four.

provided that the triangle does not differ considerably from an equilateral. Further studies of that model enabled general qualitative conditions which are compatible with the paramagnetism of closed electron shells to be determined.

One possible application of this theory to bound state molecular problems is to replace the effective potential corresponding to each occupied molecular orbital by a combination of ZRPs. Then the magnetic susceptibility can be obtained from (3.5.23). Such calculations were carried out by Rebane and Sharibdzhanov (68) for molecular hydrogen. The $H_2$ molecule serves as an example of a system where electrons are not weakly bound and where they move in long-range potentials. In these circumstances, an adequate numerical choice of parameters in the ZRPs becomes a difficult problem. A choice of $\kappa$ from the average one-electron energy does not give good results for the magnetic susceptibility. This is because the wavefunctions $\exp(-\kappa r)/r$ of the ZRP are more compact than the correct wavefunctions of the molecule. It was suggested by Rebane and Sharibdzhanov (68) that semi-empirical methods should be used to correct the ZRP results for $\chi$. One of the ways of doing this is to introduce a correcting factor into $\chi$ and then to determine this factor from the experimental value of the magnetic susceptibility for a one-center system (this approach can be compared with a choice of a normalizing factor made in Sec. 1.3). Very satisfactory results can be obtained if the simple formula (3.5.23) is retained, but the singular orbitals $\exp(-\kappa r)/r$ appearing there are replaced by the usual STOs. It was found in reference (68) that this method gives results for $\chi$ of the same quality as those obtained in quantum mechanical calculations using medium-sized expansions for the wavefunctions.

## 3.6 SOLUTION OF THE SCHRÖDINGER EQUATION WITH THE HELP OF ZRPs

Up to now we have considered problems where, in most cases, the effective potential (of an atomic or molecular system) was not known exactly and the parameter adjustments in the ZRPs had to be made using some general properties of the system such as, for instance, the scattering length. It is of interest to investigate a case where ZRPs are used to build up an approximation to the analytically known exact potential of the system. In such an approach, the ZRP method becomes an approximate method of solving the Schrödinger equation.

We shall follow Subramanyan * (73, 74) and study first a one-dimensional problem where the meaning of the replacement of the

* We also point out a paper by Nakamura (75) where separable potentials were applied to study multi-channel atomic collisions.

exact potential by a sum of δ-function wells is clearer than that in the three-dimensional case. Examples considered show that the approximate energy level in a one-dimensional problem, with a potential V(x) replaced by a sum of δ-function wells, converges to the exact energy level as the number N of ZRPs used to approximate V(x) increases. In the three-dimensional case, the approximate level also tends to the exact one as $N \to \infty$, though the pattern of convergence is more complicated than that in one-dimensional problems.

In the one-dimensional case, a bound state solution of the Schrödinger equation

$$\left( -\frac{1}{2}\frac{d^2}{dx^2} - \sum_{j=1}^{N} \alpha_j \delta(x - X_j) \right) \psi = E\psi \tag{3.6.1}$$

can be written in the form

$$\psi(x) = \sum_{j=1}^{N} c_j \exp(-\kappa|x - X_j|) \ . \tag{3.6.2}$$

We find, using the boundary conditions (1.2.17), that the coefficients $c_j$ satisfy the following system of equations:

$$\kappa c_j = \alpha_j \sum_{i=1}^{N} c_i \exp(-\kappa|X_i - X_j|) . \tag{3.6.3}$$

The parameter κ can be obtained from the secular equation for the system (3.6.3). For a symmetrical position of the ZRPs, the order of this system can be reduced if the parity classification of the states is taken into account.

The simplest case is that of two identical ZRPs separated by distance a. We find that

$$\alpha = \kappa (1 \pm \exp(-\kappa a))^{-1} \ . \tag{3.6.4}$$

For $\alpha a < 1$, there exists a single symmetrical bound state. The second antisymmetrical state only appears when $\alpha a = 1$.

For three ZRPs situated at $X = 0, \pm a$, there are two levels (first and third) of symmetrical bound states, defined by the equation

$$\alpha = 2\kappa \{2 + \exp(-2\kappa a) \pm \exp(-\kappa a)(8 + \exp(-2\kappa a)^{\frac{1}{2}}\}^{-1} \tag{3.6.5}$$

with the signs "+" and "-", respectively, and one level (second) of an antisymmetrical bound state. The latter is defined by the equation

$$\alpha = \kappa\,(1 - \exp(-2\kappa a))^{-1} . \tag{3.6.6}$$

The second level appears if $\alpha a = \frac{1}{2}$, and the third level appears if $\alpha a = \frac{3}{2}$.

Now we shall outline how a continuous potential $V(x) < 0$ can be approximated by a set of ZRPs. Let us divide the interval $-\infty < x < +\infty$ into N subintervals

$$(-\infty, \bar{x}_1),\ (\bar{x}_1, \bar{x}_2), \ldots, (\bar{x}_{N-1}, +\infty) \tag{3.6.7}$$

and introduce the notation

$$V_j^{(n)} = \int_{\bar{x}_{j-1}}^{\bar{x}_j} x^n\, V(x)\, dx. \tag{3.6.8}$$

We shall assume that the subintervals (3.6.7) are chosen in such a way that the integrals

$$V_j^{(0)}, \quad j = 1,2,\ldots, N, \tag{3.6.9}$$

do not depend on j (we have assumed in (3.6.9) that $\bar{x}_0 = -\infty$ and $\bar{x}_N = +\infty$). We shall place a ZRP inside each subinterval at the "center of mass" of the potential, i.e., at the point

$$x_j = V_j^{(1)}/V_j^{(0)} . \tag{3.6.10}$$

The identical depths of all ZRPs will be taken according to

$$\alpha = \alpha_j = V_j^{(0)} . \tag{3.6.11}$$

(It is obvious that the parameters $\alpha_j$, $j = 1,2,\ldots,N$, must assume different values if the equality of all integrals (3.6.9) is not demanded). The choice of this approximate potential appears to be natural though, of course, it is not unique.

As an example let us consider the potential

$$V(x) = -\bar{u}_0/\cosh^2 \lambda x, \tag{3.6.12}$$

for which exact solutions are known to be (54)

$$\varepsilon_n = \chi_n^2 = \{(1 + u_0)^{\frac{1}{2}} - (2n + 1)\}^2,$$
$$\varepsilon_n = -8E_n/\lambda^2, \quad u_0 = 8\,\bar{u}_0/\lambda^2 . \tag{3.6.13}$$

Table 3.1 The critical values of $u_o$ corresponding to the appearance of the nth bound state level (n = 0, 1, 2, 3, and 4) for the potential (3.6.12). N is the number of ZRPs

| n \ N | 2 | 3 | 4 | 5 | Exact solution |
|---|---|---|---|---|---|
| 0 | 0.0 | 0.0 | 0.0 | 0.0 | 0 |
| 1 | 5.77 | 6.28 | 6.68 | 6.96 | 8 |
| 2 | - | 18.85 | 18.53 | 18.83 | 24 |
| 3 | - | - | 42.44 | 40.60 | 48 |
| 4 | - | - | - | 75.13 | 80 |

According to (3.6.13), the (n + 1)th bound state level appears when $u_o^{(n)} = 4n(n + 1)$.

Table 3.1 presents numerical values of $u_o^{(n)}$ obtained in the N ZRP approximation (N = 1, 2, ..., 5) together with the exact values (3.6.13), which correspond to the appearance of the nth bound state level. It can be seen that this approximation to V(x) gives good results for the first two levels taken when N (the number of ZRPs) is small.

For the ground state (n = 0), the exact formula (3.6.13) for the eigenvalue can be written thus:

$$u_o = 2\chi_o + \chi_o^2 ,$$

whereas in the approximation of N zero-range potentials we obtain

$$u_o = 2\chi_o + a_N \chi_o^2 ,$$

$a_N$ being 0.0, 0.6932, 0.8488, and 0.9091 for N = 1, 2, 3, and 4, respectively.

Figure 3.5 shows the dependence of the energy levels upon the parameter $u_o$. The solid curves are those for the ground state and the broken curves are those for the first excited state. In both cases, the dependence derived from the exact solution of the Schrödinger equation with potential (3.6.12) is shown by the bold curve. The numbers marked above the curves are the number of the ZRPs included into the approximation. It can be asserted that, for any level, the agreement between the approximate solution and the exact

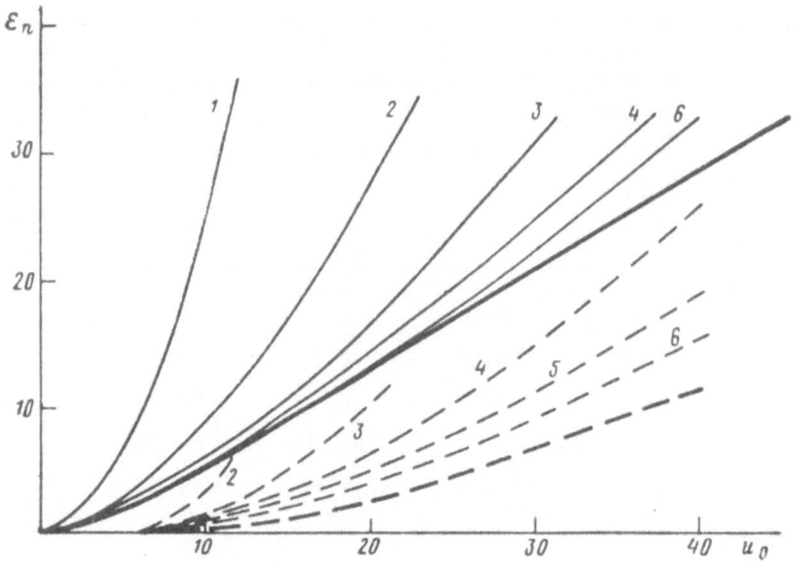


Fig. 3.5. One-dimensional potential (3.6.12) from reference (73). Dependence of eigenvalues $\varepsilon_n$ on the parameter $u_0$ for the ground state n = 0 (solid lines) and the first excited state n = 1 (broken lines). The bold curves represent the exact dependence of $\varepsilon_0$ and $\varepsilon_1$ on $u_0$. The fine curves have a similar meaning for $\varepsilon_0$ and $\varepsilon_1$ obtained in the approximation of N δ-function potentials (N is shown at each curve).

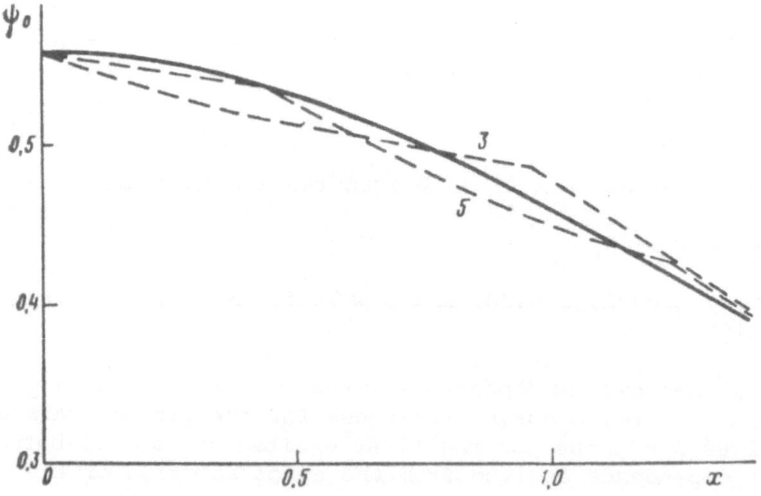


Fig. 3.6. One-dimensional potential (3.6.12). Wavefunction $\psi_0$ of the ground state (from reference (74)). Solid line: the exact wavefunction. Broken lines: approximate wavefunctions obtained using N δ-function potentials (N = 3 and N = 5).

becomes better as N increases.

Keller (76) showed that among all potentials V(x) which satisfy the condition

$$\int V(x)\,dx = D,$$

where D is a fixed real number, the δ-function leads to the lowest eigenvalue of the Schrödinger equation with the corresponding V(x), i.e., in the case of a single δ-function (N = 1), the method gives a lower bound to the exact energy. The present numerical study suggests that this conclusion remains probably true when N > 1 (we shall note that variational calculations with the exact V(x) give upper bounds to the same energy levels).

For the potential V(x) given by (3.6.12), the exact wavefunction of the ground state is

$$\psi_o(x) = c\,(\text{sech}\,x)^{\chi_o/2},$$

where c is a normalization constant. Figure 3.6 compares this exact wavefunction with those obtained in an N δ-function approximation to the potential V(x). All wavefunctions in Fig. 3.6 are normalized and correspond to the same energy level $\varepsilon_o = 1$. For different N, the parameter $u_o$ which gives this eigenvalue is close to the value $u_o = 3$ for the exact potential, (3.6.13), and depends slightly on N.

Let us turn now to the three-dimensional case. It is more difficult here to select "the most natural" way of replacing $V(\underline{r})$ by a sum of ZRPs. There is an infinite number of ways to divide the three-dimensional domain where the potential operates into N sub-domains of equal volume. This ambiguity can be reduced by adding additional conditions, for instance, by requiring the centers of mass to be evenly distributed in the region where the potential acts. The latter condition is convenient because it does not depend upon N.

As before, we divide the whole domain Ω into N sub-domains $\Omega_j$ (j = 1, 2, ..., N) in such a way that the integral

$$V_j^{(o)} = \int_{\Omega_j} V(\underline{r})\,d\underline{r},$$

which is a generalization of (3.6.8) to the three-dimensional case, has the same value independent of j. We introduce the centers of mass $\underline{R}_j$ for $V(\underline{r})$ in sub-domains $\Omega_j$ according to

$$\underline{R}_j = \underline{V}_j^{(1)} / V_j^{(o)} \tag{3.6.14}$$

and we place identical ZRPs at $\underline{R}_j$. Replacement of $V(\underline{r})$ by a pseudopotential (in Fermi's sense (1)) was discussed by Baz' et al. (19), and here we shall give only the final conclusion of that discussion. The parameter $\alpha$, which defines the depth of a three-dimensional ZRP, has to be taken as

$$\alpha = -2\pi N \left( \int_{\Omega} V(\underline{r})\, d\underline{r} \right)^{-1}, \tag{3.6.15}$$

$\Omega$ being the whole domain where the potential operates (this definition assumes that the integral (3.6.15) exists).

Subramanyan (74) considered two examples of $V(\underline{r})$. In the first case, $V(\underline{r})$ was chosen as a spherical potential well:

$$V(\underline{r}) = \begin{cases} -u_0, & \text{if } r \le 1, \\ 0, & \text{if } r > 1. \end{cases} \tag{3.6.16}$$

The sphere $r \le 1$ was divided into 2, 4, and 8 parts of equal volume by one, two, and three planes, respectively, passing through the center of the sphere. For N = 16, the sphere was divided by four planes passing through a common diameter and by one additional plane normal to them. Energy levels in the approximation of N ZRPs were obtained numerically on a computer, from the corresponding transcendental equation. The fine curves in Fig. 3.7 show the dependence of

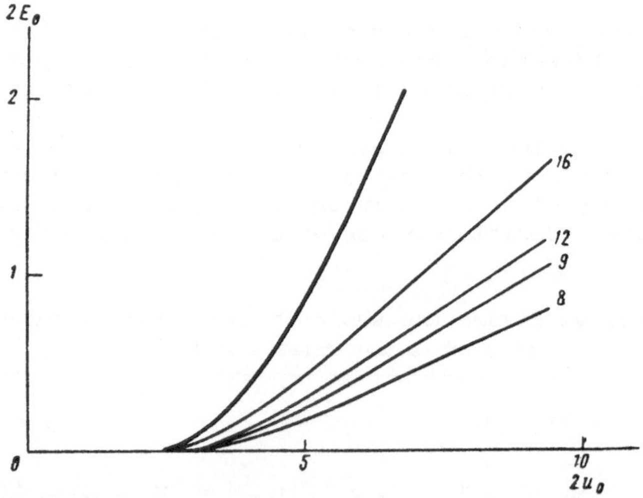


Fig. 3.7. Three-dimensional spherical well (3.6.16) from reference (73). Dependence of the ground state energy $E_0$ on the parameter $u_0$. Bold line: the exact dependence. Fine lines: approximates obtained using N $\delta$-function potentials (N is shown at each curve).

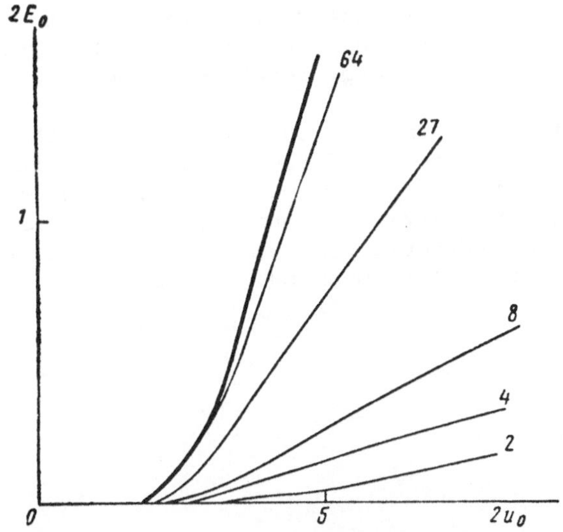


Fig. 3.8. Three-dimensional cubic potential well (3.6.17) from reference (73). Dependence of the ground state energy $E_0$ on the parameter $u_0$. Bold line: variational calculation. Fine lines: approximates obtained using N $\delta$-function potentials (N is shown at each curve).

the energy level on the parameter $u_0$ in this approximation. The bold curve gives the exact dependence of $E_0$ on $u_0$ in this potential.

The second three-dimensional potential investigated in reference (74) was

$$V(\underline{r}) = \begin{cases} -u_0 & \text{inside cube } -1 < x, y, z < 1, \\ 0 & \text{outside this cube.} \end{cases} \qquad (3.6.17)$$

Figure 3.8 presents the results obtained in the N ZRP approximation* (fine curves) together with the result obtained in a variational calculation which employed a wavefunction of the form

$$\tilde{\psi} = c \exp(-\beta^2 r^2/2).$$

Unlike the one-dimensional ZRPs discussed before, the application of three-dimensional ZRPs to approximate the potentials (3.6.16) and

* For the cubic domain (3.6.17), the method of dividing it into N parts is self-evident. Due to the symmetry of the problem the actual number of equations to be solved is greatly reduced. For instance, only four different equations need to be solved when $V(\underline{r})$ in (3.6.17) is replaced by 64 ZRPs (see also Sec. 4.4 below, where a further discussion is given).

(3.6.17) leads to energy estimates which lie higher than the corresponding exact levels. This is similar to the result of variational calculations of the energy level positions.

In the particular examples considered above, the potentials were rather simple functions and the variational method was easier to apply than the ZRP method. For more complicated potentials $V(\underline{r})$, the calculation of the matrix elements required in the variational method may become a difficult problem. In the latter case, the use of ZRPs to approximate $V(\underline{r})$ could be advantageous.

## Chapter 4

# SCATTERING BY A SYSTEM OF ZERO-RANGE POTENTIALS AND THE PARTIAL WAVE METHOD FOR A NONSPHERICAL SCATTERER

## 4.1 THE PARTIAL WAVE METHOD

Before considering the scattering by a system of ZRPs we shall investigate the general case of a scatterer described by a non-spherical potential operator $V(\underline{r})$. We shall assume that V is Hermitian and time independent and that its magnitude falls off rapidly as $r \to \infty$. Under these assumptions stationary scattering of waves or particles is described by the equation

$$\left(-\frac{1}{2}\nabla^2 - \frac{1}{2}k^2 + V\right)\psi = 0. \qquad (4.1.1)$$

In this chapter, it will often be convenient to use the dimensionless scattering amplitude $F_o(\underline{n}_o,\underline{n})$ defined by the asymptotic expression

$$\psi(\underline{r}) \underset{r\to\infty}{=} e^{ik\underline{n}_o\underline{r}} + F_o(\underline{n}_o,\underline{n})\frac{e^{ikr}}{kr} + O(r^{-2}), \qquad (4.1.2)$$

$$\underline{n} = \underline{r}/r,$$

where $\underline{n}_o$ is the unit vector specifying the direction of the incoming wave. The amplitude $F_o$ relates to the conventional amplitude $F(\underline{n}_o,\underline{n})$ according to $F_o = kF$.

The effective cross sections, that is, the differential cross section $\sigma(\underline{n}_o,\underline{n})$, the total cross section $\sigma(\underline{n}_o)$, and the total $\bar{\sigma}$ obtained by averaging $\sigma(\underline{n}_o)$ over all directions of the incoming wave, are defined in the following way:

$$\sigma(\underline{n}_o,\underline{n}) = \frac{1}{k^2}\,|F_o(\underline{n}_o,n)|^2 ,$$

$$\sigma(\underline{n}_o) = \int \sigma(\underline{n}_o,\underline{n})\, d\underline{n} , \qquad (4.1.3)$$

$$\bar{\sigma} = \frac{1}{4\pi}\int \sigma(\underline{n}_o)\, d\underline{n}_o ,$$

where integration is carried out over all directions of the unit vectors $\underline{n}$ and $\underline{n}_o$.

The case of a nonspherical scatterer is encountered in various physical problems described by equation (4.1.1) such as scattering of sound or electromagnetic waves or, for instance, scattering of electrons by a molecule.

It should be noted that the problem (4.1.1) - (4.1.2) stated above is difficult even in the simplest case of the axial symmetry (a diatomic molecule) when the operator V is chosen in such a way that equation (4.1.1) becomes separable in the elliptic coordinates (see, for instance, (77) and further references cited there).

Following Demkov and Rudakov (78), we shall consider the general problem with the help of the method of eigenstates of the S-matrix. This method is a generalization of the partial wave method for scattering by a spherically symmetrical scatterer, and it can be extended to include the treatment of nonstationary problems (79) (see also Sec. 10.2 below).

The wavefunction $\psi$ relates to an eigenfunction of the S-matrix, provided that the amplitude of the scattered wave differs from that of the incoming wave, for all directions $\underline{n}$, by a factor of unit modulus. The said condition can be written in the form

$$\psi_\lambda(\underline{r}) \sim \frac{1}{2ikr}\left[ A_\lambda(-\underline{n})e^{-ikr - i\delta_\lambda} - A_\lambda(\underline{n})e^{ikr + i\delta_\lambda}\right]. \qquad (4.1.4)$$

The functions $\psi_\lambda$ satisfying equations (4.1.3) and (4.1.4) for all k form together with the functions of the bound states a complete set. We shall call $A\lambda(n)$ the scattering eigenamplitudes and $\delta_\lambda(k)$, which are real for a Hermitian operator, the corresponding eigenphases. It is not difficult to show that the translation of the origin through vector $\underline{d}$ does not alter the phases and introduces an additional factor of $\exp(ik\,\underline{n}\cdot\underline{d})$ into the amplitudes.

Let us prove that eigenamplitudes $A_\lambda$ and $A_\mu$, related to different phases $\delta_\lambda$ and $\delta_\mu$, are orthogonal to each other. We shall use the identity

$$\int d\underline{r}\left(\psi_\mu^* \nabla^2\psi_\lambda - \psi_\lambda\nabla^2\psi_\mu^*\right) = 2\int d\underline{r}\left[\psi_\mu^* V\psi_\lambda - \psi_\lambda (V\psi_\mu)^*\right]. \qquad (4.1.5)$$

Taking a sphere of radius b as the domain of integration, we transform the left-hand side of (4.1.5) into a surface integral and demand that $b \to \infty$ so that the asymptotic formula (4.1.4) can be used for $\psi_\lambda$ and $\psi_\mu$. The right-hand side of (4.1.5) vanishes because V is Hermitian. Thus we obtain

$$\frac{1}{k}\sin(\delta_\lambda - \delta_\mu)\int d\underline{n}\; A_\mu^*(\underline{n})\; A_\lambda(\underline{n}) = 0. \qquad (4.1.6)$$

Note that the eigenphases are determined by (4.1.4) up to an arbitrary additive term multiple of $\pi$ only; hence (4.1.6) proves the orthogonality of $A_\mu$ and $A_\lambda$. If there is degeneracy so that several partial waves $\psi_\lambda$ and eigenamplitudes $A_\lambda$ correspond to the same phase $\delta_\lambda$, the amplitudes are then determined up to an arbitrary linear transformation and we can carry out the usual procedure to construct a set of orthogonal amplitudes. These amplitudes will be determined within an arbitrary unitary transformation of the subspace related to the phase $\delta_\lambda$.

We can choose the normalizing factor for $\psi_\lambda$ in such a way that eigenamplitudes $A_\lambda$ satisfy the normalization condition. The resulting functions $A_\lambda(\underline{n})$ form, for each value of k, a complete set of functions orthonormal on the unit sphere.

In the simplest case of a spherically symmetrical scatterer, the wavefunctions

$$\psi_{\ell m} = (kr)^{-1}\, u_\ell(r)\; Y_{\ell m}(\underline{n}) \qquad (4.1.7)$$

are the partial waves $\psi_\lambda$, the spherical harmonics $Y_{\ell m}$ are the eigenamplitudes $A_\lambda$, and the usual phases $\delta_\ell$ determined from the asymptotic form of the radial equation

$$\left(\frac{d^2}{dr^2} + k^2 - \frac{\ell(\ell+1)}{r^2} - 2V(r)\right) u_\ell = 0,$$
$$u_\ell(r) \sim \sin\left(kr - \frac{\ell\pi}{2} + \delta_\ell\right) \qquad (4.1.8)$$

are the eigenphases $\delta_\lambda$. In this particular case degeneracy is due to the spherical symmetry of V and the phases depend on the angular momentum $\ell$ but not on its projection m.

In the general case, if the scatterer possesses an additional group symmetry (the most evident here are various types of point groups such as axial, cubic, etc.), all partial waves may be classified with respect to representations of this group. Degeneracy exists if there are representations of dimension higher than unity. That will always be the case for a non-commutative symmetry group.

Let us express the scattering amplitude $F_0(\underline{n}_0, \underline{n})$ in terms of the eigenamplitudes $A_\lambda(\underline{n})$ and the eigenphases $\delta_\lambda(k)$. To accomplish this we shall write the asymptotic formula (4.1.2) in the form

$$\psi(\underline{r}) \underset{r \to \infty}{\sim} \frac{4\pi}{2ikr} \left[ \delta(\underline{n}_0 + \underline{n})\, e^{-ikr} - \left( \delta(\underline{n}_0 - \underline{n}) - \frac{2i}{4\pi} F(\underline{n}_0, \underline{n}) \right) e^{ikr} \right]. \tag{4.1.9}$$

Expansion of the wavefunction $\psi$ in terms of the partial waves $\psi_\lambda$ is determined by the coefficient of the incoming wave. We have

$$\delta(\underline{n}_0 + \underline{n}) = \sum_\lambda A^*_\lambda(\underline{n}_0)\, A_\lambda(-\underline{n}), \tag{4.1.10}$$

so it follows that

$$\psi(\underline{r}) = 4\pi \sum_\lambda A^*_\lambda(\underline{n}_0)\, e^{i\delta_\lambda}\, \psi_\lambda, \tag{4.1.11}$$

and

$$F(\underline{n}_0, \underline{n}) = \frac{4\pi}{2i} \sum_\lambda (e^{2i\delta_\lambda} - 1)\, A^*_\lambda(\underline{n}_0)\, A_\lambda(\underline{n}). \tag{4.1.12}$$

Finally, for the total and averaged effective cross sections we obtain

$$\sigma(\underline{n}_0) = \frac{(4\pi)^2}{k^2} \sum_\lambda |\, A_\lambda(\underline{n}_0)\, |^2 \sin^2\delta_\lambda ,$$

$$\bar{\sigma} = \frac{4\pi}{k^2} \sum_\lambda \sin^2\delta_\lambda. \tag{4.1.13}$$

All these formulae reduce, in the case of a spherically symmetrical potential, to the conventional expressions of the partial wave method. For a spherically symmetrical scatterer, the subscript $\lambda$ numbering the eigenphases $\delta_\lambda$ can be replaced by a pair of indices: $\ell$ and m having simple physical interpretation.

## 4.2 BEHAVIOR OF THE PHASES AT LOW ENERGY

The question now arises: how to classify the partial waves in the general case of a nonspherical scatterer? One possible method is the inspection of the behavior of the partial waves at low energy, i.e., as $k \to 0$. In this case, the de Broglie wavelength of the particle is greater than the effective size of the scatterer and the results are expected to become, in some respect, nearer to those derived for a spherically symmetrical scatterer. In particular, we shall see that the eigenphases can be classified by the azimuthal quantum number $\ell$, so that the phase $\delta_\lambda \sim k^{2\ell+1}$ as $k \to 0$, and the corresponding eigenamplitude approaches some spherical function $Y_\ell(\underline{n})$.

We shall establish this result by considering the limit of a partial wave $\psi_\lambda$ as $k \to 0$, assuming that the phase $\delta_\lambda$ behaves as

$$\delta_\lambda \sim k^{2\ell_o + 1} .$$

For large r, the potential V can be neglected in (4.1.1) and the function $\psi$ is the general solution of the Laplace equation,

$$\psi \sim \sum_{\ell=0}^{\infty} \sum_{m=-\ell}^{\ell} (c_{\ell m} r^{\ell} + d_{\ell m} r^{-\ell-1}) \, Y_{\ell m}(\underline{n}) . \tag{4.2.1}$$

In the region outside the scatterer, the same solution can be written, for small $k > 0$, in the form of the general solution of the Helmholtz equation

$$(\nabla^2 + k^2)\,\psi = 0 ,$$

that is,

$$\psi \sim \sum_{\ell=0}^{\infty} \sum_{m=-\ell}^{\ell} (C_{\ell m} R_\ell^{(r)}(kr) + D_{\ell m} R_\ell^{(i)}(kr)) \, Y_{\ell m}(\underline{n}) , \tag{4.2.2}$$

where $R_\ell^{(r)}$ and $R_\ell^{(i)}$ are, respectively, the regular and irregular radial parts of the solutions of the Helmholtz equation which are expressible in terms of the Bessel functions of orders $\ell + \frac{1}{2}$ and $-\ell - \frac{1}{2}$:

$$R_\ell^{(r)}(kr) = \left(\frac{\pi}{2kr}\right)^{\frac{1}{2}} J_{\ell+\frac{1}{2}}(kr)$$

and

$$R_\ell^{(i)}(kr) = \left(\frac{\pi}{2kr}\right)^{\frac{1}{2}} J_{-\ell-\frac{1}{2}}(kr) .$$

For small r,

$$R_\ell^{(r)}(kr) \sim \frac{(kr)^\ell}{(2\ell+1)!!}$$

and (4.2.3)

$$R_\ell^{(i)}(kr) \sim \frac{(kr)^{-\ell-1}}{(2\ell-1)!!},$$

whereas for large r they have the form

$$R_\ell^{(r)}(kr) \sim \sin(kr - \frac{\ell\pi}{2})/kr$$

and (4.2.4)

$$R_\ell^{(i)}(kr) \sim \cos(kr - \frac{\ell\pi}{2})/kr.$$

The amplitude $A_\lambda$ is an eigenfunction of the S-matrix if and only if the ratio

$$D_{\ell m}/C_{\ell m} = \tan\delta \tag{4.2.5}$$

does not depend upon $\ell$ and m. This follows from (4.1.4), taking into account the parity $(-1)^\ell$ of the spherical harmonics. It can be seen from formulae (4.2.1) - (4.2.5) that, if $\tan\delta = k^{2\ell_o+1}$, then in the limit $k \to 0$ we can obtain non-vanishing coefficients $c_\ell$ and $d_\ell$ only for $\ell = \ell_o$. Also, in this case, $C_{\ell_o m}$ and $D_{\ell_o m}$ behave as $k^{-\ell_o}$ and $k^{\ell_o+1}$, respectively. For all other spherical harmonics with $\ell \neq \ell_o$, $c_{\ell m} = 0$ if $\ell > \ell_o$, and $d_{\ell m} = 0$ if $\ell < \ell_o$. Thus, in the limit $k \to 0$, the wavefunction must have the following form:

$$\psi_{\ell_o} = \sum_{\ell=0}^{\ell_o} \sum_{m=-\ell}^{\ell} c_{\ell m} r^\ell Y_{\ell m}(\underline{n}) + \sum_{\ell=\ell_o}^{\infty} \sum_{m=-\ell}^{\ell} d_{\ell m} r^{-\ell-1} Y_{\ell m}(\underline{n}). \tag{4.2.6}$$

For $\ell = \ell_o$, the ratio $\gamma_\ell = d_{\ell_o m}/c_{\ell_o m}$ must be independent of m. Then

$$\tan\delta \sim \gamma_\ell k^{2\ell_o+1} + O(k^{2\ell_o+3}). \tag{4.2.7}$$

In other words, if the phase is proportional to $k^{2\ell_o+1}$, then it diminishes too fast, as $k \to 0$, for the harmonics with $\ell < \ell_o$ and in the limit $k = 0$ only the regular solution is retained. For $\ell > \ell_o$, the phase diminishes too slow, and in this limit only the irregular solution is retained. Only for $\ell = \ell_o$ are both solutions retained in the limit $k = 0$.

The inverse is also correct: if the equation $\nabla^2\psi = 2V\psi$ has a solution whose asymptotic behavior is given by (4.2.6), then there exists a partial wave, with an eigenphase $\delta \sim k^{2\ell_o+1}$, which reduces to this solution as $k \to 0$.

Let us show that such solutions of the equation

$$\nabla^2\psi = 2V\psi \tag{4.2.8}$$

do exist, the total number of them, for a given $\ell$, being, in the general case, $2\ell+1$. In order to show this we consider the solutions of (4.2.8) satisfying the integral equation

$$\Phi_{\ell m}(\underline{r}) = r^{\ell} Y_{\ell m}(\underline{n}) - \frac{1}{2\pi}\int dr' \frac{V(\underline{r}')}{|\underline{r}-\underline{r}'|} \Phi_{\ell m}(\underline{r}'). \tag{4.2.9}$$

If the operator V falls off sufficiently rapidly, the iterative process in (4.2.9) will converge and the unique solution of this equation can be constructed. Evidently, this solution has the asymptotic form

$$\Phi_{\ell_o m_o} \sim r^{\ell} Y_{\ell_o m_o}(\underline{n}) + \sum_{\ell=0}^{\infty}\sum_{m=-\ell}^{\ell} g_{\ell m}^{\ell_o m_o} r^{-\ell-1} Y_{\ell m}(\underline{n}). \tag{4.2.10}$$

We can now use these functions to construct successively functions $\phi_{\ell_o m_o}$ whose asymptotic form is given by (4.2.6). Clearly, $\Phi_{oo}$ itself satisfies the condition (4.2.6) for $\ell_o = 0$. Further we have to exclude terms with $\ell < \ell_o$ from the sum over $\ell$ in (4.2.10). We obtain

$$\phi_{oo} = \Phi_{oo},\quad \phi_{1m} = \Phi_{1m} - (g_{oo}^{1m}/g_{oo}^{oo})\,\Phi_{oo},\ \ldots . \tag{4.2.11}$$

For any $\ell_o > 0$, there always exist $\ell_o^2$ functions $\Phi_{\ell m}$ with $\ell < \ell_o$, to be used for elimination of the terms with $\ell < \ell_o$ from the expansion (4.2.10). In this way, we obtain a set of functions $\phi_{\ell m}$ which have the following asymptotic form:

$$\phi_{\ell_o m_o} \sim \sum_{\ell=0}^{\ell_o-1}\sum_{m=-\ell}^{\ell} p_{\ell m}^{\ell_o m_o} r^{\ell} Y_{\ell m}(\underline{n}) + r^{\ell_o} Y_{\ell_o m_o}(\underline{n})$$
$$+ \sum_{\ell=\ell_o}^{\infty}\sum_{m=-\ell}^{\ell} q_{\ell m}^{\ell_o m_o} r^{-\ell-1} Y_{\ell m}(\underline{n}). \tag{4.2.12}$$

These functions do not yet satisfy all conditions since the coefficients with $\ell = \ell_o$ in the regular and irregular solutions are not proportional to each other as formula (4.2.5) requires. We introduce a linear form:

$$\psi_{\ell_o\mu} = \sum_{m=-\ell_o}^{\ell_o} h_{\ell_o m}^{\ell_o\mu} \phi_{\ell_o m} \tag{4.2.13}$$

and demand this proportionality. Thus construction of the required functions has been reduced to diagonalization of the $(2\ell_o + 1)$ order matrix with elements $q_{\ell_o m}^{\ell_o m_o}$ from (4.2.12).

It is not difficult to prove that this matrix is Hermitian and that its eigenvalues, which we denote as

$$\gamma_{\ell_o\mu}, \quad \mu = -\ell_o, -\ell_o+1, \ldots, \ell_o,$$

determine $(2\ell_o+1)$ phases $\delta_{\ell_o\mu} \sim \gamma_{\ell_o\mu} k^{2\ell_o+1}$ and $(2\ell+1)$ scattering eigenamplitudes $A_{\ell_o\mu}$. As $k \to 0$, these amplitudes will tend to the linear combination of spherical harmonics,

$$A_{\ell_o\mu} = \sum_{m=-\ell_o}^{\ell_o} h_{\ell_o m}^{\ell_o\mu} Y_{\ell_o m}(\underline{n}), \tag{4.2.14}$$

whose coefficients are determined by the specific form of the potential V. (In the general case, the index $\mu$ in (4.2.14) does not admit such a simple interpretation as, for example, the projection of the angular momentum).

The classification of phases by $\ell$ in the general case of a non-spherical scatterer, proven above, allows an estimate to be obtained of the number of the phases which contribute significantly, in slow collisions, to the total cross section. If $\rho$ is the effective radius of the scatterer, it follows that $\ell_{max} = k\rho$ determines the maximum value of the angular momentum which gives an appreciable contribution to the cross section. The phases $\delta_\ell$, $\ell \leq \ell_{max}$, will be significantly different from zero, whereas the phases $\delta_\ell$, $\ell > \ell_{max}$, will be small, with their contribution to the scattering rapidly diminishing as $\ell$ increases. For $k\rho \ll 1$, the scattering is dominated by $\ell = 0$ and the problem reduces to finding a solution of (4.2.8) with the asymptotic form

$$\psi_o \sim 1 - \frac{a}{r} + O(r^{-2}) . \tag{4.2.15}$$

Alternatively, an equivalent integral equation (4.2.9) with $\ell = 0$, that is,

$$\psi_o(\underline{r}) = 1 - \frac{1}{2\pi} \int d\underline{r}' \frac{V(\underline{r}')\,\psi_o(\underline{r}')}{|\underline{r} - \underline{r}'|}, \tag{4.2.16}$$

can be used to obtain the solution.

The scattering length a,

$$a = \frac{1}{2\pi}\int d\underline{r}\, V(\underline{r})\, \psi_o(\underline{r}) , \qquad (4.2.17)$$

determines the limit of the effective cross section,

$$\sigma\Big|_{k \to 0} = 4\pi a^2, \qquad (4.2.18)$$

with the scattering becoming isotropic as $k \to 0$.

We do not consider here the special case where one of the coefficients $\gamma_\ell$ vanishes. In this case the relevant phase is, in fact, proportional to the $(2\ell + 3)$th or higher power of k. The resonance formed when a bound state moves up to the continuum boundary is another special case. The integral equation (4.2.9) has no solution because, in the latter case, the corresponding homogeneous equation has a non-trivial solution. Consequently, $\tan\delta_\ell$ tends to zero at a rate slower than $k^{2\ell+1}$.

## 4.3 THE VARIATIONAL PRINCIPLE

We shall now state the variational principles leading to a direct determination of the partial waves and phases. For the sake of simplicity we restrict ourselves to considering a local potential energy operator $V(\underline{r})$. Let us look for the solution of (4.1.1), with the boundary conditions (4.1.4), in the form

$$\psi_\lambda(\underline{r}) = \int d\underline{r}'\, \chi_\lambda(\underline{r}')\, \frac{\sin(k|\underline{r} - \underline{r}'| + \delta_\lambda)}{k|\underline{r} - \underline{r}'|} . \qquad (4.3.1)$$

Consider the asymptotic behavior of the right-hand side of (4.3.1) under the assumption that $V(\underline{r})$ falls off sufficiently rapidly as $r \to \infty$. It is easy to verify that the condition (4.1.4) is fulfilled for any function $\chi_\lambda$, the scattering amplitude being of the form

$$A_\lambda(\underline{n}) = -\int d\underline{r}'\, e^{ik\,\underline{n}\cdot\underline{r}'}\, \chi_\lambda(\underline{r}') . \qquad (4.3.2)$$

Substituting (4.3.1) into (4.1.1), we obtain an equation which has to be satisfied by $\chi_\lambda(\underline{r})$:

$$\frac{2\pi}{V(\underline{r})}\chi_\lambda(\underline{r}) + \int d\underline{r}'\, \frac{\cos(k|\underline{r} - \underline{r}'|)}{|\underline{r} - \underline{r}'|}\, \chi_\lambda(\underline{r}') =$$

$$= \cot \delta_\lambda \int d\underline{r}' \frac{\sin(k|\underline{r} - \underline{r}'|)}{|\underline{r} - \underline{r}'|} \chi_\lambda(\underline{r}'). \tag{4.3.3}$$

Making use of this integral equation and identifying $-\cot \delta_\lambda$ as eigenvalues, we obtain a functional whose stationary values are

$$-\left[\cot \delta_\lambda\right] = \frac{A}{B}, \tag{4.3.4}$$

where

$$A = 2\pi \int dr \frac{|\chi_\lambda(\underline{r})|^2}{V(\underline{r})} + \iint d\underline{r}\, d\underline{r}'\, \chi_\lambda^*(\underline{r}) \frac{\cos(k|\underline{r} - \underline{r}'|)}{|\underline{r} - \underline{r}'|} \chi_\lambda(r')$$

and

$$B = \iint d\underline{r}\, d\underline{r}'\, \chi_\lambda^*(\underline{r}) \frac{\sin(k|\underline{r} - \underline{r}'|)}{|\underline{r} - \underline{r}'|} \chi_\lambda(\underline{r}') .$$

The function $\chi_\lambda$ can be called, in accordance with (4.3.1), the source distribution function for a partial wave $\psi_\lambda$. Evidently, $\chi_\lambda$ has to vanish at the points where the potential $V(\underline{r})$ is zero, and it has to fall off sufficiently rapidly as $r \to \infty$ in order to ensure the convergence of the first integral in the numerator A of (4.3.4).

With the help of (4.3.2) the denominator B of (4.3.4) can be written in the form

$$B = \iiint d\underline{r}\, d\underline{r}'\, d\underline{n}\, \chi_\lambda^*(\underline{r})\, e^{ik\underline{n}\cdot(\underline{r} - \underline{r}')} \chi_\lambda(\underline{r}')$$

$$= \int d\underline{n}\, |A_\lambda(\underline{n})|^2 . \tag{4.3.5}$$

It follows from (4.3.5) that B is non-negative. Had the denominator B been positive-definite, the variational principle (4.3.4) would have yielded lower or upper bounds for the phase because $-\cot \delta_\lambda$ itself is lower- or upper-bounded, depending on whether $V(\underline{r})$ is, respectively, positive or negative. However, in reality, it is possible to construct such a source function $\chi(\underline{r})$ that it cancels all scattered waves and the amplitude $A(\underline{n})$ vanishes in any direction $\underline{n}$. For this it is sufficient (and necessary) to choose a function $\chi_\lambda(\underline{r})$ whose Fourier transform would vanish on a sphere of radius k in the momentum space. Therefore the operator with the kernel $\sin(k|\underline{r} - \underline{r}'|) \times |\underline{r} - \underline{r}'|^{-1}$ is similar to a projection operator and, consequently, we do not obtain, in the general case, a bounded variational principle for the phase. However, in certain special cases, the positive-definite property of B and therefore the boundedness of the variational principle (4.3.4) can be ensured, as we shall see below.

In the spherically symmetrical case, the variational principle (4.3.4) reduces to the well-known variational principle of Schwinger [35]. Notice that the potential $V(\underline{r})$ enters only the one-dimensional integral in A. This facilitates direct calculations by this method with various choices of the potential V.

Let us now formulate a variational principle analogous to that of Hulthén for a spherically symmetrical V. In order to do this we consider the functional

$$J(\Phi_1,\Phi_2) = \int d\underline{r}\ \Phi_2^* (\nabla^2 + k^2 - 2V)\ \Phi_1, \tag{4.3.6}$$

where the asymptotic form of the functions $\Phi_1$ and $\Phi_2$ is given by (4.1.4) with the normalized amplitudes $A_1(\underline{n})$, $A_2(\underline{n})$, and phases $\delta_1$, $\delta_2$, respectively. These amplitudes and phases are not required to be identical with the corresponding quantities for the exact partial waves.

We write the functions $\Phi_1$ and $\Phi_2$ in the form $\Phi_1 = \psi_\lambda + \Delta\Phi_1$, $\Phi_2 = \psi_\lambda + \Delta\Phi_2$, where $\psi_\lambda$ is the exact partial wave with amplitude $A_\lambda(\underline{n})$ and phase $\delta_\lambda$, and substitute these expressions into the functional (4.3.6). After standard manipulations, we have

$$J = \frac{1}{k}\sin(\delta_\lambda - \delta_1) \int d\underline{n}\ A_\lambda^*(\underline{n})\ A_1(\underline{n}) + \int d\underline{r}\ \Delta\Phi_2^*\ (\nabla^2 + k^2 - 2V)\ \Delta\Phi_1. \tag{4.3.7}$$

Assuming that $\Delta\Phi_1$ and $\Delta\Phi_2$ are small and setting $\delta_1 - \delta_\lambda = \Delta_1\delta$, we arrive at the Hulthén variational principle:

$$\delta J = -k^{-1}\ \Delta_1\delta\ . \tag{4.3.8}$$

The phase can be obtained as the stationary value of the functional

$$|\delta| = \delta_1 + kJ(\Phi_1,\Phi_2)\ . \tag{4.3.9}$$

Various results which follow from the Hulthén variational principle for a spherically symmetrical V can be reproduced making use of the variational principle derived above. For instance, by varying the scale, we obtain the virial theorem in a way similar to that in reference [80]:

$$\frac{d\delta}{dk} = 2\int d\underline{r}\,\psi_\lambda^* \,(2V + \underline{r}\cdot\nabla V)\,\psi_\lambda . \tag{4.3.10}$$

By adding a small perturbation $V_1$ to the potential V, we readily obtain, from the variational principle (4.3.9), the first-order perturbation theory correction to the phase,

$$\Delta^{(1)}\delta_\lambda = -2k\int d\underline{r}\,\psi_\lambda^* V_1 \psi_\lambda . \tag{4.3.11}$$

The formula (4.3.11) is correct, however, only in the absence of degeneracy. In the latter case, the matrix $V_1$ has to be diagonalized in the subspace of the degenerate functions $\psi_\lambda$, exactly in the same way as is usually done in stationary perturbation theory for energy levels.

The above result leads directly to the von Neumann-Wigner theorem for the phases; if two phases $\delta_1$ and $\delta_2$ coincide with each other (or differ by a multiple of $\pi$) at certain k, this requires an additional condition to be fulfilled. A small perturbation of a sufficiently general form will then result in the change from a phase crossing to a phase pseudocrossing. The additional condition is satisfied automatically and a crossing is possible provided that the two partial waves and phases belong to different representations of the symmetry group for the scatterer. This holds, for example, for the phases $\delta_\ell$ of a spherically symmetrical problem.

We see therefore that the method of partial waves for a nonspherical scatterer retains many features of that for a spherically symmetrical scatterer. The chief advantage of the method is that asymmetry associated with the incident plane wave is taken into account at the last stage of the calculations (for $\bar{\sigma}$, this asymmetry does not show at all). This may be used particularly to take account of the symmetry of the scatterer in the most natural way.

The actual calculation of the partial waves for a specific potential $V(\underline{r})$ is quite a difficult problem. However, the difficulties encountered exhibit the complex nature of the problem and do not imply that the method itself is poor. Calculations of partial waves are similar to those of the energy levels of bound states, and the method shows certain symmetry in the treatment of the bound states (negative energy) and scattering (positive energy). The existence of variational principles for the direct determination of the phases and partial waves also demonstrates this symmetry.

In analogy with the spherically symmetrical case there are reasons to believe that for slow collisions (i.e., where the wavelength is greater than the size of the scatterer) the method of partial waves is most convenient to use to calculate scattering by nonspherical systems.

## 4.4 SCATTERING BY A SYSTEM OF ZRPs

It is particularly easy to formulate the method of partial waves for a scatterer consisting of N zero-range potentials. We shall look for the partial wave $\psi_\lambda$ in the following form of the solution of the Schrödinger equation for a free particle:

$$\psi_\lambda = \sum_{j=1}^{N} c_j \frac{\sin(k|\underline{r} - \underline{R}_j| + \delta)}{|\underline{r} - \underline{R}_j|}, \tag{4.4.1}$$

i.e., the function $\chi$ in (4.3.1) has the form $\chi = \sum_j c_j\, \delta(\underline{r} - \underline{R}_j)$. Requiring the function (4.4.1) to satisfy the boundary conditions (1.2.7), we obtain the system of homogeneous equations for the coefficients $c_j$:

$$\sum_{\substack{i \\ i\neq j}}^{N} c_i \frac{\sin(kR_{ij} + \delta)}{R_{ij}} + c_j(k\cos\delta + \alpha_j \sin\delta) = 0. \tag{4.4.2}$$

With the notation

$$M_{ij} = \frac{\cos kR_{ij}}{R_{ij}}, \qquad L_{ij} = \frac{\sin kR_{ij}}{R_{ij}}, \quad \text{for } i \neq j,$$

and (4.4.3)

$$M_{ii} = \alpha_i, \qquad L_{ii} = k,$$

the eigenphases can be obtained from the secular equation

$$\det\| M_{ij} + \cot\delta\, L_{ij}\| = 0. \tag{4.4.4}$$

Thus there are only N non-vanishing phases (this conclusion remains valid for any separable potential of the form (1.4.2)).

The matrix $L_{ij}$ is positive-definite.* Hence it follows that $\sin\delta_\lambda$ does not vanish if $k \neq 0$. Therefore each phase $\delta_\lambda$ is defined in the interval $0 < \delta_\lambda < \pi$ (this differs, for instance, from the spherically symmetrical case, which cannot be realized with the help of a finite set of ZRPs). This conclusion is closely related to

---

* The quadratic form $\sum_{i,j} L_{ij} c_i c_j$ is the analog to the denominator B in the expression (4.3.4) for the stationary values. It is evident that, by varying the intensities of the sources $c_i$ at points $\underline{R}_i$, it is not possible to ensure the cancellation (on account of interference) of the scattered wave in all directions $\underline{n}$, and therefore the positive-definiteness does not take place in this particular case.

the Levinson theorem [22,38]; the sum of all phases, at $k = 0$, is $s\pi$, where s is the number of bound states of the system. Thus, we obtain a one-to-one correspondence between N partial phases and N possible bound states. If a particular bound state really exists, then the corresponding phase is $\pi$ at $k = 0$. If, however, the bound state does not exist, that is, the pole of the Green's function is on a non-physical sheet of the complex energy plane, then the corresponding phase vanishes at $k = 0$.

Let us transform the linear equations (4.4.2) - (4.4.4) to obtain a compact expression for the averaged total effective cross section $\overline{\sigma}$. We have

$$\begin{aligned} MC &= -LC\cot\delta, \\ L^{-1/2}ML^{-1/2}C' &= DC' = -C'\cot\delta, \\ (1 + D^2)^{-1}C' &= C'\sin^2\delta, \\ L^{-1/2}(M + iL)^{-1}L(M - iL)^{-1}L^{1/2}C' &= C'\sin^2\delta. \end{aligned} \tag{4.4.5}$$

Then

$$\overline{\sigma} = \frac{4\pi}{k^2}\,\mathrm{Sp}\left[L(M + iL)^{-1}L(M - iL)^{-1}\right]. \tag{4.4.6}$$

Formula (4.4.6) is a generalization of the simple resonance formula (1.3.5) for a single ZRP well.

Therefore the problem has been reduced to the following stages: (i) inversion of the complex symmetrical matrix P of order N,

$$P = M + iL,\quad P_{ij} = \exp(ik R_{ij})/R_{ij},\ \text{for } i \neq j,$$

and

$$P_{jj} = \alpha_j + ik, \tag{4.4.7}$$

(ii) multiplication of the matrices L, $P^{-1}$, $(P^{-1})^*$, and, finally, (iii) the calculation of the trace of the product of the matrices.

If k is large enough so that all elements $kR_{ij} \gg 1$ and $\alpha_i R_{ij} \gg 1$, then P is near to a diagonal matrix, i.e.,

$$P_{ij} \approx (\alpha_i + k)\,\delta_{ij}.$$

In this case, formula (4.4.6) gives

$$\overline{\sigma} = 4\pi \sum_{j=1}^{N} (k^2 + \alpha_j^2)^{-1}. \tag{4.4.8}$$

In other words, the scattering on all N centers is independent. Hence the multiple scattering (for details see Sec. 4.5) can be neglected, provided that the de Broglie wavelength of the incident electron is small in comparison to the distance between the scatterers. The condition $\alpha_i R_{ij} \gg 1$ for the discrete spectrum corresponds to strong interaction between the electrons and the centers so that the wavefunctions of individual electrons hardly overlap each other.

In the case of a high-order symmetry possessed by the scatterer (i.e., by the system of ZRPs) coefficients $c_j$ in (4.4.2) can be obtained immediately from symmetry considerations. As a simple example, a system of two identical ZRPs has solutions S(+1,+1) and P(+1,-1). For a system of eight identical ZRPs placed at the vertices of a cube we have one monopole solution S of the type (+1,+1,+1,+1,+1,+1,+1,+1), three dipole solutions P of the type (+1,+1,+1,+1,-1,-1,-1,-1), three quadrupole solutions D of the type (+1,+1,-1,-1,-1,-1,+1,+1), and one octopole solution F of the type (+1,-1,+1,-1,+1,-1,+1,-1). The numeration of the vertices is explained in Fig. 4.2.

If we denote the distance between two neighboring potentials by R, measure the cross section in units of $R^2$, and measure k and $\alpha$ in units $R^{-1}$, then we obtain immediately, without any intermediate calculations, the averaged cross section for two identical wells:

$$\bar{\sigma} = \frac{4\pi}{k^2}\left\{\left[1 + \left(\frac{\alpha + \cos k}{k + \sin k}\right)^2\right]^{-1} + \left[1 + \left(\frac{\alpha - \cos k}{k - \sin k}\right)\right]^{-1}\right\}. \tag{4.4.9}$$

For the cubic configuration of the ZRPs considered above, the averaged cross section is

$$\bar{\sigma} = \frac{4\pi}{k^2} \times (\sigma_s + \sigma_p + \sigma_d + \sigma_f), \tag{4.4.10}$$

where

$$\sigma_s = \left[1 + \left(\frac{\alpha + 3\cos k + (3/\sqrt{2})\cos k\sqrt{2} + (1/\sqrt{3})\cos k\sqrt{3}}{k + 3\sin k + (3/\sqrt{2})\sin k\sqrt{2} + (1/\sqrt{3})\sin k\sqrt{3}}\right)^2\right]^{-1},$$

$$\sigma_p = 3\left[1 + \left(\frac{\alpha + \cos k - (1/\sqrt{2})\cos k\sqrt{2} - (1/\sqrt{3})\cos k\sqrt{3}}{k + \sin k - (1/\sqrt{2})\sin k\sqrt{2} - (1/\sqrt{3})\sin k\sqrt{3}}\right)^2\right]^{-1},$$

$$\sigma_d = 3\left[1 + \left(\frac{\alpha - \cos k - (1/\sqrt{2})\cos k\sqrt{2} + (1/\sqrt{3})\cos k\sqrt{3}}{k - \sin k - (1/\sqrt{2})\sin k\sqrt{2} + (1/\sqrt{3})\sin k\sqrt{3}}\right)^2\right]^{-1},$$

$$\sigma_f = \left[1 + \left(\frac{\alpha - 3\cos k + (3/\sqrt{2})\cos k\sqrt{2} - (1/\sqrt{3})\cos k\sqrt{3}}{k - 3\sin k + (3/\sqrt{2})\sin k\sqrt{2} - (1/\sqrt{3})\sin k\sqrt{3}}\right)^2\right]^{-1}.$$

are the monopole, dipole, quadrupole, and octopole contributions, respectively, to the total averaged cross section. The resonance peaks in (4.4.10) are at the values of k which correspond simultaneously to a vanishing numerator and small denominator in the same bracketted fraction. Then the phase is $\pi/2$, and the contribution from the partial wave in question is the largest one. The resonances in the higher-order multipoles at small k are especially narrow. The denominators in (4.4.10) are proportional to $k^{2\ell+1}$. Therefore the width of a resonance situated at $k = k_0$ (small) is $\Delta k \sim k_0^{2\ell+1}$ similarly to that of the spherically-symmetrical case considered in Chapter 2.

The trigonometric functions in formulae (4.4.9) and (4.4.10) for the cross sections $\bar{\sigma}$ cause oscillations of the cross sections. These oscillations are related to diffraction which takes place when the distance between the potentials is a multiple of the de Broglie wavelength. Figure 4.2 shows the cross section defined by (4.4.10) for $\alpha = 0.2$. Contributions from individual partial waves are shown in Fig. 4.1.

Systems of non-overlapping scatterers (clusters of atoms) are known to be of considerable significance in solid state theory, where the muffin-tin potential method has been developed [81] as well as in the theory of molecules, where it is known as the X-$\alpha$ method [82]. John and Ziesche [83,84] have generalized the equations of the theory of partial waves (4.4.2) - (4.4.6) to the case when each center scatters not only the s-wave but also higher-momentum partial waves. Recent calculations show some promising results for the s- and p-wave electron scattering by a system of two carbon atoms [85], and they give a good illustration of general features of the theory developed above and, in particular, of the theorem stating that the phases of the amplitudes of the same symmetry never cross each other.

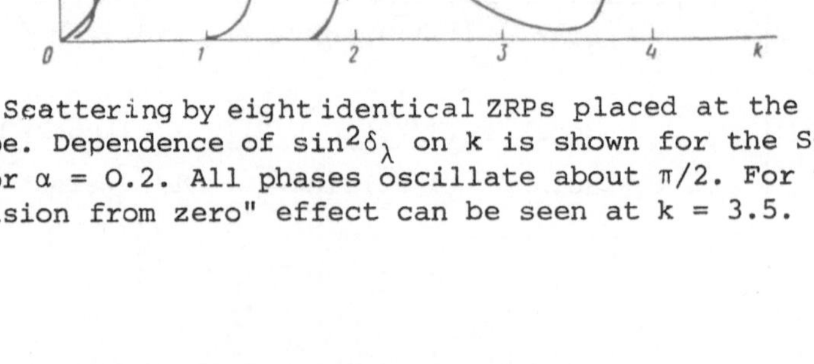

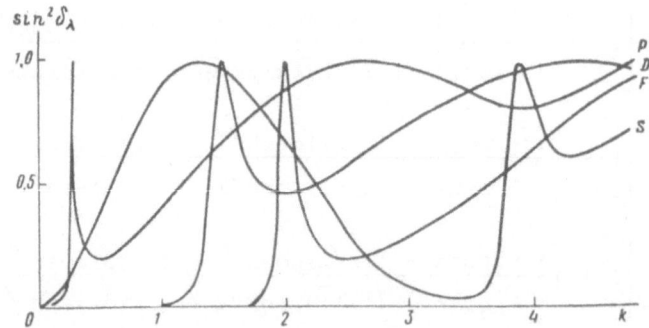

Fig. 4.1. Scattering by eight identical ZRPs placed at the vertices of a unit cube. Dependence of $\sin^2\delta_\lambda$ on k is shown for the S-, P-, D-, and F-waves for $\alpha = 0.2$. All phases oscillate about $\pi/2$. For the S-phase the "repulsion from zero" effect can be seen at k = 3.5.

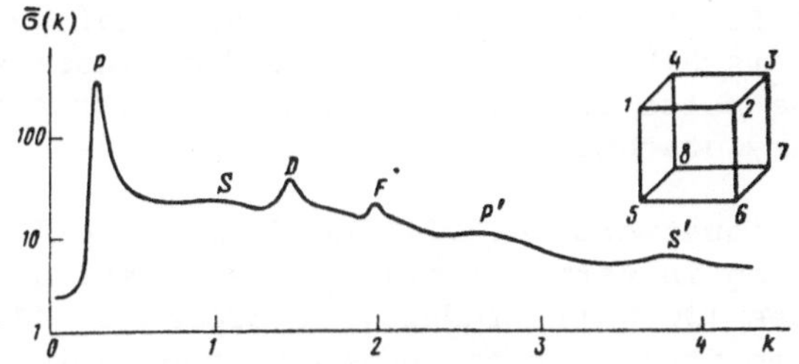


Fig. 4.2. Dependence of the total averaged scattering cross section $\bar{\sigma}(k)$ on k [78] for a cubic configuration of zero-range scatterers shown at the top of the diagram. Resonances in $\bar{\sigma}(k)$ are identified in terms of the relevant partial waves.

## 4.5 ZRPs IN THE THEORY OF MULTIPLE SCATTERING

Now we shall consider a somewhat different approach to the same problem of scattering by a system of ZRPs. Let us seek a solution of the scattering problem in the form

$$\psi = e^{ik\underline{n}_o \cdot \underline{r}} + \sum_{j=1}^{N} c_j \frac{e^{ik|\underline{r} - \underline{R}_j|}}{|\underline{r} - \underline{R}_j|} . \tag{4.5.1}$$

As usual, we shall obtain a system of equations for coefficients $c_j$ in (4.5.1) from the boundary conditions. Introducing quantities

$$F_j = c_j \exp(-ik\underline{n}_o \cdot \underline{R}_j)$$

instead of $c_j$, we obtain the following equations for $F_j$:

$$\begin{aligned} F_j &= f_j + f_j \sum_{i \neq j} F_i \frac{e^{ikR_{ij}}}{R_{ij}} e^{ik\underline{n}_o \cdot \underline{R}_{ij}} , \\ \underline{R}_{ij} &= \underline{R}_i - \underline{R}_j , \end{aligned} \tag{4.5.2}$$

where $f_j = -(\alpha_j + ik)^{-1}$ is the scattering amplitude for scattering by an isolated jth center. Considering the asymptotic form of the wavefunction, we obtain the amplitude for scattering by the whole system, in the form

$$\begin{aligned} F(\underline{k}_o,\underline{k}) &= \sum_j F_j(\underline{k}_o) e^{i(\underline{k}_o - \underline{k}) \cdot \underline{R}_j} , \\ \underline{k}_o &= k\underline{n}_o , \\ \underline{k} &= k\underline{n}. \end{aligned} \tag{4.5.3}$$

The system (4.5.3) for $F_j$ is a non-homogeneous analogy of the system (3.1.3). Both systems have the same determinant whose zeros specify the position of the poles of the S-matrix, i.e., the bound, virtual, and quasi-stationary states.

The system of equations (4.5.2) and formula (4.5.3) allow a simple interpretation in terms of multiple scattering (see, for instance, references [86-88]). Indeed, according to (4.5.3), the total scattering amplitude corresponding to the whole system is the sum of the scattering amplitudes by the individual centers. The coefficients $F_j$ include translational factors which take care of the phase difference due to the space separation between the scatterers. Equation (4.5.2) for the amplitudes $F_j$ takes into account that, for each scatterer, the total incoming wave is the sum of the primary incoming plane wave and the secondary outgoing spherical waves generated by other scatterers of the system. In the case of the ZRPs, the waves travelling between the centers of the system are free waves, and their departure from spherical waves is not essential.

The theory of multiple scattering considers various generalizations of equation (4.5.2). One of them is to consider, for each individual scatterer, apart from s-wave scattering, higher-order partial waves [88]. A generalization to include the many-channel case as well as the case where both the incoming particle and the scatterer possess spin will be discussed in the next chapter.

Let us now consider scattering by a system of separable potentials. An interesting feature of this problem is that the resulting equations do not become more complicated than the original ZRP equations (4.5.2), though the overlap between wavefunctions of different centers cannot be neglected in the case of separable potentials. Let us introduce, as before, vectors $\underline{R}_j$ describing the position of an individual scatterer with the potential $|\phi_j\rangle v_j \langle\phi_j|$, and seek the wavefunction in the form

$$|\psi\rangle = |e^{i\underline{k}_o\cdot\underline{r}}\rangle + \sum_j \frac{F_j}{\eta_j(\underline{k})} e^{i\underline{k}_o\cdot\underline{R}_j} |\chi_j\rangle . \tag{4.5.4}$$

The total scattering amplitude is given by (4.5.3), as before, but now the coefficients $F_j$ are functions of $\underline{k}$. For a re-defined $F_j$, we obtain

$$F_j = f_j + \frac{f_j}{\langle\phi_j| e^{i\underline{k}_o\cdot\underline{r}}\rangle} \sum_{i\neq j} F_i \frac{\langle\phi_j|\chi_i\rangle}{\eta_i(\underline{k})} e^{i\underline{k}_o\cdot\underline{R}_{ij}} . \tag{4.5.5}$$

As before, $f_j$ in (4.5.5) has the meaning of the scattering amplitude for an isolated jth center placed at the origin of the coordinate system. We have

$$f_j = - \frac{v_j\langle\phi_j|e^{i\underline{k}_o\cdot\underline{r}}\rangle}{1 + v_j\langle\phi_j|\chi_j\rangle} e^{-i\underline{k}_o\cdot\underline{R}_j} \eta_j(\underline{k}) . \tag{4.5.6}$$

In (4.5.5), instead of the original factor $e^{ikR_{ij}}/R_{ij}$ for the scattered spherical wave propagating from kth to jth centers, now we have the more complicated expression

$$\eta_i(\underline{k})^{-1}\langle\phi_j|\chi_i\rangle = \eta_i(\underline{k})^{-1} \langle\phi_j|(H_o - E)^{-1}|\phi_i\rangle . \tag{4.5.7}$$

The factor (4.5.7) takes into account both the finite size of the scattering centers and the overlap between them. A factor $\langle\phi_j|\exp(i\underline{k}_o\cdot\underline{r}\rangle^{-1}$ in (4.5.5) accounts for the difference of the actual wave falling on the jth center and the plane wave assumed in the ZRP theory.

In conclusion, we note that, for a proper choice of functions $\phi_j$, the equation (4.5.5) includes also the case of nonspherical separable scatterers.

# Chapter 5

# ZERO-RANGE POTENTIALS IN MULTI-CHANNEL PROBLEMS

## 5.1 ZERO-RANGE POTENTIALS FOR A MANY-COMPONENT WAVEFUNCTION

The theory developed in the previous chapter can be directly applied to describe electron scattering by molecules. Subramanyan [89] considered e + $H_2$ scattering in the approximation of a two-center ZRP model. In the simplest case, the parameter $\alpha$ for an isolated hydrogen atom was determined from the scattering length computed in the static approximation without exchange. We have already pointed out in Chapter 1 that in the approximation without exchange, the hydrogen atom has two different (singlet and triplet) scattering lengths (a similar situation arises in all atoms with non-zero total spin). Therefore the spin state of the system may influence scattering considerably. Within the ZRP model, an account of spin was first made by Smirnov and Firsov [17]. Further development of the theory was reported in [90,91] and also in [64,86,87].

A general restriction on the class of problems suitable for treatment by the ZRP method in the form presented above is that only elastic scattering can be considered. In particular, this means that the region of high energies lies beyond the scope of the theory.

In the present chapter, we shall consider a natural generalization of the method by introducing many-component wavefunctions. This allows, in particular, the restriction on the original method to be partly removed as well as a (phenomenological) description of the structure of the centers (spin, excited states, etc.) to be introduced. A similar approach to nuclear problems (however, only for one force center of spherical symmetry) was outlined by Dalitz [92].

The many-channel formulation of the theory makes it possible to consider inelastic scattering as well as to study many other aspects such as the influence of the inelastic channel on the elastic one, threshold phenomena, the transition of bound states into quasi-stationary states, the passage of the S-matrix poles from one sheet of the complex energy plane onto another, the meaning of various resonances, etc.

Let us assume that in the M-component wavefunction $\psi$,

$$\psi = \begin{pmatrix} \psi_1 \\ \psi_2 \\ \cdot \\ \cdot \\ \cdot \\ \psi_M \end{pmatrix}, \tag{5.1.1}$$

each component satisfies the Schrödinger equation for a free particle:

$$\left( -\frac{1}{2}\nabla^2 + \frac{1}{2}k_\beta{}^2 \right) \psi_\beta(\underline{r}) = 0. \tag{5.1.2}$$

The boundary conditions are imposed at points $\underline{R}_j$, $j = 1,2,\ldots,N$, and they have the form

$$\psi_\beta(\underline{r}) \Big|_{r_j \to 0} \sim \frac{c_{\beta j}}{r_j} - b_{\beta j} + O(r_j), \quad \underline{r}_j = \underline{r} - \underline{R}_j, \tag{5.1.3}$$

$$b_{\beta j} = \sum_{\gamma=1}^{M} \alpha_{\beta\gamma}^{(j)} c_{\gamma j},$$

where in the coefficients $b_{\beta j}$, $c_{\beta j}$ the first (Greek) subscript numbers the component of the wavefunction (i.e., it specifies the channel), whereas the second (Latin) subscript is used to number the scattering center. Therefore the j-th scatterer requires now an $M \times M$ matrix of parameters $\alpha_{\beta\gamma}^{(j)}$ (in a one-channel theory, only one parameter was required for each center). The energy conservation law connects the wavenumbers $k_\beta$ in different channels:

$$k_\beta{}^2 = k_1{}^2 - 2\,\varepsilon_\beta{}^{(1)},$$

where $\varepsilon_\beta^{(1)}$ are constants. The self-adjointness of the problem can be proved in a way similar to that used in Chapter 1 for a one-component wavefunction. In the present case, we should consider $\psi$ as an M-component column and $\psi^+$ as an M-component row.

The partial wave method of Chapter 4 can be extended without difficulty to the many-channel case. However, we shall restrict ourselves here only to the formulation of a multiple scattering problem. We shall seek a wavefunction with a plane incoming wave in the $\mu$-th channel, in the following form:

$$\psi_\beta(\underline{r}) = e^{i\underline{k}_{o\mu}\cdot\underline{r}}\delta_{\beta\mu} + \sum_{j=1}^{N} c_{\beta\mu j} \frac{e^{ik_\beta|\underline{r} - \underline{R}_j|}}{|\underline{r} - \underline{R}_j|}. \qquad (5.1.4)$$

Let us introduce, instead of $c_{\beta\mu j}$, new quantities $F_{\beta\mu j}$ thus:

$$F_{\beta\mu j} = c_{\beta\mu j} \exp(-i\underline{k}_{o\mu}\cdot\underline{R}_j).$$

For each permitted value of j, coefficients $F_{\beta\mu j}$ form an $M \times M$ matrix $F_j$ of the scattering amplitudes on the j-th center, from channel number $\mu$ into channel number $\beta$. Applying the boundary conditions (5.1.3) to the wavefunction (5.1.4), we obtain an inhomogeneous $M \times M$ system of algebraic equations to determine coefficients $F_{\beta\mu j}$. In matrix form this system is similar to that of (4.5.2):

$$F_j = f_j + f_j \sum_{i \neq j} F_i \frac{e^{ikR_{ij}}}{R_{ij}} e^{i\underline{k}_{o\mu}\cdot\underline{R}_{ij}}, \qquad (5.1.5)$$

where $\mathbf{f}_j = -(\alpha^{(j)} + ik)^{-1}$ is a matrix of scattering amplitudes on the isolated j-th center, and k is an $M \times M$ matrix $k_{\beta\mu} = \delta_{\beta\mu}k_\beta$. Generally speaking, matrices $F_j$ and $f_j$ do not commute with each other. The total scattering amplitude for the whole system of N centers corresponding to a transition from channel $\mu$ into channel $\beta$ is given by

$$F_{\mu\to\beta}(\underline{k}_{o\mu}, \underline{k}_\beta) = \sum_{j=1}^{N} F_{\beta\mu j}\, e^{i(\underline{k}_{o\mu} - \underline{k}_\beta)\cdot\underline{R}_j}. \qquad (5.1.6)$$

The energy of the bound, virtual, and quasi-stationary states of the particle in the many-channel problem is specified by the roots of the determinant of the system (5.1.5).

A general many-channel problem with s-waves in all channels can be reduced to the model developed above. Then the model wavefunction (5.1.4) coincides at large r with the asymptotic form of the exact wavefunction. The corresponding matrix $\alpha$ depends on the wavenumber and is identical with the matrix M introduced by Ross and Shaw in the effective range theory for many-channel problems (see reference [42]).

The model described above can be used for electron scattering by atoms if the potentials in the strong coupling equations are replaced by equivalent boundary conditions. In this way, ZRPs can be used to describe excitation of the atomic systems by electron impact. If elastic scattering of the electron by the atom is accompanied by exchange, this problem can be treated as the simplest multichannel problem with one threshold ($k_1 = k_2$).

We shall now apply this approach to consider electron scattering by the hydrogen atom in the static exchange approximation. The results can be used for different target atoms with non-zero spin provided that the target is treated as a valence electron moving in the effective field of the ion core. In the static exchange approximation, the two-electron wavefunction in a non-symmetrized form with respect to exchange of the electrons is

$$\psi(\underline{r}_1,\underline{r}_2) = \psi_1(\underline{r}_1)\,\phi_a(\underline{r}_2) + \psi_2(\underline{r}_2)\,\phi_a(\underline{r}_1)\,,$$

where $\phi_a$ is the wavefunction of the atomic electron, and the functions $\psi_1$ and $\psi_2$ satisfy the coupled equations

$$(\nabla^2 + k^2)\,\psi_1 = 2V_{11}\psi_1 + 2V_{12}\psi_2\,,$$

$$(\nabla^2 + k^2)\,\psi_2 = 2V_{21}\psi_1 + 2V_{22}\psi_2\,,$$

where the potentials $V_{pq}$ are defined by the equations

$$V_{11}(\underline{r}) = V_{22}(\underline{r}) = -\frac{1}{r} + \int \frac{\phi a(\underline{r}_1)^2}{|\underline{r} - \underline{r}_1|}\,d\underline{r}_1$$

and

$$V_{12}\,\psi(\underline{r}) = V_{21}\,\psi(\underline{r}) = \phi_a(\underline{r})\int\left(\frac{1}{|\underline{r} - \underline{r}_1|} - \frac{1}{2} - \frac{k^2}{2}\right)\phi_a(\underline{r}_1)\psi(\underline{r}_1)\,d\underline{r}_1\,.$$

Replacing the potentials $V_{\beta\gamma}$ by boundary conditions, we obtain for the functions $\psi_\beta$, $\beta = 1,2$, equations (5.1.2) with $k_1 = k_2$; also it follows from the identity $V_{11} = V_{22}$ that the elements of the matrix $\alpha$ satisfy the additional condition $\alpha_{11} = \alpha_{22}$. Solving equations (5.1.5), we obtain the relation between the matrix elements $\alpha_{11}$ and $\alpha_{12}$, on one hand, and the scattering lengths $a_s$ and $a_t$ (for the singlet and triplet states, respectively), on the other:

$$\alpha_{11} = \frac{a_s + a_t}{2a_s a_t}$$

and (5.1.7)

$$\alpha_{12} = -\frac{a_s - a_t}{2a_s a_t}\,.$$

We note that in this particular example it would be possible to take a one-component function and construct the symmetrical and anti-symmetrical wavefunctions. However, in the general case the problem requires more than one component in $\psi$.

Let us consider now the case of an electron scattering on a system of two atoms, a and b. Assuming that $\phi_a$ and $\phi_b$ are the wave-functions of the valence electrons in atoms a and b, respectively, we shall seek an approximate wavefunction of the system in the form

$$\psi(\underline{r}_1,\underline{r}_2,\underline{r}_3) = \psi_1(\underline{r}_1)\phi_a(\underline{r}_2)\phi_b(\underline{r}_3) + \psi_2(\underline{r}_2)\phi_a(\underline{r}_1)\phi_b(\underline{r}_3)$$
$$+ \psi_3(\underline{r}_3)\phi_a(\underline{r}_1)\phi_b(\underline{r}_2) + \psi_4(\underline{r}_1)\phi_a(\underline{r}_3)\phi_b(\underline{r}_2)$$
$$+ \psi_5(\underline{r}_2)\phi_a(\underline{r}_3)\phi_b(\underline{r}_1) + \psi_6(\underline{r}_3)\phi_a(\underline{r}_2)\phi_b(\underline{r}_1). \tag{5.1.8}$$

Assuming that $\phi_a$ and $\phi_b$ do not overlap and neglecting the distortion these functions experience when the atoms come closer to each other, we may replace the potentials by the boundary conditions and obtain for functions $\psi_\beta$, $\beta = 1,2,\ldots 6$, a set of equations (5.1.2) where $k_1 = k_2 = \ldots = k_6$. The boundary conditions of the form (5.1.3) with matrices $\tilde{\alpha}^a$ and $\tilde{\alpha}^b$ of order $6 \times 6$ are set at points $\underline{R}_a$ and $\underline{R}_b$. Non-zero elements of the $\alpha$-matrices are expressed in terms of the quantities $\alpha^a_{11}$, $\alpha^a_{12}$, $\alpha^b_{11}$, and $\alpha^b_{12}$ related to the individual atoms, thus:

$$\tilde{\alpha}^a_{nn} = \alpha^a_{11},$$
$$\tilde{\alpha}^a_{2n-1,2n} = \tilde{\alpha}^a_{2n,2n-1} = \alpha^a_{12},$$
$$\tilde{\alpha}^b_{nn} = \alpha^b_{11},$$
$$\tilde{\alpha}^b_{2n,2n+1} = \tilde{\alpha}^b_{2n+1,2n} = \alpha^b_{12},$$
$$\tilde{\alpha}^b_{16} = \tilde{\alpha}^b_{16} = \alpha^b_{12}\,.$$

If a unitary transformation U is applied to the vector $\psi$ in formula (5.1.1), that is,

$$\psi' = U\psi\,, \tag{5.1.9}$$

then the $\alpha$-matrices defining the boundary conditions are transformed according to

$$\tilde{\alpha}^{a'} = U\,\tilde{\alpha}^a U^{-1}, \qquad \tilde{\alpha}^{b'} = U\,\tilde{\alpha}^b U^{-1}\,. \tag{5.1.10}$$

Generally speaking, two arbitrary Hermitian matrices cannot be diagonalized by the same transformation. However, in the present case, due to the symmetry of the Hamiltonian with respect to permutations of electrons, there exists a unitary transformation U such that the matrices $\tilde{\alpha}^a$ and $\tilde{\alpha}^b$ are reduced simultaneously by (5.1.10) to a quasi-diagonal (block-wise) form. Now we shall outline how this matrix U can be constructed. First we note that the formula (5.1.8) for the total wavefunction $\psi(\underline{r}_1,\underline{r}_2,\underline{r}_3)$ can be considered as an expansion of the function $\psi$ in terms of the basis $\phi_a(\underline{r}_k)\phi_b(\underline{r}_i)$, k,i = 1,2,3, k ≠ i, with coefficients $\psi_\beta(\underline{r}_j)$, j = 1,2,3, j ≠ k,i. Clearly, any permutation of electrons results in a linear transformation of (5.1.8). For instance, the permutation of electrons 1 and 2 in (5.1.8) results in the same form but with new coefficients $\psi'_\beta$:

$$\psi'_1 = \psi_2, \qquad \psi'_2 = \psi_1, \qquad \psi'_3 = \psi_6,$$
$$\psi'_4 = \psi_5, \qquad \psi'_5 = \psi_4, \qquad \psi'_6 = \psi_3.$$

In other words, the functions $\psi_\beta$ realize a representation of the permutation group $S_3$. This leaves the matrices $\tilde{\alpha}^{a,b}$ unchanged due to the symmetry of the Hamiltonian with respect to permutations of the electron coordinates.

It is not difficult to prove that the representation of $S_3$ obtained on $\psi_\beta$ is regular. Using the method described in [93] and the matrices of irreducible representations of $S_3$ given in the monograph [94], we can write the regular representation on $\psi_\beta$ in terms of the corresponding irreducible representations. The final result, for the matrices U, $\tilde{\alpha}^{a'}$, $\tilde{\alpha}^{b'}$, is

$$U = \begin{pmatrix} \frac{1}{2} & 0 & -\frac{1}{2} & \frac{1}{2} & 0 & -\frac{1}{2} \\ \frac{1}{2\sqrt{3}} & \frac{1}{\sqrt{3}} & \frac{1}{2\sqrt{3}} & -\frac{1}{2\sqrt{3}} & -\frac{1}{\sqrt{3}} & -\frac{1}{2\sqrt{3}} \\ \frac{1}{\sqrt{3}} & -\frac{1}{2\sqrt{3}} & -\frac{1}{2\sqrt{3}} & -\frac{1}{2\sqrt{3}} & -\frac{1}{2\sqrt{3}} & \frac{1}{\sqrt{3}} \\ 0 & \frac{1}{2} & -\frac{1}{2} & -\frac{1}{2} & \frac{1}{2} & 0 \\ \frac{1}{\sqrt{6}} & -\frac{1}{\sqrt{6}} & \frac{1}{\sqrt{6}} & -\frac{1}{\sqrt{6}} & \frac{1}{\sqrt{6}} & -\frac{1}{\sqrt{6}} \\ \frac{1}{\sqrt{6}} & \frac{1}{\sqrt{6}} & \frac{1}{\sqrt{6}} & \frac{1}{\sqrt{6}} & \frac{1}{\sqrt{6}} & \frac{1}{\sqrt{6}} \end{pmatrix} \qquad (5.1.11)$$

$$\tilde{\alpha}^{a'} = \{ M_a, M_a, \alpha^a_{11} - \alpha^a_{12}, \alpha^a_{11} + \alpha^a_{12} \},$$

and

$$\tilde{\alpha}^{b'} = \{ M_b, M_b, \alpha_{11}^b - \alpha_{12}^b, \alpha_{11}^b + \alpha_{12}^b \},$$

where

$$M_a = \begin{pmatrix} \alpha_{11}^a - \frac{1}{2}\alpha_{12}^a & \frac{\sqrt{3}}{2}\alpha_{12}^a \\ \frac{\sqrt{3}}{2}\alpha_{12}^a & \alpha_{11}^a + \frac{1}{2}\alpha_{12}^a \end{pmatrix}$$

and (5.1.11)

$$M_b = \begin{pmatrix} \alpha_{11}^b - \frac{1}{2}\alpha_{12}^b & -\frac{\sqrt{3}}{2}\alpha_{12}^b \\ -\frac{\sqrt{3}}{2}\alpha_{12}^b & \alpha_{11}^b + \frac{1}{2}\alpha_{12}^b \end{pmatrix} .$$

In formulae (5.1.11), the quantities in the curly brackets are blocks of nonvanishing elements on the diagonal of the matrices $\tilde{\alpha}^{a'}$ and $\tilde{\alpha}^{b'}$.

The boundary conditions can be satisfied now by $\psi'$ only if $\psi_1'$ and $\psi_2'$ do not vanish. Then the wavefunction of the system takes the form

$$\begin{aligned} \psi(\underline{r}_1,\underline{r}_2,\underline{r}_3) = \frac{1}{2}\Bigg\{ & \psi_1'(\underline{r}_1)\left[\phi_a(\underline{r}_2)\phi_b(\underline{r}_3) + \phi_a(\underline{r}_3)\phi_b(\underline{r}_2)\right] \\ & - \psi_1'(\underline{r}_3)\left[\phi_a(\underline{r}_1)\phi_b(\underline{r}_2) + \phi_a(\underline{r}_2)\phi_b(\underline{r}_1)\right] \\ & + \frac{1}{\sqrt{3}}\psi_2'(\underline{r}_1)\left[\phi_a(\underline{r}_2)\phi_b(\underline{r}_3) - \phi_a(\underline{r}_3)\phi_b(\underline{r}_2)\right] \\ & + \frac{2}{\sqrt{3}}\psi_2'(\underline{r}_2)\left[\phi_a(\underline{r}_1)\phi_b(\underline{r}_3) - \phi_a(\underline{r}_3)\phi_b(\underline{r}_1)\right] \\ & + \frac{1}{\sqrt{3}}\psi_2'(\underline{r}_3)\left[\phi_a(\underline{r}_1)\phi_b(\underline{r}_2) - \phi_a(\underline{r}_2)\phi_b(\underline{r}_1)\right]\Bigg\} . \end{aligned} \quad (5.1.12)$$

This function corresponds to total spin $S = \frac{1}{2}$ (42). Equation (5.1.12) is a superposition of two configurations: (i) a weakly bound electron with the wavefunction $\psi_1'$ in a field of a singlet state of a quasi-molecule, and (ii) a weakly bound electron with the wavefunction $\psi_2'$ in a field of a triplet state of a quasi-molecule. The case where only $\psi_3'$ and $\psi_4'$ do not vanish also leads to $S = 1/2$. This is because the number of times that an irreducible representation is contained in the regular representation is the same as its dimension [93]. The case where only $\psi_5'$ does not vanish corresponds to total spin $S = 3/2$. The case where only $\psi_6'$ is nonvanishing corresponds to a totally symmetric wavefunction in all three electrons and, therefore, is forbidden by the Pauli principle.

This method of the simultaneous reduction of matrices to a quasi-diagonal (block-wise) form which employs splitting the regular representation into irreducible representations becomes too cumbersome if the system is composed of many atoms of spin 1/2. In this case it is easier to introduce explicitly spin functions of the electrons. In this approach, first used by Smirnov and Firsov in their earlier work [17] (see also Subramanyan [89]), the Hamiltonian is written in the form

$$H = -\frac{1}{2}\nabla^2 + U_a + U_b + V_a \underline{\sigma}\cdot\underline{\sigma}_a + V_b \underline{\sigma}\cdot\underline{\sigma}_b, \qquad (5.1.13)$$

where $\underline{\sigma}$ is the spin operator of the weakly bound electron, and $\underline{\sigma}_a$ and $\underline{\sigma}_b$ are the spin operators for the atoms. The interaction potentials $U_a$, $U_b$, $V_a$, and $V_b$ are of finite range and do not vanish in the vicinity of the corresponding atom. This method is equivalent to that considered above [39,90] but does not reveal so clearly the two-channel nature of the problem.

Another version of the theory which explicitly used spin functions but was based directly on the equation of multiple scattering was suggested by Drukarev and Yurova [91]. For scattering by an atom with spin 1/2, the matrix of boundary conditions is written thus:

$$\alpha^{(a)} = \frac{1}{a_s^{(a)}}\,\frac{1 - \underline{\sigma}\cdot\underline{\sigma}_a}{4} + \frac{1}{a_t^{(a)}}\,\frac{3 + \underline{\sigma}\cdot\underline{\sigma}_a}{4}, \qquad (5.1.14)$$

where the operators $\frac{1}{4}(1 - \underline{\sigma}\cdot\underline{\sigma}_a)$ and $\frac{1}{4}(3 + \underline{\sigma}\cdot\underline{\sigma}_a)$ are projection operators onto singlet and triplet states of the system "electron + atom." The spin function of the system "electron + molecule in a singlet state" has the form

$$\chi_o = \frac{1}{\sqrt{2}}\, u\,(u_b v_a - u_a v_b), \qquad (5.1.15)$$

where u and v are eigenfunctions of $\sigma_z$ with eigenvalues +1 and -1, respectively. For a triplet state of the molecule,

$$\chi_1 = \frac{1}{\sqrt{6}}\{2\, v\, u_a u_b - u(u_a v_b + u_b v_a)\}. \qquad (5.1.16)$$

With the help of the wavefunctions (5.1.15) and (5.1.16) we can now obtain the matrix elements:

$$\langle\chi_1|\alpha^{(a)}|\chi_o\rangle = \frac{\sqrt{3}}{4}\left(\frac{1}{a_t^{(a)}} - \frac{1}{a_s^{(a)}}\right),$$

$$\langle\chi_1|\alpha^{(b)}|\chi_o\rangle = -\frac{\sqrt{3}}{4}\left(\frac{1}{a_t^{(b)}} - \frac{1}{a_s^{(b)}}\right),$$

$$\langle\chi_o|\alpha^{(a,b)}|\chi_o\rangle = \frac{1}{4}\left(\frac{1}{a_s^{(a,b)}} - \frac{3}{a_t^{(a,b)}}\right), \qquad (5.1.17)$$

$$\langle\chi_1|\alpha^{(a,b)}|\chi_1\rangle = \frac{1}{4}\left(\frac{3}{a_s^{(a,b)}} - \frac{1}{a_t^{(a,b)}}\right).$$

It is easy to check that this result coincides with that obtained above (see equation (5.1.9)) by a different method. A general case of several scatterers of spin 1/2 was also considered in reference [91].

## 5.2 SINGLET-TRIPLET SPLITTING AND CROSS SECTIONS FOR ELASTIC AND INELASTIC SCATTERING

The equations for the functions $\psi_1'$ and $\psi_2'$ do not take into account the fact that the energies of the quasi-molecule in a singlet and in a triplet state differ from each other by $\varepsilon(R)$, depending on the internuclear separation R. This energy splitting can be accounted for, within the model used, in a phenomenological way by assuming that, in equations (5.1.2) for $\psi_1'$ and $\psi_2'$,

$$k_1^2 - k_2^2 = 2\varepsilon(R)$$

and retaining the old boundary conditions. The quantity $\varepsilon(R)$ should be taken either from exact calculations of molecular terms or from the experimental data.*

Let us consider scattering of an electron with momentum $\underline{k}_{o1}$ by a molecule. We assume that the molecule is in its ground state, which is a singlet state, so that the total spin of the system S = 1/2.

$$\psi_1' = e^{i\underline{k}_{o1}\cdot\underline{r}}\; c_1^a\,\frac{\exp(ik_1|\underline{r} - \underline{R}_a|)}{|\underline{r} - \underline{R}_a|} + c_1^b\,\frac{\exp(ik_1|\underline{r} - \underline{R}_b|)}{|\underline{r} - \underline{R}_b|},$$

---

* Introduction of the energy splitting $\varepsilon(R)$ is equivalent to adding the term $(1/4)(3 + \underline{\sigma}_a\cdot\underline{\sigma}_b)\varepsilon(R)$ to the Hamiltonian (5.1.13) of Smirnov and Firsov.

and

$$\psi_2' = c_2^a \frac{\exp(ik_2|\underline{r} - \underline{R}_a|)}{|\underline{r} - \underline{R}_a|} + c_2^b \frac{\exp(ik_2|\underline{r} - \underline{R}_b|)}{|\underline{r} - \underline{R}_b|} .$$

The boundary conditions give the following system for the coefficients $c_\beta^a$:

$$\begin{aligned} &ik_1 c_1^a + g_1 c_1^b + \exp(\tfrac{i}{2}\underline{k}_{01}\cdot\underline{R}) + (\alpha_{11}^{(a)} - \tfrac{1}{2}\alpha_{12}^{(a)})\, c_1^a + \tfrac{\sqrt{3}}{2}\alpha_{12}^{(a)} c_2^a = 0, \\ &ik_1 c_1^b + g_1 c_1^a + \exp(-\tfrac{i}{2}\underline{k}_{01}\cdot\underline{R}) + (\alpha_{11}^{(b)} - \tfrac{1}{2}\alpha_{12}^{(b)})\, c_1^b + \tfrac{\sqrt{3}}{2}\alpha_{12}^{(b)} c_2^b = 0, \\ &ik_2 c_2^a + g_2 c_2^b + (\alpha_{11}^{(a)} + \tfrac{1}{2}\alpha_{12}^{(a)})\, c_2^a + \tfrac{\sqrt{3}}{2}\alpha_{12}^{(a)} c_1^a = 0, \\ &ik_2 c_2^b + g_2 c_2^a + (\alpha_{11}^{(b)} + \tfrac{1}{2}\alpha_{12}^{(b)})\, c_2^b + \tfrac{\sqrt{3}}{2}\alpha_{12}^{(b)} c_1^b = 0, \end{aligned} \tag{5.2.1}$$

where $g_\beta = \exp(ik_\beta R)/R$. For identical atoms a and b, the system (5.2.1) splits into two systems of two equations according to parity. Let us denote

$$c_1^\pm = c_1^a \pm c_1^b , \quad c_2^\pm = c_2^a \mp c_2^b . \tag{5.2.2}$$

Then the solution of these two systems can be written as

$$c_\beta^\pm = \Phi_\beta^\pm \left(\exp(\tfrac{i}{2}\underline{k}_{01}\cdot\underline{R}) \pm \exp(-\tfrac{1}{2}\underline{k}_{01}\cdot\underline{R})\right) , \tag{5.2.3}$$

where

$$\begin{aligned} \Phi_1^\pm &= -\frac{ik_2 \mp g_2 + (\alpha_{11} + \frac{1}{2}\alpha_{12})}{\Delta_\pm} , \\ \Phi_2^\pm &= \frac{\sqrt{3}\,\alpha_{12}}{2\,\Delta_\pm} , \end{aligned} \tag{5.2.4}$$

and

$$\Delta_\pm = (ik_1 \pm g_1 + (\alpha_{11} - \tfrac{1}{2}\alpha_{12}))(ik_2 \mp g_2 + (\alpha_{11} + \tfrac{1}{2}\alpha_{12})) - \tfrac{3}{4}\alpha_{12}^2 .$$

The elastic scattering amplitude $F_e$, for a fixed position of the nuclei, is

$$\begin{aligned} F_e(\underline{k}_{01},\underline{k}_1) &= (\Phi_1^+ - \Phi_1^-)\cos(\tfrac{1}{2}(\underline{k}_{01} + \underline{k}_1)\cdot\underline{R}) \\ &+ (\Phi_1^+ + \Phi_1^-)\cos(\tfrac{1}{2}(\underline{k}_{01} - \underline{k}_1)\cdot\underline{R}) , \end{aligned} \tag{5.2.5}$$

where $\underline{k}_1$ is the momentum of the scattered electron. For $k_1 > \sqrt{2\varepsilon}$ , inelastic scattering also takes place when the electron is scattered with momentum $\underline{k}_2$ and the molecule is excited to the triplet state. The inelastic scattering amplitude $F_i$ is

$$F_i(\underline{k}_{o1},\underline{k}_2) = -i(\Phi_2^+ - \Phi_2^-)\sin\left(\frac{1}{2}(\underline{k}_{o1} + \underline{k}_2)\cdot\underline{R}\right) - i(\Phi_2^+ + \Phi_2^-)\sin\left(\frac{1}{2}(\underline{k}_{o1} - \underline{k}_2)\cdot\underline{R}\right). \qquad (5.2.6)$$

The total cross section for both the elastic and inelastic scattering processes, after averaging over all possible orientations of the molecule, are given by (90)

$$\sigma_e = 4\pi\left\{|\Phi_1^+|^2\left(1 + \frac{\sin k_1R}{k_1R}\right)^2 + |\Phi_1^-|^2\left(1 - \frac{\sin k_1R}{k_1R}\right)^2\right\}, \qquad (5.2.7)$$

$$\sigma_i = 4\pi\frac{k_2}{k_1}\left\{|\Phi_2^+|^2\left(1 + \frac{\sin k_2R}{k_2R}\right)\left(1 - \frac{\sin k_1R}{k_1R}\right) + |\Phi_2^-|^2\left(1 - \frac{\sin k_2R}{k_2R}\right)\left(1 + \frac{\sin k_1R}{k_1R}\right)\right\}. \qquad (5.2.8)$$

In (89) the singlet-triplet splitting of the terms was not taken into account, and the cross section was obtained with the help of the optical theorem. Consequently, formula (22) in reference (89) is, in fact, the total cross section $\sigma_e + \sigma_i$. This is in agreement with the present results (5.2.7) and (5.2.8) if $\varepsilon$ is set to be zero. Figure 5.1 compares $\sigma_e$ and $\sigma_i$ when $\varepsilon = 0$ (curves I and II) in the case of electron scattering by the hydrogen molecule. Parameters $\alpha_{11}$ and $\alpha_{12}$ were determined from the theoretical scattering lengths $a_s = 5.7$ and $a_t = 1.768$ (34) and the internuclear separation $R_o$ taken to be 1.4. Introducing the splitting $\varepsilon(1.4) = 0.3903$ (34), we obtain curve III for $\sigma_e$. At the threshold for inelastic scattering this curve has an infinite derivative (19,22). The insert in Fig. 5.1 is a blow-up of this part of the curve.* The maxima of curves I and III are due to the antisymmetric part of the scattering, i.e., due to the second term in (5.2.7). In order to illustrate this feature, the contribution of the first term to curve III has also been shown in Fig. 5.1 by a dotted line. These maxima will be discussed later, together with the quasi-stationary states.**

---

* A singularity in the elastic cross section at the threshold of the inelastic channel was observed experimentally by Read (96).

** It will also be shown there that a maximum of curve III at $k_1 = 0.44$ is nonphysical and has to be eliminated.

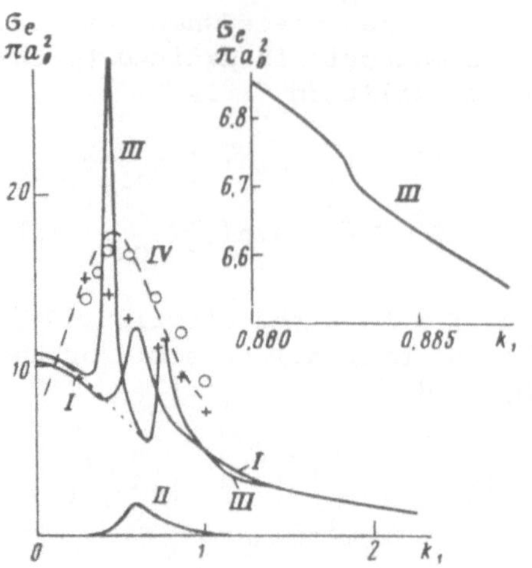


Fig. 5.1. Cross section for the elastic scattering of electrons by the hydrogen molecule. I-III: ZRP approximation (90); I - cross section for elastic scattering with $\varepsilon = 0$. II - the same for inelastic scattering. III - elastic cross section, taking into account the singlet-triplet splitting. IV: experimenal data (98); +++ calculation with exchange, Hara (101). ooo the same, taking into account both exchange and polarization. Inset: an enlarged part of the elastic cross section curve in the neighborhood of the excitation threshold for the lowest triplet state. A threshold singularity can be clearly seen.

Figures 5.1 and 5.7 also show experimental cross sections (97, 98) as well as some theoretical results obtained using much more extensive calculations involving the solution of the continuum Hartree-Fock equations (99), specially developed methods of solving equations with partial derivatives (100), and the solution of the scattering problem in spheroidal coordinates with exchange and polarization (101). The fall-off of the experimental cross section with $k_1$ decreasing in the region $k_1 < 0.2$ may indicate the significance of the polarization interaction between the electron and the molecule which has not been taken into account by the ZRP model considered above. Calculations of Hara (101) with and without the inclusion of polarization lead to the same conclusion.

In Fig. 5.2, curve I shows the excitation cross section for the lowest triplet state $^3\Sigma_u^+$ of the hydrogen molecule $H_2$ computed from (5.2.3). An earlier variational calculation (102) of this process, which is accompanied by dissociation of the molecule along the repulsive term, is also shown in Fig. 5.2. Since there is no detail of the calculations (102), it is difficult to judge how reliable they are. All other theoretical results shown in Fig. 5.2 have been obtained from various modifications of perturbation theory. The most reliable is probably curve II using the Ochkur-Rudge method [103]. We note that, in that reference, the threshold turns out to be shifted towards lower energies due to the allowance made for nuclear motion.

The quantity obtained experimentally by Corrigan [105] was the cross section for dissociation of $H_2$ by electron impact and not the cross section $\sigma_i$. The reported accuracy of the measurements was some 30 percent. Dissociation of $H_2$ takes place as a result of excitation of the molecule to a triplet state with subsequent transition to the lowest repulsive $^3\Sigma_u^+$ state. Calculations of Cartwright and Kuppermann [103], who included the three lowest triplet states of the molecule, gave a cross section for dissociation differing from the experimental one by some 40-50 per cent and with an inaccurate position of the maximum of the cross section curve.

It can be concluded from Figs. 5.1 and 5.2 that the ZRP model underestimates both $\sigma_i$ and $\sigma_e$. This result is to be expected because the real potentials of finite range lead to wavefunctions with an increased effective range. This, in turn, causes an increase in the

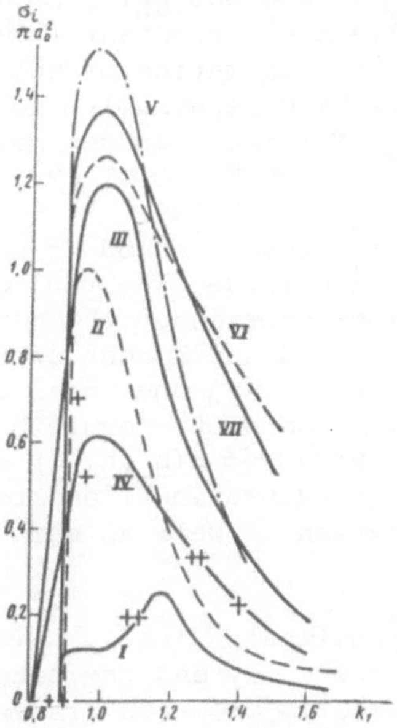


Fig. 5.2. Cross section for electron impact excitation of the lowest triplet state of the $H_2$ molecule. I - ZRP approximation [90]. II - approximation with exchange [104]. III - Ochkur's method [103]. IV - Ochkur-Rudge method [103]. V - Born-Oppenheimer approximation [104] (the cross section shown is reduced by scaling it by 1/3). VI-VII - the separable potential approximation [39]: VI - using wavefunction $\phi$ (1.4.12) and parameters (1.4.19). VII - using wavefunction (1.4.20) and parameters (1.4.21); +++ variational calculations [102].

cross sections. It is well known that perturbation theory overestimates cross sections; contrary to this, the ZRP model underestimates them. So that both methods, to some extent, complement each other.

Drukarev and Yurova [91] computed the scattering of an electron by a linear molecule $H_3$ in its ground $^2\Sigma_u^+$ state. The influence of closed channels was not taken into consideration (in a formal way this corresponded to assuming that the splitting of molecular terms $\varepsilon$ is infinite).

The ZRP method, when applied to electron scattering by molecules, involves some specific restrictions arising from the assumption that the potential wells representing the molecule are stationary [89]. This assumption is justified if the velocity of the electron is not too low so that the collision time $t_c$ of the electron is considerably shorter than the period of the molecular rotation $t_r = 2\pi/\omega_r$, i.e.,

$$t_c \ll t_r.$$

For molecular temperature T, one has $k_B T \sim M\omega_r^2 R^2$, where M is the atomic mass and $k_B$ is Boltzmann's constant. Hence $ME/k_B T \gg 1$. Stronger restrictions on the kinematics of the electron follow from the condition that $t_c$ must be considerably shorter than the period of molecular vibration $t_v$. For the hydrogen molecule, the second criterion leads to the condition $E > 0.01$ eV.

The ZRP method as well as the method of separable potentials can be used to obtain the amplitude $F(\underline{n}_0, \underline{n}, \underline{R})$ of electron scattering by a molecule for any fixed position of the nuclei, i.e., for an arbitrary vector $\underline{R}$. Only a limited amount of this information has been used in the above theory, where the total cross section is computed only for $R = R_0$, $R_0$ being the equilibrium distance between the nuclei. In fact, the amplitude $F(\underline{n}_0, \underline{n}, \underline{R})$ can also be used to calculate cross sections for vibrational and rotational transitions in the adiabatic approximation as well as spin-polarization cross sections.

In the adiabatic approximation (see, for instance, [106]) use is made of the fact that $m/M \ll 1$, and the total wavefunction of the system "molecule + incoming electron" is written as a product:

$$\Psi_A(\underline{r}, \underline{R}) = \psi(\underline{r}, \underline{R})\phi_\nu(\underline{R}), \tag{5.2.9}$$

where $\psi(\underline{r}, \underline{R})$ is the solution of the Schrödinger equation for an electron moving in the field of the fixed nuclei, and $\phi_\nu(\underline{R})$ is the wavefunction of the nuclear motion, with a set of quantum numbers $\nu$ which specify the vibrational and rotational states of the molecule. The exact formula for the amplitude for inelastic scattering is given by

$$F_{\nu\to\nu'}(\underline{n}_o,\underline{n}) = \frac{1}{2\pi}\int \phi^*_{\nu'}(\underline{R})\, e^{-i\underline{k}_{\nu'}\cdot\underline{r}}\, V(\underline{r},\underline{R})\,\Psi(\underline{r},\underline{R})\, d\underline{r}\, d\underline{R},$$

where $V(\underline{r},\underline{R})$ is the effective potential of interaction between the electron and the molecule. The adiabatic approximation for the amplitude is obtained by replacing the exact wavefunction in the expression above by an approximate formula (5.2.9).*

$$F^A_{\nu\to\nu'}(\underline{n}_o,\underline{n}) = \int \phi^*_{\nu'}(\underline{R})\; F(\underline{n}_o,\underline{n},\underline{R})\; \phi_\nu(\underline{R})\, d\underline{R}\,. \qquad (5.2.10)$$

The change $\Delta E$ in the kinetic energy E of the electron due to excitation of the nuclear motion is assumed to be negligible. More precisely the criterion for the validity of the adiabatic approximation is

$$k\,r_o\,\Delta E/E \ll 1 \quad \text{if} \quad k\,r_o \gtrsim 1,$$

$$\Delta E/E \ll 1 \quad \text{if} \quad k\,r_o \lesssim 1,$$

where $\Delta E = \max(E_{vr}, \Delta E_{vr})$, and $E = \min(E_{el}(i), E_{el}(f))$. In these expressions, $E_{vr}$ is the energy of the vibrational-rotational motion of the molecule, $\Delta E_{vr}$ is its change caused by the transition, and $E_{el}(i)$ and $E_{el}(f)$ are the initial and final kinetic energies of the scattered electron; $r_o$ is the effective range of interaction between the electron and the molecule.

If the amplitude for inelastic electron scattering by a molecule, accompanied by vibrational-rotational excitation of the molecule, can be obtained for an arbitrary position of the nuclei, the amplitude of the vibrational-rotational transitions can be found in a similar way. Finally, the wavefunction $\phi_\nu$ may belong to the continuum. In this case, the adiabatic approximation can be used to compute, for instance, the angular distribution of the nuclei in the electron impact dissociation of molecules.

We shall now turn to the calculations of cross sections for rotational transitions in the ZRP approximation. The scattering amplitude has the form (5.2.5), where all dependence upon orientation of the molecule is in the factors $\exp(\pm i(\underline{k}_o \pm \underline{k})\cdot\underline{R}_j)$. (A similar situation arises if the separable potential method is used instead of ZRPs.) Considering the two-center system, we shall put $\phi_\nu(\underline{R}) = Y_{LM}(\underline{R}/R)$ and use the plane wave expansion in terms of spherical harmonics to obtain the matrix element (5.2.10). After performing

---

* A more rigorous derivation of the adiabatic approximation for the case of a ZRP has been given by Drukarev (107) based on the Faddeev equations.

summation over all possible values M' of the projection of angular momentum of the nuclear motion in the final state, and averaging the result over all possible values M of the momentum projection in the initial state, we obtain the following differential cross section for a rotational transition:

$$\sigma_{L \to L'} = (2L' + 1) \sum_{\substack{\text{even} \\ \ell}} \left| \begin{pmatrix} L & \ell & L' \\ 0 & 0 & 0 \end{pmatrix} \right|^2 (2\ell + 1)$$

$$+ \Bigg\{ |\Phi_1^+ - \Phi_1^-|^2 \, j_\ell^2(|\underline{k}_{o1} + \underline{k}_1| \tfrac{R}{2}) + |\Phi_1^+ + \Phi_1^-|^2 \, j_\ell^2(|\underline{k}_{o1} - \underline{k}_1| \tfrac{R}{2})$$

$$+ (-1)^{\ell/2} \, 2^{-\ell} \binom{\ell}{\ell/2} \, 2\,\mathrm{Re}\, \{(\Phi_1^+ + \Phi_1^-)(\Phi_1^+ - \Phi_1^-)^*\}$$

$$\times j_\ell(|\underline{k}_{o1} + \underline{k}_1| \tfrac{R}{2}) \, j_\ell(|\underline{k}_{o1} - \underline{k}_1| \tfrac{R}{2}) \Bigg\}, \qquad (5.2.11)$$

where $j_\ell(x) = (\pi/2x)^{\frac{1}{2}} J_{\ell+\frac{1}{2}}(x)$ is the spherical Bessel function of order $\ell$. Due to the properties of 3j-symbols, (5.2.11) is a sum of a finite number of terms. Results of some calculations for $\sigma_{L \to L'}$ in the hydrogen molecule obtained using equation (5.2.11) are presented in Fig. 5.3. Agreement between the theoretical results derived here and the experimental cross sections is very good. However, for other transitions and impact energies the theory and experiment may differ from each other very considerably. Some authors (26,106,108) have suggested that rotational excitation of molecules takes place primarily due to the long-range interactions such as polarization and quadrupole interactions. This type of interaction is beyond the scope of the ZRP method developed above. On the other hand, the ZRP method is capable of taking into account multiple scattering by the short-range component of the potential. Both approaches give very often results of comparable accuracy and, at present, it is difficult to give preference to one of these two theories. Nevertheless, it appears that the total neglect of the short-range part of the potential is not justified.

Computation of the cross section for vibrational-rotational and pure rotational transitions requires integration over R, which can be performed only numerically. Such calculations were reported by Drukarev and Yurova (110,111) for $H_2$ as well as for $K_2$, $Na_2$, and $Cs_2$ with the help of the ZRP model. The influence of virtual excitation of the triplet state of the molecule was neglected because $\varepsilon$ was assumed to be large. In this case, there exists only one channel in the problem, and $\alpha$ in the boundary condition is

$$\alpha = \frac{1}{4} (1/a_s + 3/a_t). \qquad (5.2.12)$$

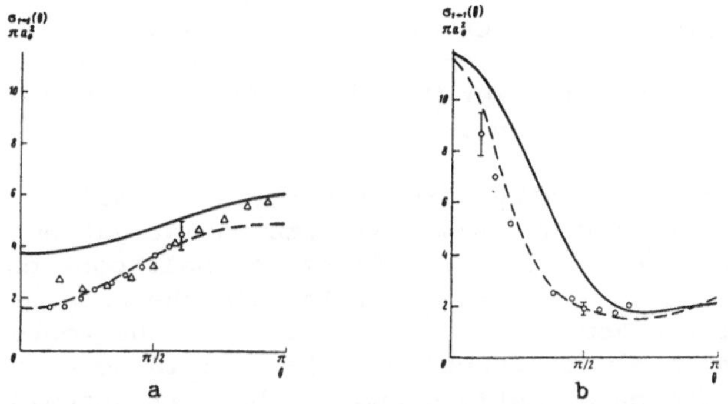

Fig. 5.3. Differential cross section $\sigma_{L \to L'}(\theta)$ for the scattering of electrons by hydrogen molecules accompanied by a transition between rotational states L and L'. The curves correspond to the case L = L' = 1 and different values of energy: (a) E = 1 eV, (b) E = 8 eV. Solid line: ZRP approximation using formula (5.2.11) with an allowance for the singlet-triplet splitting. Broken line: the strong coupling approximation without exchange, but with allowance for polarization [108]. ΔΔΔ - experimental data [97]. ooo - experimental data of Ehrhardt and Linder [109] normalized according to Lane and Geltman [108].

Alkali metals have large polarizabilities, and the selection of the correct numerical values to be used for $a_s$ and $a_t$ in formula (5.2.12) becomes a difficult problem since the polarization effects influence considerably the scattering length. Therefore the best way of determining the scattering length, for not very low energies, is to extrapolate $k \cot \delta$, assuming a linear dependence upon $k^2$, down to the point $k^2 = 0$ (see Sec. 1.4).

## 5.3 ENERGY TERMS OF THE e + $H_2$ SYSTEM AND TRAJECTORIES OF THE POLES OF THE S-MATRIX FOR A TWO-CHANNEL PROBLEM

Let us consider bound and quasi-stationary states of the system "electron + molecule" with total spin 1/2 in the case of two identical atoms. The energy levels are determined by two independent equations, for the g- and u-states, respectively:

$$\Delta_{\pm} = 0, \tag{5.3.1}$$

where $\Delta_{\pm}$ are determined by (5.2.4). Each bound state is a superposition of two configurations. For the first equation, these configurations are a $\sigma_g$-orbital for an electron moving in the field of a singlet state of the molecule, and a $\sigma_u$-orbital for an electron moving in the field of a triplet state of the molecule. For the second equation, the parities of the orbitals are to be taken opposite to those in the first equation. For this model, among the infinite number of roots of equation (5.3.1) we shall discuss only those with direct physical meaning.

If $\varepsilon(R)$ is identically zero, the roots of (5.3.1) can be obtained in analytical form, which facilitates the investigation of the general situation, i.e., when $\varepsilon(R) \neq 0$. First we shall consider the case of $\alpha_{11} > -\alpha_{12} > 0$ so that $a_s > a_t > 0$. This is the case of the hydrogen atom. We shall use Roman numerals to describe the roots $\kappa(R)$ (i.e., branches of the analytic function $E(R)$). For large R, there are two bound states of each symmetry g and u, the corresponding roots being

$$\kappa_I^{u,g} = (\alpha_{11} + \alpha_{12}) \pm \exp(-(\alpha_{11} + \alpha_{12})R)/(2R)$$

and

$$\kappa_{II}^{u,g} = (\alpha_{11} - \alpha_{12}) \mp \exp(-(\alpha_{11} - \alpha_{12})R)/(2R).$$

We note that the present term splitting happens to be only a half of that in the spinless case (3.2.5) (for details, see ref. [26]). For instance,

$$E_{Iu} - E_{Ig} = -(\alpha_{11} + \alpha_{12}) \exp(-(\alpha_{11} + \alpha_{12})R)/R. \qquad (5.3.2)$$

The function $\kappa_{II}^{u}$ reaches its minimum, $\kappa_{II}^{u} \approx \alpha_{11} - (\sqrt{3}/2)\alpha_{12}$, and the function $\kappa_{I}^{u}$ reaches its maximum, $\kappa_I^u \cong \alpha_{11} + (\sqrt{3}/2)\alpha_{12}$ at the corresponding points R determined by the equation

$$\exp((-\alpha_{11} \pm (\sqrt{3}/2)\alpha_{12})R) = -\alpha_{12}R/2.$$

The roots $\kappa_{II}^{u,g} \to +\infty$ if $R \to 0$. For large R, there exists one virtual state ($\kappa < 0$) of each symmetry, the corresponding $\kappa_{III}^{u,g} \to 0$ as $R \to \infty$. As functions of R, the roots $\kappa_I^{u,g}$ pass through the points

$$R_{cu,cg} = \left[\mp\alpha_{12}/2 + (\alpha_{11}^2 - \tfrac{3}{4}\alpha_{12}^2)^{\frac{1}{2}}\right]^{-1},$$

where $\kappa_I^{u,g}$ vanish, and then, as R decreases further, they coincide with the roots $\kappa_{III}^{u,g}$. At the points of coincidence, the roots $\kappa$ in both cases differ from zero; that is, both terms are tangent to the continuum boundary. In order to make it clear we remember that, in this model, the singlet state g and the triplet state u have the same energy if $\varepsilon(R) \equiv 0$ and, therefore, the partial wave expansion for the wavefunction of the system "electron + molecule" contains all values of the angular momentum $\ell$ ($\ell$ is even in one of the degenerate channels, and $\ell$ is odd in another one). It is evident that the u-term crosses the continuum boundary as soon as the splitting $\varepsilon(R) \neq 0$ is introduced. This, of course, corresponds to the situation in the real molecule.

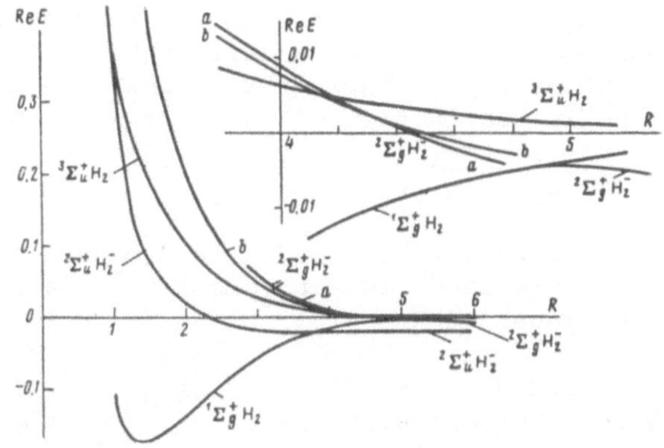


Fig. 5.4. Molecular terms of $H_2$ (variational calculations of Kolos and Wolniewicz (95)) and $H_2^-$ (ZRP approximation of Demkov and Ostrovskii (90)). Inset: Enlarged part of the graph in the neighborhood of the point where the $^2\Sigma_g^+$ term of $H_2^-$ enters the continuum.

Numerical calculations show that the introduction of the splitting $\varepsilon(R)$ attenuates the quantities $\kappa_I^{u,g}$ and $\kappa_{II}^{u,g}$ for a fixed R. Figure 5.4 shows the molecular terms of $H_2^-$ obtained by the ZRP method with the splitting $\varepsilon(R)$ deduced from the accurate variational calculations (95). The position of the $H_2^-$ term is shown relative to the $H_2$ terms (95); the latter terms are also plotted in Fig. 5.4. The $^1\Sigma_g^+$ term of the $H_2$ molecule is the continuum boundary for the molecular $H_2^-$ ion. Above it, the energy of the $H_2^-$ becomes complex (Fig. 5.4 shows the real part of the energy). The insert in Fig. 5.4 displays the behavior of the $^2\Sigma_g^+$ $H_2^-$ term near the point where it passes into the continuum. The term is tangent to the $^1\Sigma_g^+$ $H_2$ term at this point, as it ought to be according to the rigorous theory.

We have already pointed out in Section 1.4 that the ZRP model allows the existence of a $^3S$-state of $H^-$ which is forbidden by the Pauli principle. The forbidden state gives rise, in the present model, to two states of the $H_2^-$ molecule with symmetry $^2\Sigma_g^+$ and $^2\Sigma_u^+$ (roots $\kappa_{II}^g$ and $\kappa_{II}^u$). These terms lie below the real terms of $H_2^-$ and have not been shown in Fig. 5.4. One of these states (antisymmetric) passes into the continuum at R = 1.8 and causes a sharp maximum in $\sigma$ at $k_1 = 0.44$. An analogous "ghost" resonance due to a state forbidden by the Pauli principle was obtained by Herzenberg and Law (45), who carried out calculations of the elastic electron scattering by helium atoms. They found that the resonance was not present in the static approximation with exchange, but it could appear if the exchange potential was distorted. It is possible that the resonance at $k_1 = 0.44$ in the present problem is also due to distortion of the potential. In reality, such resonances do not exist so that curve III in Fig. 5.1 should be replaced, in the neighborhood of the first maximum, by the dotted curve shown there. The latter curve (see Sec. 5.2) represents the leading contribution to $\sigma$ at a given energy. It comes from a different partial wave (symmetrical). An alternative way to exclude this unwanted singularity from the scattering is subtraction of the corresponding pole from the S-matrix. In this particular case, both methods lead to the same result.

Another maximum in $\sigma_e$ (at $k_1 = 0.76$) is produced by a pole in the S-matrix which corresponds to a physical $^2\Sigma_u^+$ state. This state passes into the continuum at R = 3.8.

The results obtained in the ZRP approximation will be compared in more detail with other calculations in Sec. 5.4, where the width of the molecular terms will also be considered. *

Introduction of the term splitting $\varepsilon(R)$ into a ZRP approximation with spin (i.e., with exchange) makes evident the two-channel nature of this problem. When this feature is taken into account, it becomes possible to use the theory to describe molecular excitation by electron impact. Another physical aspect relates to the problem of dissociative attachment. In the latter case, it is required to know whether the $^2\Sigma_g^+$ term of the molecular ion $H_2^-$ lies above the $^3\Sigma_u^+$ term of the $H_2$ molecule, at the equilibrium separation between the nuclei (112). Variational calculations of molecular terms by Bardsley et al. did not take into account the two-channel nature of this problem. We shall study below some features of the movement of the S-matrix poles due to variation in R specifically related to the existence of two channels. In this respect, the present ZRP model can serve as a good illustration of the general principles of the theory in a similar way to the illustration of the role of the potential well problem in a one-channel case (52). For a two-channel problem, equation (5.3.1), which determines the position of the S-matrix poles, is much simpler than that obtained in a square well approximation.

In the two-channel case, the S-matrix is defined on a four-sheet Riemann surface of the complex E plane. We shall use in the discussion below, a two-sheet surface of the complex $k_1$ plane corresponding to the first channel. This two-sheet surface has two cuts running along the real axis from $-\infty$ to $-(2\varepsilon(R))^{1/2}$, and from $(2\varepsilon(R))^{1/2}$ to $+\infty$. On the first sheet, $\mathrm{Im}(k_2) > 0$ and on the second sheet, $\mathrm{Im}(k_2) < 0$, where $k_2$ is the momentum in the second channel. Within the two-channel theory, $\varepsilon(R) \equiv 0$ is a special case because the edges of the cuts on the $k_1$ surface join together when the splitting vanishes. Therefore any path which starts on the upper

* The width of the terms of the molecular ion $H_2^-$ found in the ZRP approximation was used by Demkov et al. (69,113) to compute various detachment reactions (also see Chapters 9 and 10 of this monograph).

half-plane of sheet I and runs downwards, avoiding the point $k_1 = 0$, arrives on the lower half-plane of sheet II. Similarly, any path which starts on the upper half-plane of sheet II and runs downwards comes to the lower half-plane of sheet I. Hence we obtain two new sheets unconnected with each other (with an exception of the common point $k_1 = 0$). Introduction of the term splitting $\varepsilon(R)$ extends the model, and in the subsequent discussion we shall not return to the case $\varepsilon(R) \equiv 0$.

The poles of the two-channel S-matrix coincide with one of the poles of the corresponding one-channel problem, in the limit of uncoupled channels. In channel one, say, each pole of the one-channel problem gives rise to two poles of the two-channel S-matrix. These two poles wil have an identical value of $k_1$ but lie on different sheets of the Riemann surface. Therefore certain qualitative correspondence will be established between poles on different sheets, the poles on the second sheet being called shadow poles.

Figure 5.5 shows the trajectories along which the S-matrix poles corresponding to the ${}^2\Sigma_g^+$ states of $H_2^-$ move as R changes. For large R, there exists, on the first channel sheet, a bound state pole lying on the positive part of the imaginary axis $k_1$. As R decreases, this pole moves downwards and coincides with a pole corresponding to a virtual state. With a further decrease in R, these two poles leave the imaginary axis $k_1$ and move along close to the real axis $k_1$ and approaching it. At $R = 2.8$, these poles arrive at the branch points at $k_1 = \pm (2\varepsilon R)^{\frac{1}{2}}$ and move through them passing onto the second sheet.* A pair of shadow poles on the second sheet move in a similar way, but they cannot pass through the cut (this would contradict the unitarity of the S-matrix) and, therefore, they always remain on the second sheet. The real parts of the complex E of these two pairs of poles are shown as curves a and b in Fig. 5.4.

For large $r_1$, the asymptotic form of the quasi-stationary state ${}^2\Sigma_g^+$ is

$$\sqrt{3}\,\Phi(\underline{r}_1,\sigma_{1z};\ \underline{r}_2,\sigma_{2z};\ \underline{r}_3,\sigma_{3z})\Big|_{r_1 \to \infty} \sim$$

$$\phi_1(\underline{r}_2,\sigma_{2z};\ \underline{r}_3,\sigma_{3z})\ q_1(\theta_1,\phi_1)\ u_1\ \frac{e^{ik_1 r_1}}{r_1} + \phi_3^{(1)}(\underline{r}_2,\sigma_{2z};\ \underline{r}_3,\sigma_{3z}) \times$$

---

*As a consequence the molecular system has, at $R = 2.8$ (that is, for $k_1 = 0.502$), a "transparence energy." For this energy, there is a continuum wavefunction which has only a converging wave in the first channel, and only a diverging wave in the second channel.

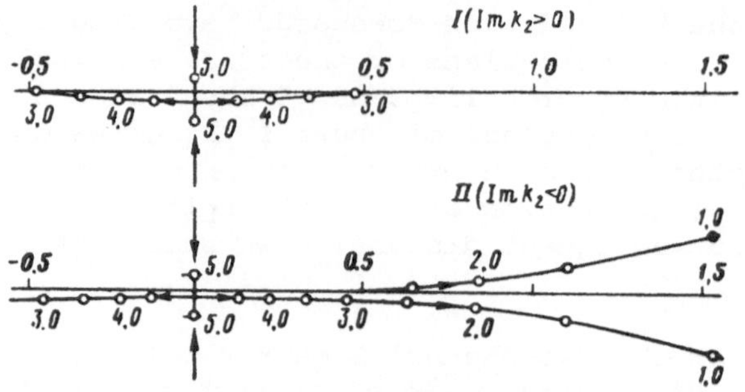


Fig. 5.5. Trajectories of the poles of the S-matrix for the $^2\Sigma_g^+$ states of the negative ion $H_2^-$ on the two-sheet surface of the complex momentum plane in the lowest channel $k_1$ (90). The numbers in the figure are values of R. The cuts have not been shown because the position of the branching points $k_{1b} = \pm\,(2\varepsilon(R))^{1/2}$ depends on the parameter R.

$$\times\ (2/3)^{\frac{1}{2}}\ q_2(\theta_1,\phi_1)\ v_1\ \frac{e^{ik_2r_1}}{r_1}\ +\ \phi_3^{(o)}(\underline{r}_2,\sigma_{2z};\ \underline{r}_3,\sigma_{3z})$$

$$\times\ (1/3)^{\frac{1}{2}}\ q_2(\theta_1,\phi_1)\ u_1\ \frac{e^{ik_2r_1}}{r_1}\ . \tag{5.3.3}$$

In (5.3.3), the wavefunction depends on both space- and spin-coordinates of the electrons; $u_1$ and $v_1$ are spin functions of the first electron and $r_1$, $\hat{\theta}_1$, and $\hat{\phi}_1$ are its spherical coordinates. The wavefunctions $\phi_S$ in equation (5.3.3) correspond to the following molecular states:

$$\phi_1 \rightarrow\ {}^1\Sigma_g\ ,$$

$$\phi_3^{(o)} \rightarrow\ {}^3\Sigma_u\quad (S_z = 0),$$

$$\phi_3^{(1)} \rightarrow\ {}^3\Sigma_u\quad (S_z = 1).$$

These functions as well as the angular momentum functions $q_1$ and $q_2$ of the escaping electron, which all depend on R, are assumed to be normalized. The quantity $k_1^{\,2}(R)/2$ is the difference between the complex energy of the $^2\Sigma_g^+$ $H_2^-$ term and the energy of the $^1\Sigma_g^+$ $H_2$ term; $k_2^{\,2}(R)/2 = k_1^2(R)/2 - \varepsilon(R)$.

If the real part of the energy of the $^2\Sigma_g^+$ $H_2^-$ term lies between the $^3\Sigma_u^+$ and $^1\Sigma_g^+$ terms of $H_2$, the negative ion $H_2^-$ can decay only to the ground state of the hydrogen molecule. In this case, the wavefunction has an exponentially increasing asymptotic form only in the first channel and it corresponds to the first pair of poles (see curve a in Fig. 5.4). If the ${}^2\Sigma_g^+$ term of $H_2^-$ lies above the ${}^3\Sigma_u^+$ term of $H_2$, there are two open channels to decay to: (i) the ground ${}^1\Sigma_g^+$ state and (ii) the excited ${}^3\Sigma_u^+$ state of the hydrogen molecule. In the latter case, the aymptotic form increases in both channels as $r \to \infty$. The second pair of poles (curve b of Fig. 5.4) describes this resonance, and it leads to a peak in the excitation cross section $\sigma_i$ (curve I in Fig. 5.2). Indeed, resonances in the scattering amplitude are due to those poles of the S-matrix which are close to the real E axis and lie on the sheet reached by starting in the upper half-plane of the physical sheet of the complex E plane and moving downwards. These poles (i.e., below the real $k_1$ axis) are on sheet I if the energy is below threshold, and on sheet II if the energy is above threshold. Thus, each of the two pairs of S-matrix poles related to the ${}^2\Sigma_g^+$ state describes resonances in its own region of R. Between these two R-regions there is an intermediate domain of R, where certain quasi-crossing in the complex plane takes place. The existence of this intermediate region may be of interest in the theory of dissociative recombination.

The quasi-stationary state ${}^2\Sigma_u^+$ is always below the excitation threshold in the second channel (see Fig. 5.4), provided that $R > 1$ (it may not be true for $R < 1$; however, in this case, the model itself is not applicable). Consequently, in this case the poles do not pass from one sheet to another. The general form of the trajectories of both pairs is the same as that of the second pair of poles for the ${}^2\Sigma_g^+$ states.

In addition to the total width of the terms, the relative probability $\Lambda(R)$ of disintegration via the two channels

$$H_2^-({}^2\Sigma_g^+) \to H_2({}^1\Sigma_g^+) + e \tag{5.3.4}$$

and

$$H_2^-({}^2\Sigma_g^+) \to H_2({}^3\Sigma_u^+) + e \tag{5.3.5}$$

is of importance for the theory of disintegration. If we take into account the usual physical interpretation (see reference [19] for instance) of the wavefunction of a quasi-stationary state (5.3.3) which behaves exponentially for large R, the following formula for $\Lambda(R)$ can be derived:

$$\Lambda(R) = \frac{\mathrm{Re}\, k_1(R)}{\mathrm{Re}\, k_2(R)} \; \frac{\int |q_1(\theta,\phi)|^2 \, d\Omega}{\int |q_2(\theta,\phi)|^2 \, d\Omega} \tag{5.3.6}$$

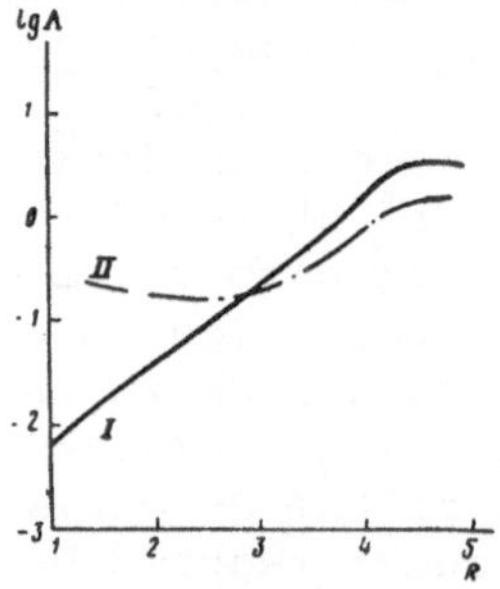


Fig. 5.6. Disintegration of the negative hydrogen ion $H_2^-$ in the state $^2\Sigma_g^+$. Dependence of log Λ upon R showing the ratio of the probabilities of disintegration via channels (5.3.4) and (5.3.5) from reference (39). I - ZRP approximation; II - separable potential approximation.

and

$$\Lambda(R) = \frac{\mathrm{Re}\, k_1}{\mathrm{Re}\, k_2} \left|\frac{c_1^+}{c_2^-}\right|^2 \times \frac{\dfrac{\mathrm{sh}(2\mathrm{Im}\, k_1 R)}{\mathrm{Im}\, k_1 R} + \dfrac{\sin(2\mathrm{Re}\, k_1 R)}{\mathrm{Re}\, k_1 R}}{\dfrac{\mathrm{sh}(2\mathrm{Im}\, k_2 R)}{\mathrm{Im}\, k_2 R} - \dfrac{\sin(2\mathrm{Re}\, k_2 R)}{\mathrm{Re}\, k_2 R}} . \tag{5.3.7}$$

The function Λ(R) computed from these expressions is shown above, in Fig. 5.6, together with the results obtained with the help of a model employing separable potentials.

## 5.4 ELECTRON SCATTERING BY MOLECULES IN THE SEPARABLE POTENTIAL APPROXIMATION

We shall now consider a many-component wavefunction (5.1.1), assuming that each component $\psi_\beta$ of the total wavefunction Ψ satisfies the equation

$$(-\frac{1}{2}\nabla^2 + \frac{1}{2}k_\beta^2)|\psi_\beta\rangle + \sum_{\gamma=1}^{M} |\phi_{\beta\gamma}\rangle\, v_{\beta\gamma}\, \langle\phi_{\beta\gamma}|\psi_\gamma\rangle = 0 . \tag{5.4.1}$$

Equations (5.4.1) represent a generalization of the theory for ZRPs developed in Sec. 5.1. These equations can be solved by a simple extension of the methods of Sec. 4.5 and Sec. 5.1, but we shall not concentrate on that here. Instead we shall turn directly to the case of an electron moving in a field of two atoms each of spin 1/2. The total spin of the system e + A may be either 0 or 1, and separable potentials $\hat{V}^a_s$ and $\hat{V}^a_t$, describing the interaction between the electron and atom in these two spin states, are

$$\hat{V}^a_s = |\phi_s> v_s <\phi_s|$$

and

$$\hat{V}^a_t = |\phi_t> v_t <\phi_t|,$$

respectively. Let us introduce the potentials

$$\hat{V}^a_{11} = \frac{1}{2}(\hat{V}^a_s + \hat{V}^a_t)$$

and

$$\hat{V}^a_{12} = \frac{1}{2}(\hat{V}^a_s - \hat{V}^a_t) ,$$

which are similar to the coefficients $\alpha^a_{11}$ and $\alpha^a_{12}$ in the boundary conditions stated in Sec. 5.1 for the case of ZRPs. Further, treatment of the two models of the system $AB^-$ will be identical. This becomes evident on noticing that both the separable potential matrix $V_{ij}$ in (5.4.1) and in the matrix of coefficients $\alpha_{ij}$ appearing in the ZRP model transform according to (5.1.10) under the unitary transformation (5.1.9).

In particular, when the total spin of the system $AB^-$, S, is S = 1/2, the coordinate function $\psi(\underline{r}_1,\underline{r}_2,\underline{r}_3)$ has the old form (5.1.12). However, now the one-electron functions $\psi_1'$ and $\psi_2'$ in the expression for $\psi$ have to satisfy the equations

$$(-\frac{1}{2}\nabla^2 - \frac{1}{2}k_1^2)\psi_1' + (\hat{V}^a_{11} - \frac{1}{2}\hat{V}^a_{12} + \hat{V}^b_{11} - \frac{1}{2}\hat{V}^b_{12})\psi_1'$$
$$+(\sqrt{3}/2)(-\hat{V}^a_{12} + \hat{V}^b_{12})\psi_2' = 0,$$
$$(-\frac{1}{2}\nabla^2 - \frac{1}{2}k_2^2)\psi_2' + (\hat{V}^a_{11} + \frac{1}{2}\hat{V}^a_{12} + \hat{V}^b_{11} + \frac{1}{2}\hat{V}^b_{12})\psi_2'$$
$$+(\sqrt{3}/2)(-\hat{V}^a_{12} + \hat{V}^b_{12})\psi_1' = 0. \quad (5.4.2)$$

The term splitting $\varepsilon(R)$ has been taken into account in equations (5.4.2). The scattering problem is reduced, in the usual way, to a system of linear algebraic equations. The (transcendental) equation for the energies of bound and quasi-stationary states of $AB^-$ is then obtained by equating the determinant of the system to zero. For identical atoms, A and B, the states can be classified according to parity and the resulting system of algebraic equations split into two systems of lower order.

We turn now to the molecular ion $H_2^-$ and assume that $\phi_s = \phi_t = \phi$ (this assumption is not necessary and is made for the sake of some simplifications only). Then the problem is reduced to two systems of algebraic equations of order two each. The potentials $\hat{V}_s$ and $\hat{V}_t$ take the form

$$\hat{V}_s = |\phi\rangle\, v_s\, \langle\phi|$$

and

$$\hat{V}_t = |\phi\rangle\, v_t\, \langle\phi|,$$

whereas the potentials $V_{11}$ and $V_{12}$ become of a one-term form similar to that of $\hat{V}_s$ and $\hat{V}_t$, but with different coefficients, namely,

$$V_{11}^{a,b} = |\phi_{a,b}\rangle\, v_{11}\, \langle\phi_{a,b}|$$

and

$$V_{12}^{a,b} = |\phi_{a,b}\rangle\, v_{12}\, \langle\phi_{a,b}|,$$

where

$$\phi_{a,b} = \phi(\underline{r} - \underline{R}_{a,b})$$

and

$$v_{11} = \frac{1}{2}(v_t + v_s), \quad v_{12} = \frac{1}{2}(v_t - v_s).$$

Electron scattering on the singlet state of the molecule is described by the functions

$$\psi_1' = e^{i\underline{k}_o\cdot\underline{R}} + c_1^a \chi_a^1 + c_1^b \chi_b^1,$$

$$\psi_2' = c_2^a \chi_a^2 + c_2^b \chi_b^2,$$

where $\chi_{a,b}^{\beta} = (-\tfrac{1}{2}\nabla^2 - \tfrac{1}{2}k_\beta^2)^{-1}|\phi_{a,b}\rangle$. The coefficients $c_1^{\pm}$ and $c_2^{\pm}$ defined by (5.2.2) satisfy the system of equations

$$(v_{11} - \frac{1}{2}v_{12})\left(\langle\phi_a|\chi_a^1\rangle \pm \langle\phi_a|\chi_b^1\rangle\right) c_1^{\pm} + c_1^{\pm}$$

$$- (\sqrt{3}/2)\, v_{12}\left(\langle\phi_a|\chi_a^2\rangle \mp \langle\phi_a|\chi_b^2\rangle\right) c_2^{\pm}$$

$$= -(v_{11} - \frac{1}{2}v_{12})\langle\phi|e^{ik_1 z}\rangle\left(e^{-\frac{1}{2}i\underline{k}_{o1}\cdot\underline{R}} \pm e^{\frac{1}{2}i\underline{k}_{o1}\cdot\underline{R}}\right),$$

and

$$(v_{11} + \frac{1}{2} v_{12}) \left(<\phi_a|\chi_a^2> \mp <\phi_a|\chi_b^2>\right) c_2^{\pm} + c_2^{\pm}$$
$$- (\sqrt{3}/2)\, v_{12} \left(<\phi_a|\chi_a^1> \pm <\phi_a|\chi_b^1>\right) c_1^{\pm}$$
$$= (\sqrt{3}/2)\, v_{12} <\phi|e^{ik_1 z}> \left(e^{-\frac{1}{2}i\, \underline{k}_{01}\cdot\underline{R}} \pm e^{\frac{1}{2}i\, \underline{k}_{01}\cdot\underline{R}}\right). \qquad (5.4.3)$$

The solution of this system can be written in the form

$$c_\beta^{\pm} = \eta^{-1}(k_\beta)\, \Phi_\beta^{\pm} \left(e^{\frac{1}{2}i\, \underline{k}_{01}\cdot\underline{R}} \pm e^{-\frac{1}{2}i\underline{k}_{01}\cdot\underline{R}}\right),$$

where the quantities $\Phi_\beta^{\pm}$ are defined by

$$\Phi_1^{\pm} = \mp (\Delta_\pm)^{-1} <\phi|e^{ik_1 z}> \eta(k_1) \Big[(v_{11}^2 - v_{12}^2)$$
$$\times \{<\phi_a|\chi_b^2> \mp <\phi_a|\chi_b^2>\} + (v_{11} - \frac{1}{2} v_{12})\Big],$$

$$\Phi_2^{\pm} = \pm (\Delta_\pm)^{-1} <\phi|e^{ik_1 z}> \eta(k_2)\, (\sqrt{3}/2)\, v_{12},$$

with $\Delta_\pm$ being the determinant of the system (5.4.3). The way the coefficients $\Phi_\beta^{\pm}$ have been defined above ensures that the expressions for the amplitudes and cross sections take the same form in the case of separable potentials as they do in the case of ZRPs, i.e., (5.2.7) and (5.2.8). The positions of the S-matrix poles in both cases are determined by the same equation (5.3.1).

Electron scattering by the hydrogen molecule and the energy terms of the molecular ion $H_2^-$ were obtained in the separable potential approximation in references [39] and [47]. Numerical studies by Ostrovskii [39, 47] showed that the two methods of choosing the potential (described in Sec. 1.4 of the present monograph), i.e., with the wavefunction (1.4.12) and parameters (1.4.19) or, alternatively, with the wavefunction (1.4.20) and parameters (1.4.21), give very similar results.

The separable potential approximation does not show any tendency to underestimate cross sections. In this approximation, cross sections both for elastic and inelastic scattering are larger than those of the ZRP method (see Figs. 5.7 and 5.2). In general, the elastic cross sections are in better agreement with more accurate calculations. For inelastic scattering with excitation of the triplet state of $H_2$, the separable potential approximation probably leads to an overestimation of the cross section (see Fig. 5.2). The result

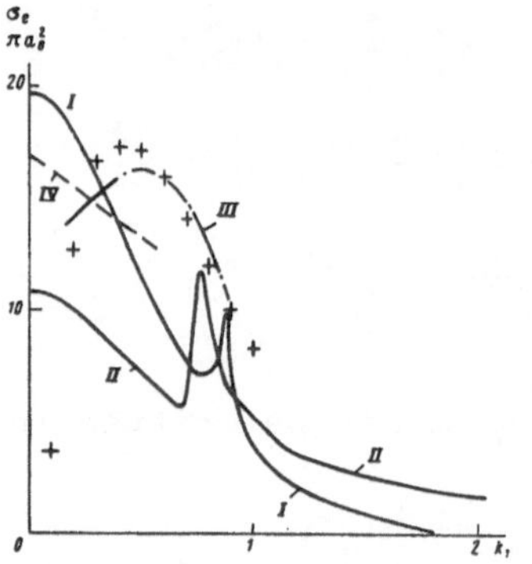


Fig. 5.7. Cross section of elastic electron scattering by the hydrogen molecule. I - separable potential approximation (47); II - ZRP approximation (non-physical resonance has been eliminated following the method of Sec. 5.3); III - experimental data (97); IV - calculation of Tully and Berry (100); +++ calculation of Wilkins and Taylor (99).

is much better than that given by the Born-Oppenheimer approximation (the latter greatly overestimates cross sections) but is slightly higher than that of Ochkur's method which, in turn, gives cross sections which are larger by a factor of two compared with the more accurate Ochkur-Rudge method.

The qualitative features of the movement of the S-matrix poles established in Sec. 5.4 for the ZRP approximation remain qualitatively the same for separable potentials. The computed positions and widths are shown in Figs. 5.8 and 5.9. We note that these calculations, as well as calculations in the ZRP approximation, show that in the neighborhood of the point of intersection between the $^3\Sigma_u^+$ term of $H_2$ and the real part of the $^2\Sigma_g^+$ term of $H_2^-$ the pair of poles lying on different sheets of the complex energy have nearly equal widths. There are two different pairs of poles corresponding to the physical quasi-stationary state depending on whether the state is above or below the disassociation threshold of the molecular ion $H_2$ via channel (5.3.5). It is clear that the transition from one pair of poles to another will be accompanied by smooth changes in energy and width. It remains to be seen if this is a specific property of the $H_2^-$ ion or if this is a more general result.

The positions of the crossing points into the continuum obtained for molecular terms in different approximations such as ZRP, separable potentials, or potentials of finite range (26), all agree with each other, suggesting that the results are trustworthy. On the other hand, variational calculations of the $^2\Sigma_g^+$ terms of $H_2^-$ give too large a value for the crossing point $R_c$ into the continuum. (According to Ostrovskii (47) and Demkov et al. (69), $R_{cu} \simeq 6$ a.u.). For large separations $R > R_c$, the term comes very near to the continuum boundary so that even a small error in the variationally determined binding energy may influence considerably the position of $R_c$. This is a possible explanation of the failure of the variational value for $R_c$. The introduction of the splitting $\varepsilon(R)$ shifts the crossing points of the terms towards larger distances to new positions which seem to be more accurate.

In the range $2 < R < 4.2$, the quasi-stationary $^2\Sigma_g^+$ term of the negative molecular ion $H_2^-$ lies above the $^3\Sigma_u^+$ term of the hydrogen molecule $H_2$ in the separable potential approximation and lies much nearer to the $^3\Sigma_u^+$ $H_2$ term than according to calculations using ZRPs. This is also in agreement with variational calculations of Bardsley et al. (112) (provided that the $H_2$ terms and $H_2^-$ terms have been obtained with the same accuracy).

For $R < 2$, the $^2\Sigma_g^+$ term of the molecular ion $H_2^-$ lies below the $^3\Sigma_u^+$ term of the hydrogen molecule $H_2$. This agrees with the relative position of the terms in the united atom limit (corresponding to the $^2S$ resonance of $He^-$ and the (1s2p) $2^3P$ state of He; see reference (112)). In the ZRP approximation, all S-matrix poles on the Riemann surface of the complex energy plane move to infinity as $R \to 0$. This distorts considerably the behavior of the terms, for small R. The separable potential approximation is free from these defects.

The reliability of the approximations described above worsens as R decreases. However, the introduction of the singlet-triplet splitting $\varepsilon(R)$ enables the disintegration of the negative molecular ion $H_2^-$ via different channels to be described as well as improving the theory near the crossing point of the $^2\Sigma_u^+$ $H_2^-$ term into the

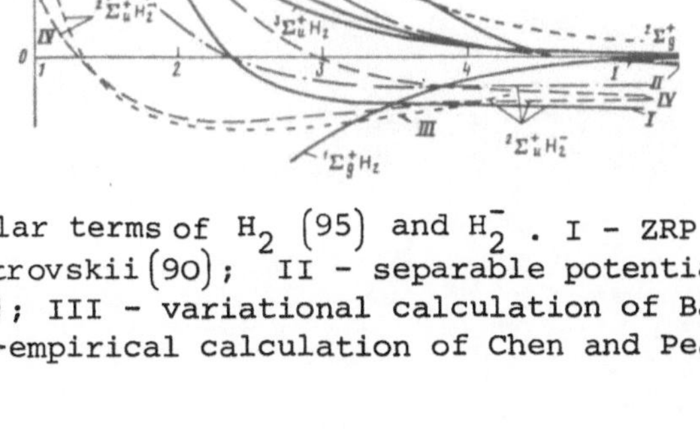

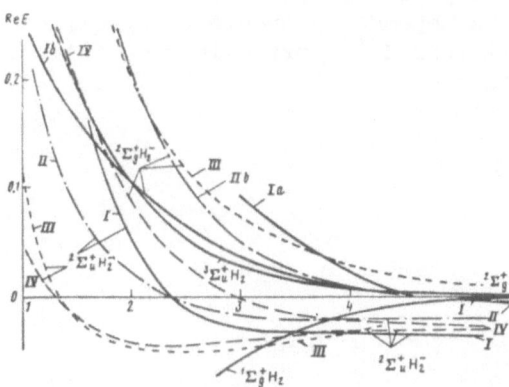

Fig. 5.8. Molecular terms of $H_2$ (95) and $H_2^-$. I - ZRP approximation of Demkov and Ostrovskii (90); II - separable potential approximation of Ostrovskii (47); III - variational calculation of Bardsley et al. (112); IV - semi-empirical calculation of Chen and Peacher (114).

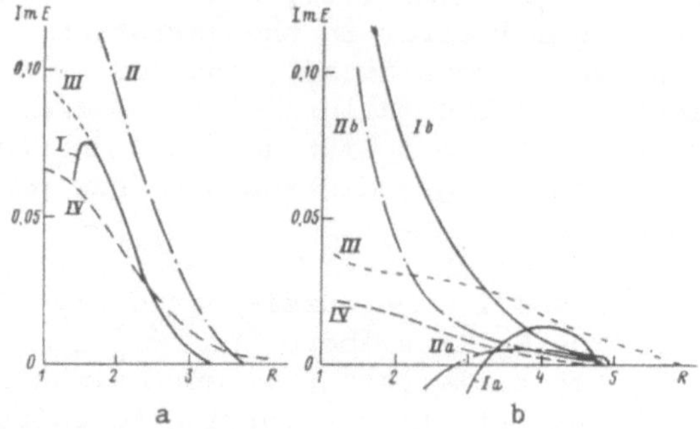


Fig. 5.9. Widths of molecular terms: (a) $^2\Sigma_u^+$ $H_2^-$ and (b) $^2\Sigma_g^+$ $H_2^-$. (The notation is the same as in Fig. 5.8.)

continuum. Allowing for the finite range of the atomic potentials and their superpositions improves the results in the range R = 2-3. In particular, the relative positions of the terms of $H_2$ and $H_2^-$ are improved.

There is another physical phenomenon which gains specific importance when the separation R becomes small, that is, the mutual distortion of the electron shells of the atom. Partly, this distortion is taken into account by introducing the singlet-triplet splitting $\varepsilon(R)$.

Variations in atomic parameters (for example, in the parameter $\alpha$ for the boundary conditions) create effects which are, in a certain sense, of next higher order, and they are important only for small R. The best way of introducing the R dependence of the atomic parameters is probably by using a semi-empirical approach. In other words, this dependence should be specified in such a way that the model will reproduce correctly some reliable experimental or theoretical data.

## Chapter 6

# MOTION OF A PARTICLE IN A PERIODIC FIELD OF ZERO-RANGE POTENTIALS

## 6.1 ONE-DIMENSIONAL LATTICE IN A THREE-DIMENSIONAL SPACE. BOUND STATES

In all applications considered in the previous chapters of this monograph, it was assumed that the particle was in a field of a finite number of ZRPs. Now we shall consider an infinite number of potential wells. This will require the calculation of infinite sums extended over all force centers. A physically important model is an infinite number of identical wells forming a regular (periodic) structure similar to that of a crystal. The regular feature of this model facilitates the calculation of the sums. It can be considered as a generalization of the well-known Kronig-Penney model (81). In the latter, a particle moves in the field of a one-dimensional periodic lattice in a one-dimensional space. In this and the next section, we shall consider a one-dimensional lattice in a three-dimensional space (a model of an electron moving in the field of a polymer molecule), where the result of the summation can be written in an analytical form. In Sec. 6.3 and Sec. 6.4 we shall study the case of two- and three-dimensional lattices and discuss various approximate methods of calculating the sums.

Following Demkov and Subramanyan (115), we consider an infinite number of identical ZRPs placed on the x-axis at points $x = na$, where $n = 0, \pm 1, \pm 2, \ldots$. According to Bloch's theorem, the solution of the Schrödinger equation with the periodical potential is

$$\psi_q(\underline{r}) = \exp(iq\,\underline{\nu}\cdot\underline{r})\, u_q(\underline{r}) \,, \qquad (6.1.1)$$

where $\underline{\nu}$ is a unit vector directed along the x-axis, q is the quasi-momentum of the particle, and $u_q(\underline{r})$ is a periodic function

$$u_q(\underline{r} + m\underline{a}) = u_q(\underline{r}), \qquad \underline{a} = \nu\,\underline{a}, \tag{6.1.2}$$

and m is an integer. For a lattice formed by ZRPs, the solution of the Schrödinger equation which satisfies Bloch's theorem is

$$\psi(\underline{r}) = c \sum_n \frac{\exp(inqa - \kappa|\underline{r} - n\underline{a}|)}{|\underline{r} - n\underline{a}|} . \tag{6.1.3}$$

Making use of the boundary condition (1.2.7), we obtain the relation between q and $\kappa$, that is, between the quasi-momentum and the energy of the particle ($E = -\kappa^2/2$):

$$\alpha = \kappa - S(q,\kappa), \tag{6.1.4}$$

where

$$S = \sum_{n=-\infty}^{\infty}{}' \frac{1}{na} \exp(-|n|\kappa a + iqna)$$

$$= \sum_{n=1}^{\infty} \frac{1}{na} \exp(-n\kappa a) \cos(nqa). \tag{6.1.5}$$

The sum $\Sigma'$ in the first line of (6.1.5) excludes the term n = 0. After summation of the series for $S(q,\kappa)$, we obtain

$$\cosh \kappa a = \frac{1}{2} \{\exp(\alpha a) + 2 \cos(qa)\}. \tag{6.1.6}$$

For a positive energy $E = k^2/2$, we find, in a similar way, that

$$\alpha a = \ln\{2 \cos(ka) - 2 \cos(qa)\}. \tag{6.1.7}$$

Figure 6.1 shows the energy of the particle as a function of $\alpha a$ for three different values of the quasi-momentum q. For any value of $\alpha a$, there exists one and only one zone of negative energy (a discussion about zones with positive energy will be given in Sec. 6.2).

Interaction between potential wells is small for $\alpha a >> 1$ and the entire energy zone lies, in this case, near the position of the energy level of the isolated well, i.e., $\alpha \approx \kappa$. This result follows directly from (6.1.6) if cos(qa) is neglected in comparison with cosh ($\kappa$a), and cosh ($\kappa$a) is replaced by exp ($\kappa$a)/2. In this limiting case, it is easy to estimate the width of the energy zone. For the upper bound of the zone ($qa = \pi$), and for the lower bound of the zone (qa = 0), we have

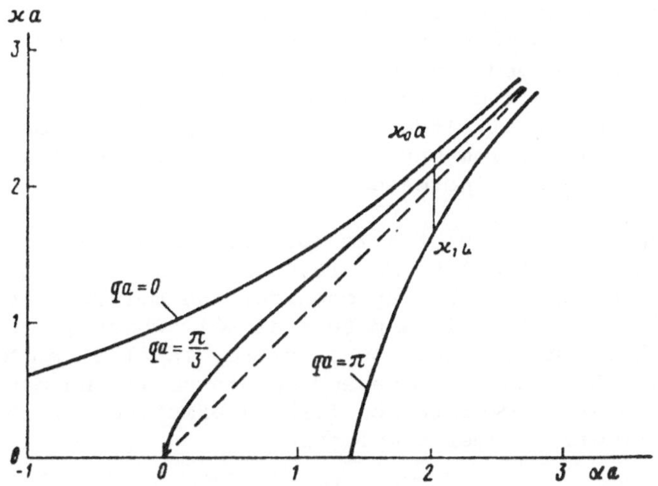


Fig. 6.1. One-dimensional periodic structure of ZRPs. Dependence of $\kappa a$, where $\kappa = (2|E|)^{1/2}$, on the parameter $\alpha a$ for different values of quasi-momentum q (115) (solid lines). Broken line: the energy level in an isolated well; $\kappa_0$ and $\kappa_1$ are the bottom bound and the upper bound of the zone, respectively.

$$\kappa_1 a \approx \ln\{\exp(\alpha a) - 2\} \approx \alpha a - 2\exp(-\alpha a),$$

and (6.1.8)

$$\kappa_0 a \approx \ln\{\exp(\alpha a) + 2\} \approx \alpha a + 2\exp(-\alpha a),$$

respectively. Then it follows from (6.1.8) that, for $\alpha a >> 1$, the width of the zone is

$$\Delta E = (\kappa_0^2 - \kappa_1^2)/2 \approx 4\alpha \exp(-\alpha a)/a. \tag{6.1.9}$$

Hence $\Delta E$ falls off exponentially as $\alpha a$ increases.

If $\alpha a$ decreases, the zone widens in such a way that the energy level of the isolated well always remains inside the zone. The latter is true due to the inequality

$$\frac{1}{2}(\exp|x| + 2) > \cosh(x) > \frac{1}{2}(\exp|x| - 2). \tag{6.1.10}$$

At $\alpha a = \ln 4$, the upper bound of the zone ($qa = \pi$) coincides with the continuum boundary ($E = 0$). For smaller $\alpha a$, including $\alpha a < 0$, the continuum boundary is reached at $qa < \pi$. However, the lower bound of the zone always remained in the region of negative energies. This follows from the fact that for $q = 0$, equation (6.1.6) has a solution at any value of $\alpha a$. This is similar to a two-dimensional motion of a particle in a field of a single potential well where there always exists a bound state.

One of the consequences of this theory is the conclusion that a linear polymeric molecule modelled by a one-dimensional lattice of ZRPs always has bound states if the molecule is sufficiently long. We note that this conclusion does not depend on a particular choice of this potential. Indeed, the ZRP approximation becomes exact, in the limit of low energies, for any potential which falls off rapidly enough as r increases.

For large negative $\alpha a$, it is possible to get an estimate of the critical number of atoms, $N_o$, in such a molecule required for a bound state to appear. We can consider the molecule as an N-link structure where each link is a potential well which is too shallow to bind the electron. For the lowest energy level in an N-link molecule, $Nqa \approx \pi$. Assuming that qa and ka are small and making use of (6.1.6), we obtain the criterion for the existence of the bound state in the linear polymeric molecule.

$$\frac{\pi^2}{2N_o^2a^2} < \frac{1}{2a^2}\exp(\alpha a),$$

or (6.1.11)

$$N_o > \pi\exp(-\alpha a/2).$$

A linear polymeric molecule which consists of $nN_o$ "links" will have n bound states.

Using this theory, we can also obtain an expression for the effective mass of a particle moving in the field of a one-dimensional lattice of ZRPs.

Making use of equation (6.1.6) and the relation between E and $\kappa$, we obtain, after simple calculations,

$$\left.\frac{d^2E}{dq^2}\right|_{q=0} = -\left(\kappa\frac{d^2\kappa}{dq^2}\right)_{\kappa=0} = \frac{\kappa_o a}{\sinh(\kappa_o a)}, \tag{6.1.12}$$

where $\kappa_o$ is the value $\kappa$ takes at $q = 0$. For the effective mass of the particle, $m^*$, we find

$$m^* = 1\Big/\left(\frac{d^2E}{dq^2}\right)_{q=0} = \frac{\sinh(\kappa_o a)}{\kappa_o a}. \tag{6.1.13}$$

With the help of equation (6.1.6) the effective mass of the particle can be written thus

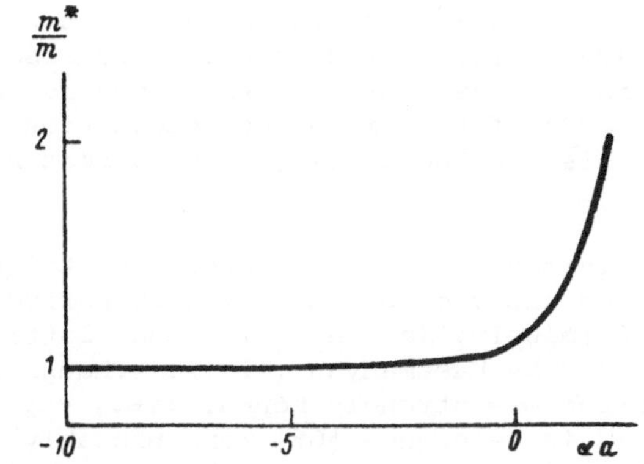


Fig. 6.2. Ratio $m^*/m$ as a function of the parameter $\alpha a$ (115). m is the mass of a free electron; m* is the effective mass of an electron in one-dimensional periodic structure of ZRPs.

$$m^* = \frac{(\Lambda^2 - 1)^{1/2}}{\ln\{\Lambda + (\Lambda^2 - 1)^{1/2}\}}, \tag{6.1.14}$$

where $\Lambda = \exp(\alpha a)/2 + 1$.

The $\alpha a$-dependence of $m^*/m$ is shown in Fig. 6.2. As $\alpha a$ increases (the well deepens), the effective mass $m^*$ also increases. For large negative $\alpha a$ (very shallow wells), $m \approx m^*$; that is, the effective mass is close to the real mass. The latter corresponds to free motion of the particle.

This model works correctly if the energy of the electron moving in a field of the molecule is low. For higher energies, the modelling of the electron-atom interaction by a single ZRP may be too crude. Another defect of this theory is the assumption that all atoms in the molecule are identical. Both defects can be rectified and better results can be obtained if a ZRP in each "link" of the molecule is replaced by several wells of, generally, different depth. This extension of the present theory will be considered in more detail in Sec. 6.4 for the general case of a three-dimensional periodic structure.

An interesting problem concerned with such periodic structures arises in connection with a hypothesis of Saxon and Hatner. Let us consider two periodic structures of the same period, one being realized by atoms of type A and another by atoms of type B. Let us further assume that E is a particular value of energy which lies in the forbidden zones of these structures. Then, according to the hypothesis, this E will lie in a forbidden zone for any alloy obtained by a partial replacement of some atoms A by atoms B. This hypothesis was tested, in the past, in one-dimensional models. Bhagwat and Subramanyan [116] showed that it held also in three-dimensional space, for a linear lattice which consisted of equidistant alternating ZRPs of two different intensities. However, no general proof that this hypothesis is always correct has been given up to now.

Dalidchick and Ivanov [66] investigated the states of an electron moving in the field of a finite linear periodic structure which had centers of impurity (a one-dimensional lattice with impurities was considered by Kasamanyan [117]). Though it was assumed there that the electron was strongly bound, i.e., $\exp(-\alpha a) \ll 1$, many results obtained in reference [66] are probably valid under more general conditions.

## 6.2 ELECTRON SCATTERING BY LONG LINEAR MOLECULES

We shall now consider electron scattering by one-dimensional periodic structures (polymeric molecules) following Demkov et al. [118]. This type of scattering process involves translational symmetry, which is a rather unusual property in scattering theory. This symmetry is not exact because of the finite size of the molecule. Nevertheless, its existence leads to some singular features in the differential and total cross sections. A similar situation arises in the theory of x-ray scattering [119].

First we have to formulate the scattering problem where the scatterers are infinitely long linear periodic structures. We shall start with a periodic structure of a finite length. In this case, two asymptotic regions ought to be considered. The first region corresponds to distances much smaller than the size of the structure itself. The second (external) region is that where the wavefunction can be written as a sum of the initial plane wave and the scattered wave. For the infinitely long structure, the second region does not exist, and in the first one the wavefunction is close to that for the case of a long finite size scatterer.

The infinitely long periodic structure possesses exact translational symmetry. It scatters incident electrons in the direction of the generating lines of the cones whose axis coincides with the axis of the structure and the vertices are at the scattering centers of the structure. (For simplicity we have shown in Fig. 6.3 only one of these cones.) For cone number m, the momentum k of the scattered electron satisfies the equation

$$\underline{K} \cdot \underline{a} = 2\pi m, \quad m = 0, \pm 1, \pm 2, \ldots, \tag{6.2.1}$$

where $\underline{K} = \underline{k} - \underline{k}_o$ is the transferred momentum. In other words, the component of the transferred momentum along the structure must be a multiple of the period of the reciprocal structure. This interference condition ensures that waves scattered by different centers add together coherently to give the total amplitude.

In reality, of course, we can observe scattering by scatterers of finite size, i.e., by long linear molecules consisting of N identical links (polymeric molecules). Under normal conditions, these molecules are randomly oriented, and therefore we shall be interested in cross sections averaged over all possible orientations of the molecules (this is analogous to diffraction in x-ray structure analysis).

Let us introduce the critical angle of scattering, $\theta_m$, for cone number m:

$$\theta_m = 2 \arcsin (m\pi/ka), \quad m = 1,2,\ldots. \tag{6.2.2}$$

In Fig. 6.3(b), the relative orientation of vectors $\underline{k}_o$ and $\underline{a}$ is such that there exists scattering through angle $\theta_m$ (as well as through angles $\theta > \theta_m$ along other generating lines of the cone). For any other orientation of the molecule with respect to $\underline{k}_o$, the least angle of scattering is greater than $\theta_m$ (Fig. 6.3(a), (c)). Therefore as a result of averaging, the m-th cone will contribute only to scattering through angles larger than $\theta_m$.

Now we shall investigate the singularity of the averaged differential cross section at $\theta = \theta_m$. Let us consider the integral form of the Schrödinger equation for the scattered electron:

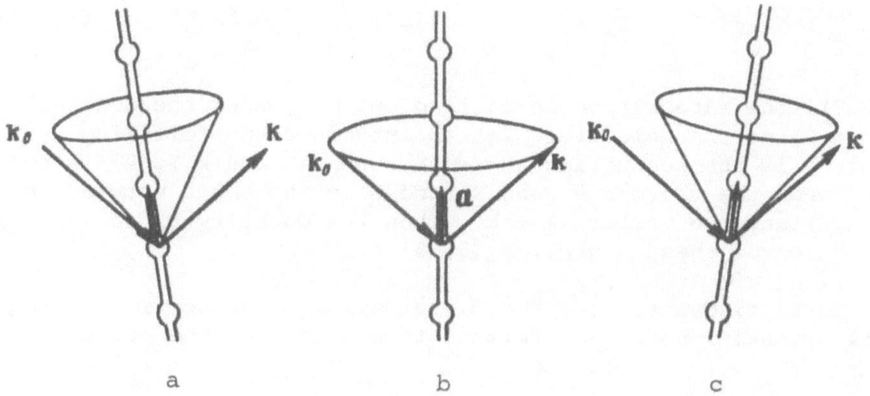


Fig. 6.3. Scattering of electrons by a long linear structure takes place in the direction of the cone surfaces. $\theta_m$ is an angle between $\underline{k}_o$ and $\underline{k}$. The diagrams a - c show that for any relative orientation of $\underline{k}_o$ and $\underline{a}$, the cone contributes to scattering only through angles $\theta > \theta_m$.

$$\psi_{\underline{k}_o}(\underline{r}) = e^{i\underline{k}_o \cdot \underline{r}} - \int \frac{e^{ik|\underline{r} - \underline{r}'|}}{2\pi\, |\underline{r} - \underline{r}'|} V(\underline{r}')\psi_{\underline{k}_o}(\underline{r}')d\underline{r}', \quad (6.2.3)$$

where $V(\underline{r})$ is the interaction potential between the electron and a linear periodic structure of finite length. We assume that the wavefunction $\psi_{\underline{k}_o}(\underline{r})$ satisfies approximately Bloch's theorem

$$\psi_{\underline{k}_o}(\underline{r} + m\underline{a}) = e^{imqa}\, \psi_{\underline{k}_o}(\underline{r}) \quad (6.2.4)$$

provided that both vectors $\underline{r}$ and $\underline{r} + m\underline{a}$ are in the region where the potential $V(\underline{r})$ differs from zero. The quasi-momentum $q$ in (6.2.4) must be chosen so that the first term on the right-hand side of (6.2.3), that is, the plane wave, also satisfies Bloch's theorem (6.2.4). This gives

$$qa = \underline{k}_o \cdot \underline{a}. \quad (6.2.5)$$

The scattering amplitude $F$ is obtained in the usual way by considering the asymptotic expression for the right-hand side of (6.2.3) when $r \gg Na$, that is, in the external asymptotic region. Making use of (6.2.4) and (6.2.5) as well as of the periodicity of the potential $V(\underline{r})$, we find, after simple manipulations, that

$$F(\underline{k}_o, \underline{k}, \underline{a}) = \frac{\sin(NKa/2)}{\sin(Ka/2)} B_N(\underline{k}_o, \underline{k}, \underline{a}), \quad (6.2.6)$$

where

$$B_N(\underline{k}_o, \underline{k}, \underline{a}) = -\frac{1}{2\pi} \int e^{-i\underline{k}\cdot\underline{r}'} V(r')\psi_{\underline{k}_o}(\underline{r}')d\underline{r}'. \quad (6.2.7)$$

In (6.2.7), the integration is carried out only over the central cell ($m = 0$) of the molecule. Let us introduce the following notation: $\theta$ is the scattering angle (between $\underline{k}$ and $\underline{k}_o$), $\theta'$ is the angle between the vectors $\underline{a}$ and $\underline{K}$, and $\phi$ is the angle between the plane $(\underline{a}, \underline{K})$ and the vector $\underline{k} + \underline{k}_o$. Then the quantity $B_N$ in (6.2.7) is a function of these angles, $B_N(\theta, \theta', \phi)$.

We shall also introduce the differential cross section averaged over all orientations $\underline{a}$ and related to one cell of the structure:

$$\tilde{\sigma}_N(\theta) = \frac{1}{4\pi N} \int_0^{\pi} \sin\theta'\, d\theta' \int_0^{2\pi} d\phi\, |F(\underline{k}_o, \underline{k}, \underline{a})|^2. \quad (6.2.8)$$

For a further discussion, it is convenient to introduce the derivative of the cross section (6.2.8), $\tilde{\sigma}'_N(\theta)$. Making use of (6.2.6) and (6.2.8) and integrating by parts, we obtain the following result:

$$\begin{aligned}\tilde{\sigma}'_N(\theta) &= \frac{1}{8\pi N}\cot\frac{\theta}{2}\,\frac{\sin^2(NKa/2)}{\sin^2(Ka/2)}\\ &\times\int_0^{2\pi} d\phi\{|B_N(\theta,\,0,\,\phi)|^2 + |B_N(\theta,\,\pi,\,\phi)|^2\}\\ &+ \frac{1}{4\pi N}\int_0^{\pi}\sin\theta' d\theta'\int_0^{2\pi} d\phi\,\frac{\sin^2(NKa/2)}{\sin^2(Ka/2)}\\ &\times\left[\frac{\partial}{\partial\theta}|B_N|^2 + \frac{1}{2\sin\theta'}\cot\frac{\theta}{2}\,\frac{\partial}{\partial\theta'}(\cos\theta'\,|B_N|^2)\right].\end{aligned} \tag{6.2.9}$$

The second term in (6.2.9) is finite as $N \to \infty$, at $\theta \neq 0$, whereas the first term displays singular behavior as N increases. Indeed, the function

$$\delta_p(x,\,N) = \sin^2(Nx)/(\pi N\sin^2 x)$$

has a $\delta$-function type of behavior, for large N, with maxima at the points $x = m\pi$ $(m = 0,\,\pm1,\,\pm2,\,\ldots)$; thus

$$\delta_p(m\pi,\,N) = N/\pi \tag{6.2.10}$$

and

$$\lim_{N\to\infty}\delta_p(x,N) = \delta_p(x) = \sum_{m=-\infty}^{\infty}\delta(x + m\pi). \tag{6.2.11}$$

This shows that the cross section $\tilde{\sigma}_\infty(\theta)$ for an infinitely long periodic structure is discontinuous at points $\theta_m$, equation (6.2.2). As $\theta$ increases, the increments at these points (see also [119]) are given by

$$\begin{aligned}(\Delta\tilde{\sigma})_m &= \tilde{\sigma}_\infty(\theta_m + 0) - \tilde{\sigma}_\infty(\theta_m - 0)\\ &= \frac{1}{2m}\left[|B_\infty(\theta_m,\,0,\,\phi)|^2 + |B_\infty(\theta_m,\,\pi,\,\phi)|^2\right].\end{aligned} \tag{6.2.12}$$

In (6.2.12), $|B_\infty|^2$ do not actually depend on $\phi$ because the values 0 and $\pi$ assumed there for angle $\theta$ correspond to the poles of the sphere. Generally, the increment $(\Delta\tilde{\sigma})_m$ at $\theta = \theta_m$ is of the same order of magnitude as the cross section itself.

In the case of large finite N, the cross section $\tilde{\sigma}_N(\theta)$ rises sharply by a quantity $(\Delta\tilde{\sigma})_m$ in the vicinity of the point $\theta_m$. The derivative $\tilde{\sigma}'_N(\theta_m)$ at this point increases linearly as N increases (see equations (6.2.9) and (6.2.10) above):

$$\tilde{\sigma}'_N(\theta_m)\Big|_{N\to\infty} = N\frac{ka}{2\pi}\cos\frac{\theta_m}{2}(\Delta\tilde{\sigma})_m + O(1). \qquad (6.2.13)$$

For the total scattering cross section, its derivative with respect to momentum is discontinuous at points $k_m = m\pi/a$ as shown by Demkov et al. [118].

The condition (6.2.4) is exactly satisfied only as $N \to \infty$. For finite N, there is a departure from (6.2.4), the maximum error occuring at the ends of the linear molecular structure. Bloch's theorem holds exactly within the Born approximation. On the other hand, as N increases, the value of $\psi_{ko}$ inside the potential range approaches that of $\psi_{k_o}$ for an infinite structure, and the latter satisfies the condition of Bloch's theorem exactly.

In order to judge at what N the singular features in the cross section become noticeable, it is convenient to consider scattering by a ZRP linear lattice. Some numerical results for differential cross sections are shown graphically in Fig. 6.4 (see also reference [119]). The arrows in the diagrams mark the scattering angles $\theta_m$ at which the general theory, for $N \to \infty$, predicts a discontinuity of the cross section. The linear dependence of the derivative $\tilde{\sigma}'_N(\theta_m)$ upon N given by (6.2.13) is clearly confirmed by the model calculation.

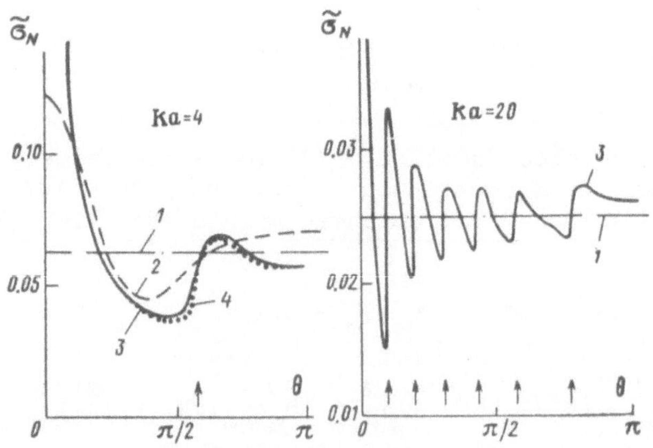


Fig. 6.4. Electron scattering by a linear structure of ZRPs (for $\alpha a = 0.01$). Differential cross sections $\tilde{\sigma}_N$, for different N and ka [118], are averaged over all orientations and given in units of $a^2$. 1 - N = 1; 2 - N = 6; 3 - N = 10; 4 - N = 16.

Vibrational motion of the nuclei distorts the periodicity of the molecular potential and it may influence scattering. The order of magnitude of the displacement from the equilibrium position of the nuclei may be estimated from the formula

$$\Delta \sim (k_B TN)^{\frac{1}{2}}/(\omega_o M^{\frac{1}{2}}),$$

where T is the Kelvin temperature of the gaseous target, $k_B$ is the Boltzmann constant, $\omega_o$ is the vibrational frequency for a pair of neighboring atoms in the molecule, and M is the mass of the atom. One must demand

$$k\Delta << \pi \tag{6.2.14}$$

in order to ensure that the interference picture is not distorted. On the other hand, the singular features become observable if

$$ka > \pi. \tag{6.2.15}$$

Model calculations with ZRPs show that the interference effects are clearly developed, even for not very large N, if $\Delta/a << 1$. In this case it is possible to reconcile the criteria (6.2.14) and (6.2.15) and, by selecting the energy E of incident electrons ($E \gtrsim 150$ eV), to observe nearly singular features of differential cross sections.

For an infinitely long periodic lattice of ZRPs, $B_\infty$ can be found in an analytical form:

$$B_\infty = a\{\ln(2\cos ka - 2\cos qa) - \alpha a\}^{-1}. \tag{6.2.16}$$

Consider $B_\infty$ for a fixed value of q as a function in the complex k plane. Then $B_\infty$ has an infinite number of logarithmic branch points:

$$k_b^{(n)} = 2\pi n/a \pm q, \quad n = 0, \pm 1, \pm 2, \ldots.$$

We shall make cuts in the complex k plane which start from the branch points $k_b^{(n)}$ and are directed downwards parallel to the imaginary k axis. The selection of the physical sheet k out of the infinite number of sheets of the Riemann surface is determined by the requirement that the bound states given as roots of equation (6.1.6) have to be the poles of the scattering amplitude and, therefore, of the function $B_\infty$.

Thus, if k lies on the positive imaginary axis, the principal value of the many-valued function ln(x) ought to be taken in equation (6.2.16). The path along the real k axis must avoid the branch points $k_b^{(n)}$ by means of small semicircles in the upper half-plane. Therefore if k is real and there are m branch points in the interval (0,k), then

$$B_\infty = a\{\ln|2\cos ka - 2\cos qa| - im\pi - \alpha a\}^{-1}. \qquad (6.2.17)$$

The same result can be obtained from (6.2.16) if the periodicity of the infinite sums arising there is accurately taken into account.

Now we can turn our attention to the properties of positive energy zones which exist according to equation (6.1.7). This is an interesting problem because these zones do not correspond to bound states of the electron (the wavefunctions fall off too slowly in the direction normal to the linear structure). One might have expected these zones to lead to some singularities in the scattering amplitude. The above discussion shows that this will not happen because the corresponding poles of the scattering amplitude lie on remote nonphysical sheets of the complex surface. By a special selection of $\alpha$ it is possible, of course, to ensure that function $B_\infty$ has a pole on the physical sheet in the interval $0 < k < q$. In the problem we are considering, $qa = \underline{k_0}\cdot\underline{a} \ll ka$, which follows from (6.2.5), and this special pole will correspond, at most, to a virtual state. Numerical calculations [118] confirm the validity of the formula (6.2.17).

## 6.3 TWO-DIMENSIONAL LATTICE IN THREE-DIMENSIONAL SPACE

Let us consider a two-dimensional square lattice of ZRPs with period a. For a particle moving in the field of such a structure, we obtain, similarly to Sec. 6.1, the following equation connecting the energy of the particle and its quasi-momentum (the latter has now two components $q_1$ and $q_2$ corresponding to two coordinate axes in the lattice):

$$\alpha = \kappa - S(q_1, q_2, \kappa). \qquad (6.3.1)$$

In (6.3.1),

$$S(q_1, q_2, \kappa) = \sum_{m,n=-\infty}^{\infty}{}' \frac{e^{-\kappa a(n^2+m^2)^{1/2}}}{a(m^2+n^2)^{1/2}} e^{i(q_1 m + q_2 n)a} \qquad (6.3.2)$$

where the prime means that the center $n = 0$, $m = 0$, where we have placed the origin of the coordinates, has to be excluded from summation. Following Subramanyan [74], we shall write (6.3.2) as a sum of two terms $S = S_1 + S_2$, where

$$S_1 = 2\sum_{n=1}^{\infty} \frac{e^{-\kappa an}}{an} \{\cos(q_1 an) + \cos(q_2 an)\} \qquad (6.3.3)$$

and

$$S_2 = 4 \sum_{n=1}^{\infty} \sum_{m=1}^{\infty} \frac{e^{-\kappa a(m^2+n^2)^{1/2}}}{a(m^2+n^2)^{1/2}} \cos(q_1 am) \cos(q_2 an). \quad (6.3.4)$$

The first sum $S_1$ runs over centers $\underline{R}_{nm}$ lying on the x- and y-axes, whereas the second sum $S_2$ includes all other centers of the lattice. According to Sec. 6.1, $S_1$ can be obtained in an analytical form:

$$S_1 = -\ln\{4(\cosh \kappa a - \cos q_1 a)(\cosh \kappa a - \cos q_2 a)\}/a.$$

In order to derive a more convenient expression for $S_2$, we first use the formula

$$\frac{e^{-\kappa a(m^2+n^2)^{1/2}}}{(m^2+n^2)^{1/2}} = \frac{2}{\pi} \int_0^{\infty} K_0\left(m\sqrt{(\kappa a)^2 + x^2}\right) \cos nx\, dx \quad (6.3.5)$$

to rewrite $S_2$ as follows:

$$S_2 = \frac{1}{a} \int_0^{\infty} dx \sum_{m=1}^{\infty} \sum_{n=1}^{\infty} K_0\left(m\sqrt{x^2 + (\kappa a)^2}\right)$$

$$\times \cos nx \cos mq_1 a \cos nq_2 a. \quad (6.3.6)$$

For the partial sum over m in (6.3.6), which has the form

$$\Lambda(a, b) = \sum_{m=1}^{\infty} K_0(ma) \cos mb,$$

we use the integral representation for the Bessel function $K_0$:

$$K_0(ma) = \int_0^{\infty} e^{-ma\cosh t} dt,$$

and obtain the following integral formula for the partial sum $\Lambda$:

$$\Lambda(a, b) = \int_0^{\infty} dt\, \{e^{-a\cosh t} \cos b - e^{-2a\cosh t}\}$$

$$\times \{1 - 2e^{-a\cosh t} + e^{-2a\cosh t}\}^{-1}.$$

For the second partial sum (over n) in (6.3.6), one can find with

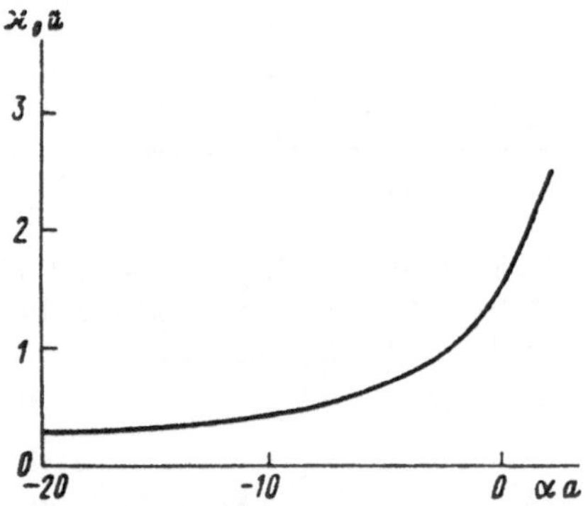


Fig. 6.5. Positions of the bottom of the energy zone for an electron in a two-dimensional square lattice formed by identical ZRPs. Dependence of $\kappa_o a$ on the parameter $\alpha a$ (74).

the help of (6.2.11) that

$$\sum_{n=1}^{\infty} \cos nx \cos nq_2 a = -\frac{1}{2} + \pi \sum_{m=-\infty}^{\infty} \{ \delta(x + q_2 a - 2m\pi) + \delta(x - q_2 a - 2m\pi)\} .$$

Finally, combining the two results above, we find that

$$S_2 = -\frac{1}{2a}\int_0^{\infty} dx \int_0^{\infty} dt \frac{B\cos q_1 a - B^2}{1 - 2B\cos q_1 a + B^2} + \frac{\pi}{a}\sum_{m=-\infty}^{\infty}\int_0^{\infty} dt \frac{b_m \cos q_1 a - b_m^2}{1 - 2b_m \cos q_1 a + b_m^2} ,$$

where

$$B(x, t) = \exp\{-\sqrt{x^2 + (\kappa a)^2}\cosh t\},$$

and

$$b_m(t) = \exp\{-\sqrt{(2m\pi + q_2 a)^2 + (\kappa a)^2}\cosh t\}.$$

In the general case, this expression cannot be simplified. However, even in its present form, it is convenient enough for various numerical calculations because the exponential factors in $S_2$ fall off very rapidly if the energy parameter $\kappa$ is sufficiently large. Figure 6.5 shows the position of the bottom of the zone ($\kappa = \kappa_0$) as a function of $\alpha a$. We note that an alternative way of computing the sum S is to use the generalized Ewald's method. This method will be described in detail in the next section.

## 6.4 THREE-DIMENSIONAL LATTICE AND THE METHOD OF EWALD

We shall now consider a three-dimensional cubic lattice of identical ZRPs. We start with the states of positive energy $E = k^2/2$ of a particle moving in the field of the lattice. As before, the equation for an energy zone is

$$\alpha = -ik - S'(\underline{q}, k), \tag{6.4.1}$$

$$S'(\underline{q}, k) = \sum_{\underline{R}_n}{}' \exp(i\,\underline{q}\cdot\underline{R}_n)\exp(ikR_n)/R_n. \tag{6.4.2}$$

In (6.4.2), $R_n$ is the position vector of the n-th center in the lattice, and the summation in (6.4.2) excludes the center n = 0, where the origin of the coordinate system has been placed. It will be shown below that

$$\operatorname{Im} S'(\underline{q}, k) = -k \tag{6.4.3}$$

so that the right-hand side of (6.4.1) is real. Equation (6.4.1) is valid for both positive and negative energies, that is, for real and imaginary k, respectively. Let us introduce the function

$$S(\underline{q}, k, \underline{r}) = \sum_{\underline{R}_n} \frac{\exp\{-i\underline{q}\cdot(\underline{R}_n - \underline{r}) + ik|\underline{r} - \underline{R}_n|\}}{|\underline{r} - \underline{R}_n|}. \tag{6.4.4}$$

S is a periodic function of $\underline{r}$ whose period is the same as that of the lattice. Applying the Fourier transform, this function can be written in the form of a sum over vectors $\underline{p}$ of the reciprocal lattice:

$$S(\underline{q}, k, \underline{r}) = \frac{4\pi}{v_c}\sum_{\underline{p}} \frac{\exp(i\underline{p}\cdot\underline{r})}{P^2 - k^2}, \tag{6.4.5}$$

where $\underline{P} = \underline{p} + \underline{q}$ and $v_c$ is the volume of a unit cell. From equations (6.4.2), (6.4.4), and (6.4.5), we obtain

$$S'(\underline{q}, k) = \lim_{r \to 0} \{\exp(i\,\underline{p}\cdot\underline{r})\, S(\underline{q}, k, \underline{r}) - \exp(ikr)/r\} =$$

$$= \lim_{r \to 0} \left( \frac{4\pi}{v_c} \sum_{\underline{p}} \frac{e^{i\underline{P}\cdot\underline{r}}}{P^2 - k^2} - \frac{e^{ikr}}{r} \right) \tag{6.4.6}$$

It follows from (6.4.6) that (6.4.3) holds; hence

$$S'(\underline{q},k) = \text{Re}\, S'(\underline{q},k) - ik. \tag{6.4.7}$$

On account of (6.4.6), equation (6.4.1) can be rewritten in the form

$$\alpha = \frac{4\pi}{v_c} \left( \frac{v_c}{(2\pi)^3} \int \frac{d\underline{p}}{p^2} - \sum_{\underline{p}} \frac{1}{P^2 - k^2} \right), \tag{6.4.8}$$

where both the integral and the sum in (6.4.8) diverge as $\underline{p} \to \infty$. Replacing, for large $\underline{p}$, the sum above by an integral, it is easy to see that the two divergent terms cancel each other so that $\alpha$ given by (6.4.8) is, in fact, finite.

The equation for the energy of a zone in the momentum representation, (6.4.8), was first obtained by Goldberger and Seitz (6), who also considered the motion of a nearly free particle in a three-dimensional lattice (the weak coupling approximation in zone theory) in order to describe neutrons moving in crystals. It is preferable to use the impulse approximation in this problem because $q \approx k$, and therefore the main contribution to the sum (6.4.8) comes from the term with $\underline{p} = 0$. The inclusion of this term corresponds to replacing the lattice by a constant pseudopotential of the Fermi type. We have

$$q^2 - k^2 = 4\pi n_o/\alpha ,$$

where $n_o = 1/v_c$ is the density of the medium, so that the refractive index of the medium, $\mu = q/k$, is given by

$$\mu = \left( 1 + 4\pi n_o/(\alpha k^2) \right)^{\frac{1}{2}}. \tag{6.4.9}$$

However, this formula is not applicable when $\underline{q}$ is such that the corresponding vector $\underline{p}$ of the reciprocal lattice is non-zero and the quantities $\underline{q}^2$ and $(\underline{q} + \underline{p})^2$ differ only slightly from each other (the latter is, in fact, Bragg's condition for reflection from crystal planes). Then a term in the sum (6.4.8) corresponding to this particular vector $\underline{p}$ also gives a significant contribution. An account of this effect leads (as it always does, in fact, in the weak coupling approximation) to the zone energy spectrum and to forbidden zones (see Goldberger and Seitz (6) for details).

Propagation of a plane wave through a slab whose internal structure is an assembly of ZRPs has been considered by Baryshevskii et al. (120). They have found that for a regular distribution of ZRPs in the slab (that is, for a crystal) the refractive index $\mu$ is real and given by (6.4.9). For random distributions of identical ZRPs, on the other hand, the refractive index is

$$\mu = \{1 + 4\pi n_o f/k^2\}^{\frac{1}{2}} , \tag{6.4.10}$$

where $f = -(\alpha + ik)^{-1}$ is the scattering amplitude on one center. In the second case, the non-zero imaginary part of $\mu$ is due to the incoherent scattering inside the slab causing damping of the incident wave. The same authors have also studied scatterers with spin.

In the opposite case of a strong interaction considered by Maleev (65), it is more convenient to use the equations written in the coordinate representation, (6.4.2) and (6.4.3). Because $ik \approx -\alpha$ and $\alpha a >> 1$, it is possible to retain in the sum only those terms which correspond to the nearest neighbors.

An intermediate case is more difficult to handle because neither the series over the direct lattice nor that over the reciprocal lattice have fast convergence. Summation of such series must be carried out in many problems of solid state physics, for instance, in calculations of the Madelung constant, i.e., the total electrostatic energy of ionic crystals, for a given structure of the lattice. These electrostatic sums are a particular case of (6.4.2) for $k = 0$ and for a specific value of $\underline{q}$ chosen in such a way that the signs alternate. One of the methods of performing summation in the electrostatic case suggested in (81) is to construct a sequence of spherical shells each enclosing a total electrical charge of zero value.

There exists another, very elegant method of summation proposed by Ewald for conditionally convergent lattice sums. In essence, this method achieves acceleration of the convergence by a suitable replacement of the sum being computed by two other sums, one over vectors of the direct lattice and another over vectors of the reciprocal lattice. Thus both the slow convergence of the direct lattice sum (6.4.4) due to the behavior of the terms at large r, and the slow convergence of the reciprocal lattice sum (6.4.5) due to the singularities arising as $|\underline{r} - \underline{R}_n| \to 0$, are taken into account. Below we shall consider a generalization of the Ewald method, proposed by Subramanyan (74) for sums of the form (6.4.4). Using the well-known identity

$$\frac{1}{r} = \frac{2}{\sqrt{\pi}} \int_0^\infty e^{-r^2\rho^2} d\rho ,$$

we transform $S(\underline{q}, k, \underline{r})$ to the form

$$S(\underline{q}, k, \underline{r}) = \int_0^\infty F(\underline{q},k,\underline{r},\rho)\, d\rho , \tag{6.4.11}$$

where

$$F(\underline{q},k,\underline{r},\rho) = \frac{2}{\sqrt{\pi}} \sum_{\underline{R}_n} \exp\{-i\underline{q}\cdot(\underline{r} - \underline{R}_n) + ik|\underline{r} - \underline{R}_n| - |\underline{r} - \underline{R}_n|^2 \rho^2\}. \tag{6.4.12}$$

Making use of the periodicity of F, we write

$$F(\underline{q},k,\underline{r},\rho) = \sum_{\underline{p}} \Phi(\underline{q},k,\underline{p},\rho) \exp(i\,\underline{p}\cdot\underline{r}), \tag{6.4.13}$$

where, on account of $\exp(i\,\underline{p}\cdot\underline{R}_n) = 1$, the function $\Phi$ in (6.4.13) is given by

$$\Phi(\underline{q},k,\underline{p},\rho) = \frac{1}{v_c} \int_{\Omega_c} F(\underline{q},k,\underline{r},\rho)\, e^{-i\,\underline{p}\cdot\underline{r}}\, d\underline{r}$$

$$= \frac{2}{v_c\sqrt{\pi}} \sum_{\underline{R}_n} \int_{\Omega_c} \exp\{-i\underline{p}\cdot(\underline{r} - \underline{R}_n) + ik|\underline{r} - \underline{R}_n| - \rho^2|\underline{r} - \underline{R}_n|^2\}\, d\underline{r},$$

where the integration is carried out over the volume of a unit cell, $\Omega_c$. By extending integration to the volume of the total lattice, $\Omega$, we obtain thus:

$$\Phi(\underline{q},k,\underline{p},\rho) = \frac{2}{v_c\sqrt{\pi}} \int_{\Omega} \exp(-i\,\underline{P}\cdot\underline{r} + ikr - \rho^2 r^2)\, d\underline{r} .$$

Writing $\Phi = \Phi_1 + i\Phi_2$, where $\Phi_1 = \mathrm{Re}\,\Phi$ and $\Phi_2 = \mathrm{Im}\,\Phi$, we obtain

$$\Phi_1 = \frac{\pi}{P\rho^3 v_c} \left((P+k)e^{-(P+k)^2/4\rho^2} + (P-k)e^{-(P-k)^2/4\rho^2}\right),$$

where we note that only $\Phi_1$ will be needed below. Let us divide the range of integration in (6.4.11) into two subintervals thus:

$$S(\underline{q}, k, \underline{r}) = \int_0^\eta F(\underline{q},k,\underline{r},\rho)\, d\rho + \int_\eta^\infty F(\underline{q},k,\underline{r},\rho)\, d\rho, \tag{6.4.14}$$

where the parameter $\eta$ in the integration limits will be specified later. On substituting the sum (6.4.12) into the second integral in (6.4.14), we obtain

$$\int_{\eta}^{\infty} F(\underline{q},k,\underline{r},\rho)\, d\rho = \sum_{\underline{R}_n} e^{-i\underline{q}\cdot(\underline{r}-\underline{R}_n)} \frac{e^{ik|\underline{r}-\underline{R}_n|}}{|\underline{r}-\underline{R}_n|} \operatorname{erfc}(\eta|\underline{r}-\underline{R}_n|), \tag{6.4.15}$$

where the integral

$$\frac{2}{\sqrt{\pi}} \int_{\eta}^{\infty} e^{-r^2\rho^2}\, d\rho = \frac{1}{r} \operatorname{erfc}(\eta r) \tag{6.4.16}$$

has been used, erfc(x) being the complementary error function. With the help of (6.4.13) we can rewrite the first integral in (6.4.14) in the form

$$\int_{0}^{\eta} F(\underline{q},k,\underline{r},\rho)\, d\rho = \sum_{\underline{p}} e^{i\underline{p}\cdot\underline{r}} \int_{0}^{\eta} (\Phi_1 + i\Phi_2)\, d\rho, \tag{6.4.17}$$

where

$$\int_{0}^{\eta} \Phi_1\, d\rho = \frac{2\pi}{Pv_c} \left( \frac{\exp\{-(P+k)^2/4\eta^2\}}{P+k} + \frac{\exp\{-(P-k)^2/4\eta^2\}}{P-k} \right). \tag{6.4.18}$$

On account of (6.4.6) and (6.4.7), we find that

$$s'(\underline{q},k) + ik = \operatorname{Re} s'(\underline{q},k) = \lim_{r\to 0} \operatorname{Re}\{e^{i\underline{q}\cdot\underline{r}} S(\underline{q},k,\underline{r}) - \frac{\cos kr}{r}\}.$$

Then using (6.4.14) - (6.4.18) we obtain the final expression

$$s'(\underline{q},k) + ik = \frac{2\pi}{v_c} \sum_{\underline{p}} \left( \frac{\exp\{-(P+k)^2/4\eta^2\}}{P(P+k)} + \frac{\exp\{-(P-k)^2/4\eta^2\}}{P(P-k)} \right) + \sum_{\underline{R}_n}{}' \operatorname{erfc}(\eta R_n) \frac{\cos(\underline{q}\cdot\underline{R}_n + kR_n)}{R_n} - \frac{2\eta}{\sqrt{\pi}}. \tag{6.4.19}$$

In the last sum in (6.4.19), the prime means that the term with $\underline{R}_o = 0$ must be excluded from the summation (for discussion, see reference [81]). For $\underline{p} = \underline{q} = 0$, $P = 0$ and therefore the expression in the brackets in the first sum (6.4.19) has to be replaced by its limiting form, i.e., $2\exp(-k^2/4\eta^2)/k^2$. The choice of the parameter $\eta$ is determined by the condition that the rate of convergence is approximately the same for both sums in (6.4.19). Numerical calculations [74] show that this condition is satisfied if $\eta \approx 1$.

For negative energies, the generalized Ewald method requires some further modifications (74). However, it is often possible to perform a direct summation in (6.4.2) without difficulty because in the case of negative energies the terms there fall off exponentially.

We note that the series (6.4.4) considered here is closely related to the "structural Green's function" used in calculations of the electronic states in reference (81). However, the final form of equation (6.4.1) in the ZRP approximation is of much simpler form than the analogous relation between the energy and quasi-momentum derived in the Green's function method.

The investigation of the zonal structure is considerably facilitated by the sum (6.4.2) having the same value for all equivalent points $\underline{q}$ of the reciprocal lattice (the latter is a consequence of the well-known fact that the symmetry of surfaces of equal energy in a Brillouin zone is the same as that of the lattice itself). Subramanyan carried out numerical calculations of the energy zones in a cubic lattice (74) with the help of equation (6.4.1) and using the generalized summation method of Ewald. For certain values of $\underline{q}$, the sum $S(\underline{q},0)$ is identical with the Madelung constants for various periodic structures. For instance, the sum S gives the Madelung constant for a structure of the NaCl type if $qa = (\pi, \pi, \pi)$, where a is the period of the lattice. These properties of the sum $S(\underline{q},0)$ were used in reference (74) to test the accuracy of calculations performed there.

Figure 6.6 shows $\kappa a = \sqrt{2|E|}\,a$ as a function of $\alpha a$, for both positive and negative energies, at $\underline{q}a = (0,0,0)$ and $\underline{q}a = (0,0,\pi)$. The two different branches of each curve for a given $\underline{q}a$ correspond to two different energy zones.* The curve for $\underline{q}a = (0,0,0)$ gives the position of the bottom of the zone $\kappa_0 a$ as a function of $\alpha a$. For large $\alpha a$ (that is, $\alpha a >> 1$), $\kappa_0 \approx \alpha$ and the curve shows that, in the region of negative energies, there always exists only one zone (similarly to the case of the one-dimensional structure considered earlier). The width of the zone decreases with $\alpha a$ increasing. The same can be seen in Fig. 6.7, which also shows how the energy of the particle depends on the wave vector changing along the (001) direction, i.e., along one of the axes of the reciprocal lattice.

It is also of interest to inspect the surfaces of equal energy. Figure 6.8 shows projections of these surfaces onto the plane $q_1 = 0$ (in the first Brillouin zone) for $ka = 2$ and $ka = 3$ when $\alpha a = 10$. As the diagram shows, in this case, the surfaces are nearly spherically symmetric.

---

* The discontinuity points in Fig. 6.6 correspond to the case where Bragg's condition is satisfied.

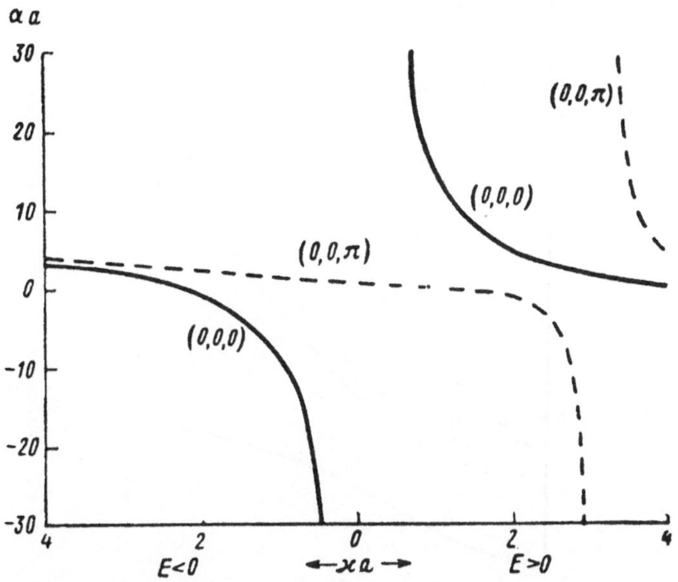


Fig. 6.6. Simple cubic lattice formed by identical ZRPs. Dependence of $\kappa a = (2|E|)^{\frac{1}{2}}a$ on the parameter $\alpha a$ for different values of quasi-momentum of the electron $\underline{q}$ [74]. Solid line: $\underline{q}a = (0,0,0)$; broken line: $\underline{q}a = (0,0,\pi)$.

The effective mass of an electron $m^*$, moving in the three-dimensional cubic lattice of equal ZRPs is given by

$$m^* = \left(\frac{\partial^2 E}{\partial q_1^2}\right)^{-1}_{\underline{q}=0} = \left(\kappa_0 \frac{\partial^2 \kappa}{\partial q_1^2}\right)^{-1}_{\underline{q}=0}, \qquad (6.4.20)$$

where $q_1$ is a component of vector $\underline{q}$ along an axis of the reciprocal lattice, $\kappa = ik$, and $\kappa_0$ is the value of $\kappa(\underline{q})$ at $\underline{q} = 0$. Making use of the fact that $\kappa(\underline{q})$ satisfies the equation (6.4.2), and that

$$\nabla_{\underline{q}}\kappa\Big|_{\underline{q}=0} = 0,$$

we find that

$$\frac{\partial^2 \kappa}{\partial q_1^2}\Bigg|_{\underline{q}=0} = - \frac{\nabla^2_{\underline{q}} S'(\underline{q},\kappa)}{3\frac{\partial}{\partial \kappa}\left(S'(\underline{q},\kappa) + \kappa\right)}\Bigg|_{\underline{q}=0}$$

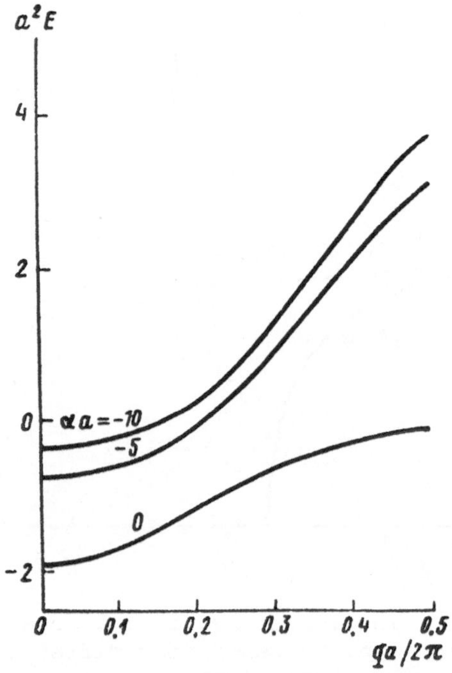


Fig. 6.7. Simple cubic lattice formed by identical ZRPs. Dependence of $a^2E$ on qa. Quasi-momentum of electron, $\underline{q} = (0,0,q)$ (74).

After finding the partial derivatives of S' in the formula above, we obtain the final expression for the effective mass:

$$m^* = \frac{3 \sum_{\underline{R}_n} e^{-\kappa_o R_n}}{\kappa_o \sum_{\underline{R}_n} R_n \, e^{-\kappa_o R_n}}, \qquad (6.4.21)$$

where the term with n = 0 has been included in the summation and the origin has been placed at the center n = 0. The series in (6.4.21) converges rapidly if $\kappa_o$ is large enough.

For large values of $\alpha$ ($a\alpha \gg 1$), $\kappa_o$ can be replaced, on the right-hand side of equation (6.4.21), by $\alpha$. For $\alpha a \to \infty$, $m^* \to \infty$. For small (and negative) values of $\alpha a$, $\kappa_o$ corresponding to $\underline{q} = 0$, in (6.4.21), can be obtained from Fig. 6.5. The dependence of the ratio $m^*/m$ on $\alpha a$ is shown in Fig. 6.9. These diagrams show that $\kappa_o$ tends to zero and the effective mass $m^*$ tends to m provided that $\alpha \to -\infty$.

Table 6.1 The effective mass of an electron moving in a lattice formed by atoms of an inert gas

| Element | Temperature, K | Density (solid phase; Ref. 121), $g/cm^3$ | Effect. period of latt., a, Å | Scatter. length $1/\alpha$, a.u. | $\alpha a$, a.u. | $m^*/m$ (Ref. 74) |
|---|---|---|---|---|---|---|
| Ar | 40 | 1.65 | 3.44 | -1.77 | -3.83 | 1.3 |
| Kr | 116 | 2.82 | 3.68 | -3.7 | -1.88 | 1.5 |
| Xe | 133 | 2.70 | 4.34 | -6.5 | -1.26 | 1.6 |

The model described here is most applicable to the motion of an electron through crystal structures formed by atoms of an inert gas. In such structures, the interaction between the atoms is weak and the intensity of the well in the ZRP approximation can be determined from the electron scattering length of free atoms. Subramanyan [74] replaced the real crystal structure by an effective cubic lattice with the same number of centers in a unit volume. The structure constant obtained in this way was used together with the experimentally known scattering length for determining the effective mass of the electron from Fig. 6.9. Results for Ar, Kr, and Xe are displayed in Table 6.1 above.

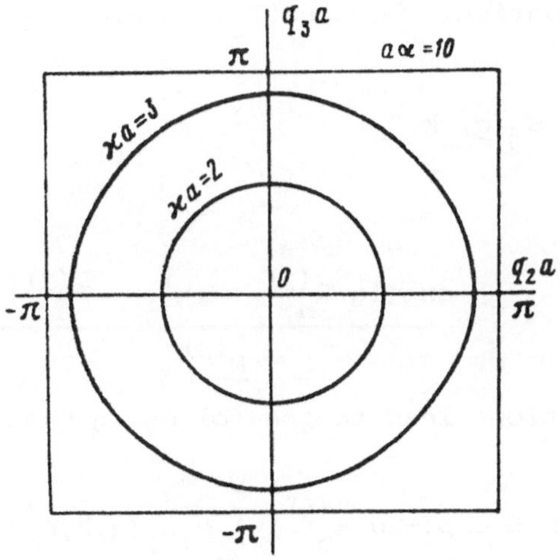


Fig. 6.8. Projection of equi-energy surfaces on the plane $q_1 = 0$ for a simple cubic lattice with $\alpha a = 10$. (From Subramanyan, Ref. 74.)

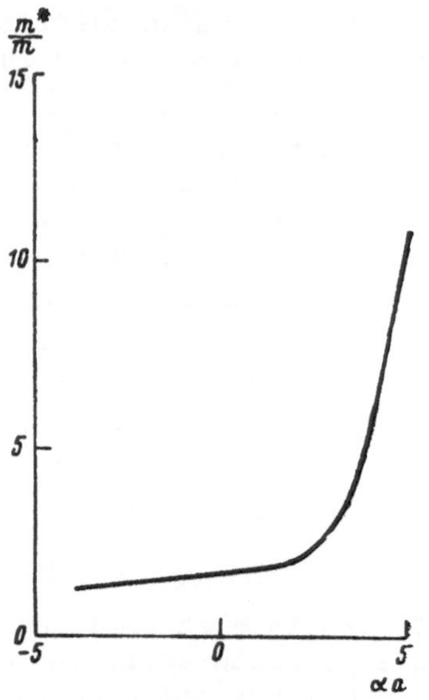


Fig. 6.9. Ratio m*/m as a function of the parameter αa. m is the mass of a free electron; m* is the effective mass of the electron in a simple cubic lattice formed by ZRPs (from Ref. 74).

We have already pointed out in Sec. 6.1 that in order to improve this theory and extend the scope of its applications, it is necessary to replace each single ZRP, in an elementary cell, by a basis of $\nu$ different ZRPs of depth $\alpha_j$ positioned at points $\underline{R}_n^{(j)}$, $j = 1,2,\ldots,\nu$, inside the cell number n. Then the solution of the Schrödinger equation, which satisfies Bloch's theorem, is sought in the form

$$\psi = \sum_{j=1}^{\nu} c_j \tilde{S}_j(\underline{q}, k, \underline{r}),$$

where

$$\tilde{S}_j(\underline{q}, k, \underline{r}) = \sum_n \frac{\exp(i\underline{q}\cdot\underline{R}_n^{(j)} + ik|\underline{r} - \underline{R}_n^{(j)}|)}{|\underline{r} - \underline{R}_n^{(j)}|} .$$

The boundary conditions lead to the following system of equations:

$$c_j(\alpha_j + ik + S'(\underline{q},k)) + \exp(-i\underline{q}\cdot\underline{R}_o^{(j)}) \sum_{i\neq j}^{\nu} c_i\tilde{S}_i(\underline{q},k,\underline{R}_o^{(j)}) = 0. \quad (6.4.22)$$

In (6.4.22), the function $S'(\underline{q},k)$ given by expression (6.4.2) is entirely determined by the "empty" Bravais lattice.

In the usual way, the equations connecting the energy of the particle with its quasi-momentum are obtained from equating the determinant of the system (6.4.22) to zero. The matrix elements of the system can be calculated with the help of the Ewald method, and then the roots of the secular equation can be found numerically using one of the existing computer codes. We note that the atomic properties of the lattice enter the determinant through the parameters $\alpha_j$ and $\underline{R}_o^{(j)}$.

Equations (6.4.22) were previously obtained by several authors [6,65,74]. It is instructive to compare these equations with those of Korringa-Kohn-Rostoker [81]. In the latter case, the atomic and structural properties are also separated in the elements of the determinant. However, the elements themselves have a more complicated form than those in the ZRP method considered above.

Chapter 7

# WEAKLY BOUND SYSTEMS IN ELECTRIC AND MAGNETIC FIELDS

## 7.1 WEAKLY BOUND SYSTEMS IN A HOMOGENEOUS ELECTRIC FIELD

In this chapter we shall consider the motion of an electrically charged particle in a combined field of a ZRP and an electric or magnetic field. In contrast to the approach used in Chapter 3, the external (electric or magnetic) field is not assumed here to be weak so that perturbation theory will not be used. This allows the inclusion of some new and interesting phenomena which could not be treated within the perturbational framework. The results derived in this chapter will be applicable to weakly bound systems such as negative ions or analogous systems in solids.

From the mathematical point of view, the problem is to construct the Green's function $G(\underline{r},\underline{r}',E)$ for a Schrödinger equation with external electric or magnetic fields and to study its properties. As formula (3.1.9) shows, the energy of a bound (or quasi-stationary) state of a particle in the resulting field (that includes the ZRP assumed to be at the point $\underline{R} = 0$) is a pole of the Green's function $G(\underline{r},0,E)$ satisfying the equation

$$\alpha + 2\pi \frac{\partial}{\partial r}\Big[r\,G(\underline{r},0,E)\Big]_{\underline{r}=0} = 0. \qquad (7.1.1)$$

The lower the binding energy, the easier for an external electric field to destroy the system. The system disintegrates by tunneling through a barrier, and the main theoretical difficulty is that there is no standard semiclassical method in the case of a three-dimensional barrier. A hydrogen-like atom in an external electric field is a rare example where the variables can be separated, by using a parabolic system of coordinates, and the problem is thus reduced to a one-dimensional form (54). In the general case of non-separable variables it was proposed in the past to average the one-dimensional barrier factor over all possible trajectories of the emitted particles. However, this averaging procedure is not uniquely defined, and that may considerably influence the result of calculations.

An analogous problem with a ZRP and an electric field admits a solution in a closed analytical form (29) which can be used as a basis for further improvements of the theory. This approximation also gives an upper limit for the lifetime of the system and for its polarizability. Indeed, if the repulsive forces and barriers are ignored, the resulting model has a state with the most compressed wavefunction. Therefore its polarizability and the rate of decay in the external electric field will be minimal among all systems with potentials $V(\underline{r}) \leq 0$ and a fixed value of the binding energy $E_o$.

Let us apply a simple scale transformation $\underline{r} \to \lambda\,\underline{r}$ to the Schrödinger equation, assuming that $\psi(\underline{r})$ is a solution which satisfies all boundary conditions for the potential $V(\underline{r})$, for the homogeneous field $-\mathcal{E}z$ and for the energy $E$. Then the function $\psi(\lambda\underline{r})$ will be a solution of the same equation for the potential $\lambda^2 V(\lambda\underline{r})$, for the homogeneous field $\lambda^3\mathcal{E}z$, and for the energy $\lambda^2 E$. Hence it follows for the whole family of potentials that the functional form of the eigenenergy $E(\lambda,\mathcal{E})$ must be

$$E(\lambda,\mathcal{E}) = \lambda^2 f(\mathcal{E}/\lambda^3),$$

where f is a universal function of arbitrary parameters $\lambda$ and $\mathcal{E}$ for all potentials under consideration. For instance, the second-order correction in $\mathcal{E}$ to the energy E is thus:

$$E_2 \sim \mathcal{E}^2/\lambda^4, \quad E_2 = -D\,\mathcal{E}^2/2E_o^{\,2} \tag{7.1.2}$$

so that polarizability is $D/E_o^2$ being inversely proportional to the square of the binding energy $E_o$. The dimensionless proportionality coefficient D in (7.1.2) depends upon the particular choice of the potential. For the ground state in the Coulomb field, $D = 9/8$, and for the extreme case of a ZRP, $D = 1/16$. The latter is the least value of D among all nonpositive potentials. This shows that the transition from a long-range Coulomb potential with the wavefunction $\psi = \exp(-\kappa r)$ to a short (zero) range potential with the wavefunction $\psi = \exp(-\kappa r)/r$ may result, at the same binding energy, in a considerable (by a factor of 18) change in the polarizability of the system. The decay constant $\Gamma$, whose functional form is

$$\Gamma = E_o\,\Phi(E_o^{\,3/2}/\mathcal{E}), \tag{7.1.3}$$

can be expected to exhibit even stronger dependence.

For a one-dimensional barrier of a triangular form, the semi-classical approximation gives

$$\Gamma \sim \exp\left(-2\int_0^{z_0}\sqrt{2(E_0 - \mathcal{E}z)}\,dz\right) = \exp\left(-\frac{2^{5/2}E_0^{\,3/2}}{3\mathcal{E}}\right) \quad (7.1.4)$$

This exponential dependence remains the same for any three-dimensional barrier. However, a pre-exponential factor contained in the exact expression for $\Gamma$ may vary by several orders of magnitude, as we shall see below, and therefore it is important to have a more precise estimate for $\Gamma$ in the case of short-range potentials. (We note that a similar formula for a Coulomb field has been given in the book [54] by Landau and Lifshitz.)

First we shall discuss the analytic properties of the scattering. In the limiting case where the external field $\mathcal{E} \to 0$, there exists a bound state of the system which lies in the continuum energy spectrum (the latter is extended from $-\infty < E < \infty$ when the field exists). There is a pole of the S-matrix lying on the real axis k which corresponds to this bound state. By switching the interaction on (i.e., by increasing $\mathcal{E}$) we shall move the pole to the lower half-plane so that its new position will give both the energy shift and width of the level. The difference between the present case and the one considered in Chapter 2 is that, in the latter case, the trajectory of the pole was from the upper half-plane to the lower half-plane along the imaginary axis, that is, through the origin. In the present situation, the pole is usually far from the origin and the imaginary axis but always near to the real axis.

The Green's function for a particle moving in a uniform electric field satisfies the equation

$$\left(-\frac{1}{2}\nabla^2 + \kappa^2/2 - \frac{1}{2}Fz\right)G = \delta(\underline{r}), \quad F = 2\mathcal{E}. \quad (7.1.5)$$

Following general rules of constructing the Green's functions and using a cylindrical system of coordinates, we can write G in the form

$$G = \frac{1}{\pi}\int_\kappa^\infty J_0(\rho\sqrt{p^2-\kappa^2})\, g(p,z)\, p\,dp , \quad (7.1.6)$$

where g is the one-dimensional Green's function for a particle moving in a uniform electrical field. This function falls off as $z \to -\infty$, has the form of an outgoing wave as $z \to \infty$, and satisfies the equation

$$d^2g/dz^2 + (Fz - p^2)\, g = \delta(z).$$

Making use of Fock's notation [122], we can express g in terms of the Airy functions u and v. Putting

$$\xi = (p^2 - Fz)\, F^{-2/3}$$

and

$$\xi_o = p^2 F^{-2/3},$$

we obtain

$$g = \begin{cases} F^{-1/3}\, v(\xi)\, \{u(\xi_o) + 2iv(\xi_o)\}, & \xi > \xi_o, \\ F^{-1/3}\, v(\xi_o)\, \{u(\xi) + 2iv(\xi)\}, & \xi < \xi_o. \end{cases} \qquad (7.1.7)$$

In the subsequent discussion we shall need asymptotic expansions for the Airy functions for large positive values of the argument given in the reference [122]:

$$u = \xi^{-1/4}\, e^{x}\, \phi_2(x), \quad v = 2^{-1}\, \xi^{-1/4}\, e^{-x}\, \phi_1(x), \qquad (7.1.8)$$

where

$$x = (2/3)\, \xi^{3/2},$$

and

$$\phi_2(x) = \phi_1(-x) = 1 + \frac{5}{72}\, x^{-1} + \frac{5\cdot 7\cdot 11}{2\cdot 72^2}\, x^{-2} + \ldots .$$

Let us now write equation (7.1.1) in the form

$$\left[ 2\pi G(\underline{r}, 0, -\kappa^2/2) - \frac{1}{r} \right]_{r=0} = -\alpha. \qquad (7.1.9)$$

Making use of the integral representation

$$\frac{1}{r} = \int_{\kappa}^{\infty} e^{-pr}\, dr + \kappa + O(r), \quad r \to 0,$$

and taking into account that in the boundary condition we can demand that $\underline{r} \to \infty$ along any path, including the z-axis, we obtain the following expression for (7.1.9):

$$\alpha = \kappa - \lim_{z \to +0}\ \lim_{p' \to \infty} \int_{\kappa}^{p'} \left[ 2\, g\, (p,z) - \frac{e^{-p|z|}}{p} \right] pdp. \qquad (7.1.10).$$

Interchanging the order of the two limits in (7.1.10), taking into account (7.1.7), and using a well-known property of the indefinite integral

$$\int UVdx = xUV - U'V', \tag{7.1.11}$$

U and V being any two solutions of the Airy equation, we shall reduce (7.1.10) to the following form:

$$\alpha = F^{1/3}\Big[\xi_1 v(\xi_1)\{u(\xi_1) + 2i\,v(\xi_1)\} - v'(\xi_1)\{u'(\xi_1) + 2i\,v'(\xi_1)\}\Big], \tag{7.1.12}$$

where

$$\xi_1 = \kappa^2 F^{-2/3}.$$

We assume that the electric field is weak, $\xi_1 >> 1$. It is easy to see that this is the only case of practical interest. Consider, for instance, $H^-$ where $E_0 = 0.75$ eV and put $\mathcal{E} = 2.5 \times 10^6$ V/cm. Then $1/\xi_1 = 0.18$. Another reason why we should not consider stronger fields is that the ZRP approximation itself becomes inaccurate in this case. From the experimental point of view, weak external fields are favorable since they do not produce wide level broadening and the system (for instance, in semiconductors) can be studied using spectroscopy. With the help of asymptotic expressions for the Airy functions and their derivatives we obtain, from (7.1.12), for $\kappa - \alpha$ and for E the following expressions:

$$\kappa - \alpha = -F^2/32\alpha^5 + i(F/8\alpha^2)\exp(-4\alpha^3/3F) \tag{7.1.13}$$

or, in the usual units,

$$E = E_0 - \frac{\hbar^2 e^2}{32\,m}\frac{\mathcal{E}^2}{E_0^2} + \frac{i}{2}\Gamma, \tag{7.1.14}$$

where

$$\Gamma = (\hbar e\mathcal{E}/2^{3/2}\,m^{1/2}\,E_0^{1/2})\exp(-2^{5/2}\,m^{1/2}\,E_0^{3/2}/3\hbar e\mathcal{E})$$

in accordance with formulae (7.1.2) - (7.1.4).

These formulae give us the polarizability and the decay constant in the ZRP approximation. A correction to the real part of E is proportional to $\mathcal{E}^2$ (and higher powers of $\mathcal{E}$) and can be derived without difficulty from perturbation theory (see Sec. 3.4 and Sec. 1.3). However, the imaginary part of E cannot be obtained in the same way since the perturbation series diverges as $\mathcal{E} \to 0$, and E as a function of $\mathcal{E}$ has an essential singularity at $\mathcal{E} = 0$, where an infinitesimal change in $E$ causes a complete change in the energy spectrum of the system. This situation is similar in a certain respect to that arising in quantum field theory if the interaction constant there tends to zero.

Our result is that, in a three-dimensional problem with a ZRP, the pre-exponential factor for the decay probability (7.1.4) is proportional to $E$. In a one-dimensional problem, this factor does not depend on $E$, and in the three-dimensional problem with a Coulomb potential (ionization of hydrogen [54]), it is inversely proportional to $E$. In other words, the decay constant contains a factor of the form $\mathcal{E}_0^{\gamma}$, where $\mathcal{E}_0 = 2(2mE_0)^{3/2}/(3me\hbar E)$ is a dimensionless parameter and $\gamma$ is an integer index depending upon the particular choice of the potential.

In real problems, this parameter $\mathcal{E}_0 \simeq 20$, and one can expect that by replacing the Coulomb potential by a ZRP, at fixed $E$ and $E_0$, it is possible to change the pre-exponential factor in (7.1.4) and, therefore, the decay constant $\Gamma$ itself, by at least two orders of magnitude.

Assuming that the average lifetime $\tau = \hbar/\Gamma$ is measured in seconds, the strength of the field $E$ in MeV/cm, and the electron affinity $E_0$ in eV, we obtain the following formula:

$$\log\tau = -13.17 + \frac{1}{2}\log E_0 - \log E + 29.67 E_0^{3/2}/E. \qquad (7.1.15)$$

For the negative hydrogen ion $H^-$, the lifetime given by (7.1.15) can be compared with the experimental data as well as with some other theoretical calculations. In Fig. 7.1 curve (a) has been computed in reference [29] using (7.1.15); curve (b) by Khuri; curve (c) by Kho; curve (d) by Hiskes; and curve (e) by Darewich and Neamtan. Kho used the rectangular barrier, and in the approximation of Khuri, the active electron was assumed to be in the field of an effective Coulomb potential and only one-dimensional motion, along the z-axis, was considered. We pointed out earlier that such an approximation considerably underestimates the lifetime and, indeed, Fig. 7.1 shows that curve (b) lies below all other curves. In the estimates of $\tau$ reported by Hiskes and by Darewich and Neamtan, the electron was assumed to move in the static field of the hydrogen atom. A considerable difference between (d) and (e) is probably due to different methods of averaging the three-dimensional barrier used in these two works. Another uncertainty is due to a value for the frequency of the orbital revolution of the electron which had to be assumed from some qualitative arguments. On the other hand, derivation of formula (7.1.15) does not require these assumptions and is universal for any system with a weak binding. Figure 7.1 shows that the ZRP approximation overestimates the lifetime (actually, it gives an upper bound on the lifetime).

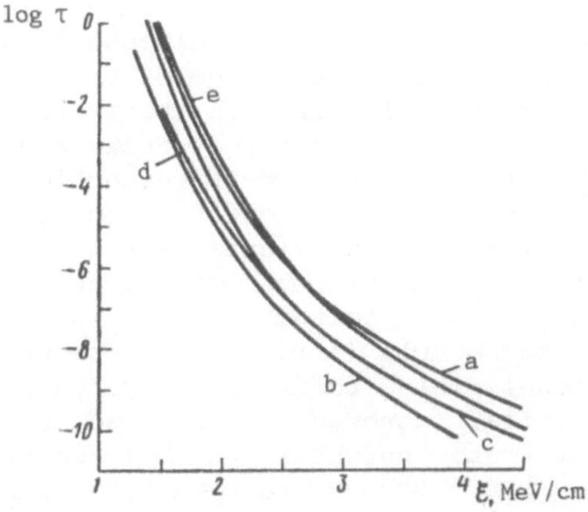


Fig. 7.1. Theoretical dependence of log τ, where τ is the average lifetime of the negative ion $H^-$, on the external electric field $\mathcal{E}$. (a) ZRP approximation; (b) N.N.Khuri (1956) and (c) T.K.Kho (1962) quoted in S.N.Kaplan, G.Paulikas, and R.V.Pyle, Phys. Rev., 131, 2574 (1963); (d) J.R.Hiskes, Nucl. Fusion, 2, 38 (1962); (e) G.Darewich and S.Neamtan, Nucl. Instr. Methods, 21, 247 (1963).

The most important simplification adopted in (7.1.15) is the neglect of the finite range of the potential. The small parameter $\alpha b$ is close to unity for $H^-$ and $\alpha b \sim 0.3$ for other negative ions. The accurate allowance for the finite range of the potential is very difficult, and we have to develop an approximate way of including it. One can expect that, without introducing any significant error, the triangular barrier factor in (7.1.14) can be replaced by a real factor of the form

$$\exp\left( - \frac{2}{\hbar} \int_{z_1}^{z_2} \{E_o + U(z) - \mathcal{E}z\}^{\frac{1}{2}} dz \right),$$

where $U(z)$ is the effective potential well for the active electron. This would reduce the lifetime by a factor of $\exp(\alpha b)$, that is, by a factor of 2-3 for the negative hydrogen ion $H^-$. If our formula (7.1.15) is corrected, the resulting curve would lie below curve (d) computed by Hiskes, the difference being particularly noticeable at the lower end of the $\mathcal{E}$ range.

The same correction can be obtained in a different way by reducing $E_o$ in formula (7.1.14) by a quantity of order $b\mathcal{E}$, in other words, by measuring the energy from the saddle point of the potential barrier. Assuming this correction to be small and expanding the quantity $(E_o - b\mathcal{E})^{3/2}$, we obtain the same result as before.

Finally, there is one more way to introduce this correction in the case of the hydrogen ion $H^-$. It was discussed in Chapter 1 that the main effect caused by the finite range of the potential was an increase in the normalization factor B in the expression for the wavefunction at $r > b$. The probability of decay and polarizability both increase approximately as $B^2$. The wavefunction for $r < b$ also changes as a result of the finite range of the potential. However, this change does not affect in any significant way either the probability of decay or the polarizability. It follows that for $H^-$, the lifetime must be reduced by a factor of 2.4. There are other possible corrections to formula (7.1.15): an account of a shift of $E_0$ due to an additional field (polarizability) which leads to an increase of the lifetime; introduction of an adiabatic potential instead of the electrostatic potential which leads to a decrease of the lifetime $\tau$; an account of the fact that $\xi_1$ cannot be considered as large if the external field is strong; and so on. However, all these corrections are small, and there is no need to take them into account unless the experimental accuracy is considerably increased.

We shall now define more precisely what is meant by the polarizability of a negative ion. Actually, we have to consider the polarizability in a time-dependent field oscillating with frequency $\Omega$ which is much lower than the electron frequency in the state with $E_0$. At the same time, $\Omega$ must be high enough if we wish to ignore the polarizability of the ion as a whole, that is, $1/M\Omega^2$ (where M is the ionic mass), in comparison with the polarizability of the electron shell of the ion, and to disregard the motion of the nucleus.

If the masses of a neutral particle $M_0$ and a charged particle $M_1$, which together form a weakly bound system, are comparable in magnitude, then the mass m in (7.1.14) must be replaced by the reduced mass $M_0M_1/(M_0 + M_1)$ and the external field $\mathcal{E}$ must be replaced by the reduced external field $\mathcal{E}M_0/(M_0 + M_1)$. Indeed, the effect of $\mathcal{E}$ is that the field accelerates the system as a whole and, at the same time, pulls the components of the molecular system apart.

In the form presented above, the results are valid for negative ions with a weakly bound s-electron. One of the ways to treat the $\ell \neq 0$ case is to introduce several ZRPs or to use the approximation of separable potentials. Smirnov and Chibisov [26] obtained a quasi-classical expression for the probability of disintegration, which coincides, for $\ell = 0$, with formula (7.1.14) above. Ivanov [123] calculated the polarizability of negative ions for the general case of $\ell \neq 0$, using a model with a potential of finite range.

One of the important applications of the theory of disintegration is the determination of the binding energy of a negative ion from the experimentally obtained probability of this process (see paper [124] and references therein).

The wavefunction considered above contained only an outgoing (diverging) wave, that is, it corresponded to a pole of the S-matrix. This type of solution for the continuum is used, in particular, in the theory of the light absorption by impurity centers in solids in the presence of an external uniform electrical field. In this theory all states of the continuum with energy $E < \mu_F$, where $\mu_F$ is the Fermi level, are assumed to be filled and the problem is reduced to calculating matrix elements of the dipole operator between the filled and vacant states (125). We shall consider below a general method of constructing the required eigenfunctions which has been used in solving other problems in the ZRP approximation.

In the general case, an eigenfunction of the continuum can be expressed in terms of the Green's function $G(\underline{r},\underline{R},E)$ (see equation (1.2.1)) thus:

$$\psi_E(\underline{r}) = \psi_E^{(o)}(\underline{r}) + c\,G(\underline{r},\underline{R},E), \tag{7.1.16}$$

where $\underline{R}$ specifies the position of the ZRP, and $\psi_E^{(o)}$ is the eigenfunction of the continuum with an external field but without the ZRP. From the boundary condition, we obtain (compare with Sec. 3.1)

$$c = \frac{2\pi\psi_E^{(o)}(\underline{R})}{\alpha + 2\pi\left[\frac{\partial}{\partial|\underline{r}-\underline{R}|}|\underline{r}-\underline{R}|\,G(\underline{r},\underline{R},E)\right]_{\underline{r}=\underline{R}}}. \tag{7.1.17}$$

An obvious consequence of this result is that the ZRP does not influence those eigenstates $\psi_E^{(o)}$ which vanish at the position of the ZRP, that is, if $\psi_E^{(o)}(\underline{R}) = 0$. States $\psi_E^{(o)}$ are degenerate (in our case, the order of degeneracy is infinite), and it is convenient to construct an orthonormal system of functions $\psi_{jE}^{(o)}$, $j = 1, 2, 3...$ belonging to the same energy E, such that only one of them, with $j = 1$, say, is nonvanishing at $\underline{r} = \underline{R}$ and, therefore, will experience the perturbing effect of the ZRP. Let us expand the functions $\psi_{jE}^{(o)}$ in terms of $\psi_E^{(o)}$. Evidently, $\psi_{1E}^{(o)}$ is the projection of $\delta(\underline{r}-R)$ onto the subspace in question. In other respects, the orthonormal system $\psi_{jE}^{(o)}$ can be constructed in an arbitrary way, for instance, with the help of the orthogonalization procedure.*

Specifically, for the case of an electric field,

$$\psi_{Emp}^{(o)}(\underline{r}) = J_{|m|}(\sqrt{p^2-\kappa^2}\,\rho)\,v((p^2 - Fz)F^{-2/3})\,e^{im\phi},\quad E = -\frac{\kappa^2}{2},$$

*This approach was adopted by Eganova and Shirokov (126, 127), who considered scattering on a system of ZRPs.

where m is the projection of the momentum of the particle on the direction of the electric field, and $(p^2 - \kappa^2)/2$ is the kinetic energy of the particle associated with its movement in the direction normal to the field. Then

$$\psi_{jE}^{(o)}(\underline{r}) = \int_{\kappa}^{\infty} \chi_j(p)\, \psi_{EOp}^{(o)}(\underline{r})\, dp .$$

Clearly, the ZRP does not perturb states with $m \neq 0$. The functions $\chi_j(p)$ form an orthonormal system, thus:

$$\int_{\kappa}^{\infty} \chi_j(p)\, \chi_{j'}(p)\, dp = \delta_{jj'} .$$

To within a normalization factor $g_1$, the wavefunction $\chi_1$ is given by

$$\chi_1(p) = g_1 \int \delta(\underline{r})\, \psi_{EOp}^{(o)}(\underline{r})\, d\underline{r} = g_1\, v(p^2 F^{-2/3}) .$$

The rest of the functions $\chi_j$ can be selected, for instance, in such a way as to ensure that states $\psi_{jE}$ are reduced to spherical functions if $F \to 0$ (125) .

## 7.2 WEAKLY BOUND SYSTEMS IN A HOMOGENEOUS MAGNETIC FIELD

First we shall discuss qualitatively the state of a charged particle in a three-dimensional well in the presence of an external magnetic field. We assume that the well is so shallow that there exists no bound states provided that the magnetic field is switched off. If the magnetic field is then switched on, it restrains the motion of the particle in the direction normal to the field, and the problem becomes virtually one-dimensional. The latter is known to have always a bound state even if the well is very shallow. Therefore we may expect (it will be proved later) that the introduction of an external magnetic field, however weak, will always lead to the appearance of a bound state in the three-dimensional well no matter how shallow it may be. To the first nonvanishing order, the binding energy is proportional to the square of the strength of the magnetic field. In this respect, the phenomenon is similar to diamagnetism but acts in the opposite direction and therefore can be called antidiamagnetism.

The magnetic field exerts a stabilizing effect on the particle. Under real microscopic conditions magnetic fields are too weak for this effect to be observed in free atoms and electrons. In semiconductors at liquid helium temperatures and for low effective masses of the electron and the hole, the binding energy can be of the same order of magnitude as $k_BT$, as we shall see, and the bound states produced by the external magnetic field may be observed experimentally.

Apart from the antidiamagnetic shift of energy downwards, which is proportional to the square of the magnetic field $H$, there is another shift, in the opposite direction, of the boundary of the continuum spectrum. This shift is linear in the magnetic field, $\Delta E = \omega_L$, where $\omega_L = H/2c$ is the Larmor frequency, and is caused by the properties of the quantum motion of a spinless particle. Such a particle cannot penetrate inside the region of a strong magnetic field by moving along the force lines, if its energy is small enough. (This is at variance with the properties of classical motion.) The energy $\Delta E$, which is the zero energy of a two-dimensional oscillator with frequency $\omega_L$, plays the role of a potential barrier. This term is due to the uncertainty principle for the coordinate and momentum of the particle in the direction normal to the magnetic field. $\Delta E$ increases the total energy of the system, and it is analogous to the paramagnetic shift, though it acts in the opposite way (antiparamagnetism), causing the system to be squeezed out of the region where the magnetic field is strong. For a particle with spin directed along the magnetic field the situation will be different. For instance, for an electron, the antiparamagnetic and paramagnetic terms completely compensate each other due to the anomalous gyromagnetic ratio of the electron, so that, in effect, only a quadratic antidiamagnetic term is retained. This results in the electron being sucked into the region where the magnetic field is strong.

In a system with a deep potential well, a bound state exists even in the absence of a magnetic field. Then there is an ordinary diamagnetic effect which increases the total energy by a term proportional to $H^2$. Correspondingly, the system will be squeezed out of the region where the magnetic field operates. Here the linear shift of the boundary of the continuous spectrum also takes place, but it does not influence the total energy. Despite the diamagnetic shift upwards, the binding energy increases as the field increases so that, in this case again, the magnetic field produces a stabilizing effect on the particle.

Consider a charged spinless particle moving in a combined field of a ZRP and a homogeneous magnetic field $H$ (128). For such a particle in a state with a zero projection of momentum on the direction of the magnetic field, the Green's function satisfies the following equation:

$$\frac{\partial^2 G}{\partial z^2} + \frac{\partial^2 G}{\partial \rho^2} + \frac{1}{\rho}\frac{\partial G}{\partial \rho} + (\varepsilon^2 - \lambda^2\rho^2)\, G = -2\,\delta(\underline{r}), \qquad (7.2.1)$$

where

$$\varepsilon = 2E \quad \text{and} \quad \lambda = H/2c.$$

Due to the condition $\underline{A}(0) = 0$ imposed on the vector $\underline{A}$, the modified boundary conditions discussed in Section 3.5 are reduced here to the ordinary ones. The Green's function G in (7.2.1) can be constructed in the form

$$G(\rho,z) = 2\sum_n \phi_n(\rho)\,\phi_n(0)\,g_n(z). \qquad (7.2.2)$$

In (7.2.2), $\phi_n$ is a regular normalized solution of the equation

$$\frac{d^2\phi_n}{d\rho^2} + \frac{1}{\rho}\frac{d\phi_n}{d\rho} + (\beta_n^2 - \lambda^2\rho^2)\,\phi_n = 0; \qquad (7.2.3)$$

that is, $\phi_n$ are eigenfunctions of a two-dimensional isotropic oscillator,

$$\phi_n = (\lambda/\pi)^{\frac{1}{2}} \exp(-\lambda\rho^2/2)\, L_n(\lambda\rho^2), \qquad (7.2.4)$$

where $L_n$ are Laguerre polynomials. The eigenvalues $\beta_n$ are given by the relation

$$\beta_n = 4\lambda\,(n + \frac{1}{2}).$$

The functions $g_n$ in (7.2.2) are one-dimensional Green's functions which satisfy the equations

$$\frac{d^2 g_n}{dz^2} - a_n^2\, g_n = -\delta(z),$$

where

$$a_n^2 = \beta_n - \varepsilon,$$

and they fall off as $|z| \to \infty$ (at the moment we are considering the bound states only). The functions $g_n$ are of the form

$$g_n(z) = \frac{1}{2\alpha_n}\exp(-\alpha_n|z|).$$

Therefore the function G(ρ,z) given by equation (7.2.2) can be written

$$G(\rho,z) = \frac{\lambda}{\pi} \sum_n \exp(-\lambda\rho^2/2)\, L_n(\lambda\rho^2)\, \frac{1}{\alpha_n} \exp(-\alpha_n|z|) .$$

For the purposes of further discussions, it is sufficient to consider the function G at ρ = 0 and on the z-axis. We have

$$G = (\lambda/\pi) \sum_{n=0}^{\infty} \exp(-\alpha_n|z|)\,/\alpha_n , \qquad (7.2.5)$$

where

$$\alpha_n = (\lambda(4n + \xi))^{\frac{1}{2}}, \quad \xi = 2 - \varepsilon/\lambda .$$

Making use of the formula

$$\int_0^{\infty} \exp(-bt^2 - c/t^2)\, dt = \frac{1}{2} (\pi/b)^{\frac{1}{2}} \exp(-2(bc)^{\frac{1}{2}})$$

where we shall put $c = \lambda z^2$ and $b = n + \xi/4$, the sum (7.2.5) can be written as a closed analytical expression. In order to achieve this, we interchange the order of summation and integration in (7.2.5) and sum the geometrical progression of the integrand. Then we obtain

$$G(0,z) = \frac{\lambda^{\frac{1}{2}}}{\pi^{3/2}} \int_0^{\infty} \frac{\exp(-\xi t^2/4 - \lambda z^2/t^2)}{1 - \exp(-t^2)}\, dt .$$

At the lower limit, this integral diverges as z → 0. Isolating the singularity, we have

$$G(0,z) = \frac{1}{2\pi|z|} \exp(-(\lambda\xi)^{\frac{1}{2}}|z|) + \frac{\lambda^{\frac{1}{2}}}{\pi^{3/2}} \int_0^{\infty} \Big( \frac{1}{2}\coth(t^2/2) + \frac{1}{2} - 1/t^2 \Big) \exp(-\xi t^2/4 - \lambda z^2/t^2)\, dt . \qquad (7.2.6)$$

For large z, we can replace the first factor in the integrand by unity and obtain

$$G(0,z) = \frac{1}{2\pi} (\lambda/\xi)^{\frac{1}{2}} \exp\big(-(\lambda\xi)^{\frac{1}{2}}|z|\big) \Big[ 1 + O(|z|^{-1}) \Big] . \qquad (7.2.7)$$

This equation shows that the asymptotic behavior of G at large z is determined by the first term in the sum (7.2.5) and is the same as in the one-dimensional case. It follows also that the continuum boundary is reached if $\xi = 0$, which is identical with the condition $E = \omega_L$.

An equation for the energy is obtained on substituting (7.2.7) into (7.1.1.):

$$\alpha/\lambda^{\frac{1}{2}} = \xi^{\frac{1}{2}} - \frac{2}{\sqrt{\pi}} \int_0^\infty \left(\frac{1}{2}\coth(t^2/2) + \frac{1}{2} - 1/t^2\right) \exp\left[-\frac{\xi t^2}{4}\right] dt. \quad (7.2.8)$$

The right-hand side of (7.2.8) can be expressed [128] in terms of the generalized Riemann zeta function $\zeta$. Thus:

$$\alpha/\lambda^{\frac{1}{2}} = -\zeta(1/2,\ \xi/4) = F(\xi). \quad (7.2.9)$$

The latter expression can be written in the form of a Laurent-type expansion. The three-term formula

$$F(\xi) \approx (\xi + 4)^{\frac{1}{2}} - 2\xi^{-\frac{1}{2}} - (\xi + 4)^{-\frac{1}{2}} - \frac{1}{3}(\xi + 4)^{-3/2} \quad (7.2.10)$$

gives a very good approximation to $F(\xi)$.

Finally, for $\xi \ll 1$ and for $\xi \gg 1$, we obtain

$$\alpha/\lambda^{\frac{1}{2}} = -2\xi^{-\frac{1}{2}} - \zeta(1/2) + O(\xi), \quad \xi \ll 1, \quad (7.2.11)$$

and

$$\alpha/\lambda^{\frac{1}{2}} = \xi^{\frac{1}{2}} - \xi^{-\frac{1}{2}} - \frac{1}{3}\xi^{-3/2} + O(\xi^{-7/2}), \quad \xi \gg 1, \quad (7.2.12)$$

where $\zeta(s)$ is the ordinary Riemann zeta function.

The case $\xi \gg 1$ is realized if $\alpha > 0$, that is, if the well is deep enough to have a bound state when there is no magnetic field. The second limiting case, $\xi \ll 1$, is realized if $\alpha < 0$, that is, for a well so shallow that no bound state exists until a magnetic field has been applied. In both cases, the right-hand side of the equation takes large values and, therefore, $\lambda^{\frac{1}{2}}\alpha \ll 1$. In other words, the radius of the first Larmor orbit is much larger than the characteristic size of the electron cloud of the atom in the absence of the magnetic field. Depending on the magnitude of $\lambda^{\frac{1}{2}}\alpha$, we shall classify the external magnetic field as either weak (if $\lambda^{\frac{1}{2}}\alpha \ll 1$), or strong (if $\lambda^{\frac{1}{2}}\alpha \gg 1$).

We shall now consider various limiting situations which may arise under the real conditions.

1. <u>$\alpha > 0$; weak field.</u> Making use of equation (7.2.12), we obtain

$$\varepsilon/\alpha^2 = -1 + \frac{1}{3}\lambda^2/\alpha^4 + O(\lambda^4/\alpha^8), \tag{7.2.13}$$

or in terms of the usual notation

$$E = E_o + \frac{e^2 H^2}{24mc^2\alpha^2}. \tag{7.2.14}$$

On taking account of the relation

$$< r^2 > -1/(2\alpha^2), \tag{7.2.15}$$

where $r^2$ is averaged over the bound states without a magnetic field, we obtain

$$E - E_o = \frac{eH^2}{12\ mc^2} < r^2 > , \tag{7.2.16}$$

which is a well-known result for the diamagnetic shift of the energy level.

Advancing arguments similar to those of Section 7.2 for the electric field, we come to the conclusion that the diamagnetic susceptibility

$$\chi = \frac{e^2 E_o}{6\ c^2} = \left.\frac{\partial^2 E}{\partial H^2}\right|_{H = 0} \tag{7.2.17}$$

will be a minimum for the ZRP approximation, in comparison with all other finite-range potentials, provided that the binding energy $E_o$ is fixed and there are no potential barriers. As an example, we note that $<r^2>$ and, therefore, the magnetic susceptibility are larger by a factor of six for a long-range Coulomb field than for a ZRP. It is also interesting to note that the corresponding change of the polarizability in an electric field is a factor of eighteen. Thus the magnetic susceptibility is less sensitive to variations in the form of the potential well than the electric polarizability.

2. <u>$\alpha < 0$; weak field.</u> Making use of equation (7.2.11), we obtain

$$2\lambda - \varepsilon = 4\lambda^2\alpha^{-2}\ \{1 + \lambda^{\frac{1}{2}}\alpha^{-1}\ \zeta(1/2)\}^{-2} + O(\lambda^{7/2}\alpha^{-7}). \tag{7.2.18}$$

As was pointed out above, $\varepsilon = 2\lambda$ is the continuum boundary so that the difference $2\lambda - \varepsilon$ is the binding energy $E_o$. In terms of the usual notation

$$E_o = \frac{e^2 H}{2mc^2\alpha^2}\{1 + 1.46\,|\alpha|^{-1}\left(\frac{eH}{2\hbar c}\right)^{\frac{1}{2}}\}^{-2} . \qquad (7.2.19)$$

The energy shift directed downwards from the continuum boundary can be called antidiamagnetic, as we have already pointed out above. The first (quadratic in $H$) term in formula (7.2.19) is identical with the total result derived by Bychkov (129), who considered electron scattering by a force center in a magnetic field. It is not difficult to see, however, that the next term in (7.2.19) is also important. Even for $\lambda\alpha^{-2} = 0.01$, it contributes some thirty percent to the total result.

3. $\alpha = 0$; field of an arbitrary strength. This is a special case where there is no bound state without a magnetic field. The introduction of the field, however weak, deepens the potential well and gives rise to a bound state. In this case, the left-hand side of equation (7.2.9) is identically zero, and we have to find $\xi = \xi_o$ such that the right-hand side of the equation vanishes. With the help of an approximate formula (7.2.10) it is not difficult to obtain $\xi_o = 1.21$. Therefore, in this special case, the binding energy $E_o$ is a linear function of the magnetic field:

$$2\lambda - \varepsilon = \xi_o\lambda , \quad E_o = 0.30\, e\hbar H/mc. \qquad (7.2.20)$$

4. An arbitrary $\alpha$; strong field. Expand $F(\xi)$ in the series around the point $\xi = \xi_o$:

$$F(\xi) = F'(\xi_o)(\xi - \xi_o) + \frac{1}{2}F''(\xi_o)(\xi - \xi_o)^2 + \dots$$

and make use of the approximate expression (7.2.10). Then

$$F'(\xi_o) = 1.02 \quad \text{and} \quad F''(\xi_o) = -0.95.$$

With the help of this result, the total energy of a bound state in a magnetic field can be expressed in the following way:

$$\varepsilon\alpha^{-2} = (2 - \xi_o)\,\lambda\alpha^{-2} - \frac{\lambda^{\frac{1}{2}}}{\alpha F'(\xi_o)} + \frac{F''(\xi_o)}{2\{F'(\xi_o)\}^3} + O(\lambda^{-\frac{1}{2}}\alpha)$$

$$= 0.79\,\lambda\alpha^{-2} - 0.98\,\lambda^{\frac{1}{2}}\alpha^{-1} - 0.45 + O(\lambda^{-\frac{1}{2}}\alpha). \qquad (7.2.21)$$

Figure 7.2 shows the dependence of energy $\varepsilon\alpha^{-2}$ upon the field $\lambda\alpha^{-2}$ (in dimensionless units) for $\alpha > 0$ and for $\alpha < 0$. The graph also shows the continuum boundary $\varepsilon = 2\lambda$ and the straight line $\varepsilon = (2 - \xi_o)\lambda$ which corresponds to the case $\alpha = 0$. The two curves $\alpha > 0$ and $\alpha < 0$ are branches of a parabola-like curve which is symmetric, to within the numerical accuracy, with respect to a

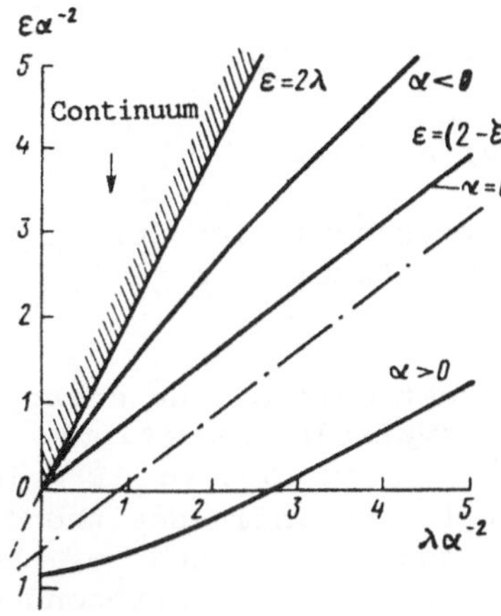


Fig. 7.2. Dependence of the energy level in a magnetic field on the field strength $H$ (in dimensionless units) for $\alpha > 0$, $\alpha < 0$, and $\alpha = 0$ [128]. The broken line is the line of symmetry (within the accuracy of the calculations) for two branches $\alpha > 0$ and $\alpha < 0$. The hatched area corresponds to the continuum.

broken line also shown in Fig. 7.2. The line $\alpha = 0$ gives the asymptotic limit, as $\lambda \to \infty$, for the gradient of both branches $\alpha > 0$ and $\alpha < 0$. A continuation of the $\alpha < 0$ branch into the non-physical region $\lambda\alpha^{-2} < 0$ is shown by a dotted line.

Now we shall formulate two conditions under which the approximation of this section is valid.

(A) Similarly to the usual condition $\alpha b \ll 1$, where b denotes the radius of the potential well, let us now write

$$\{ 2m E_0(H) \}^{1/2} b/\hbar \ll 1$$

for the bound ($\alpha < 0$) state with the binding energy $E_0(H)$ which appears as a result of the application of field $H$. Using the first term in (7.2.19) for $E_0(H)$, we find that

$$eHb/\alpha\hbar c \ll 1.$$

In reality, this criterion is always fulfilled. Take, as an example, $a = \alpha^{-1} = 10^{-7}$ cm and $b = 10^{-8}$ cm. Then the condition reads $H \ll 10^8$ Oe. The magnetic field will always create a bound state near the continuum boundary provided that the scattering length is negative. That will be the case even if the well is deep enough to have other (low-lying) bound states.

(B) The radius of the first Larmor orbit must be greater than the effective size of the potential well, i.e.,

$$b\lambda^{\frac{1}{2}} << 1, \quad (eH/2\hbar c)^{\frac{1}{2}} b << 1.$$

This condition will be always satisfied for practically realized fields and not too large b. Putting $b = 10^{-7}$ cm, we find that $H << 10^7$ Oe.

Let us now obtain an estimate of the energy of a bound state in a semiconductor under favorable conditions. Putting $m^* = 0.01\, m_e$, $a = 10^{-7}$ cm, and $H = 10^5$ Oe, we obtain $E = 10^{-3}$ eV. At a temperature of several °K, these levels can influence the concentration of electrons or holes in the conduction band of the semiconductor, leading, for instance, to a fall of electroconductivity along the applied magnetic field.

If the magnetic field is directed along the principal axis of an anisotropic crystal, there are two effective masses $m^*_{\parallel}$ and $m^*_{\perp}$, and the formula has to be slightly modified; in the final expressions, $m^*$ must be replaced by $m^*_{\perp}$ and all terms must be multiplied by a factor of $(m^*_{\parallel}/m^*_{\perp})^{\frac{1}{2}}$.

The problem of electron scattering on a ZRP in the presence of a magnetic field solved in this section has many important applications in the theory of the electroconductivity of metals in strong magnetic fields (129 - 131). In this case, the plane wave can propagate along the direction of the magnetic field only, that is, along the z-axis. Therefore we have to put, in the general formulae (7.1.16) and (7.1.17),

$$\psi^{(o)}_{n_o} = \exp(-\lambda\rho^2/2)\, L_{n_o}(\lambda\rho^2)\, \exp(ik_{n_o}z).$$

Here $k_{n_o}$ is the initial momentum, and the quantum number $n_o$ describes the state of the particle with respect to its motion normal to the field. Making use of the results for the Green's function obtained above, we can derive the following expression for the transition amplitude $n_o \to n$:

$$F_{n_o \to n} = \frac{\lambda}{2\pi k_n} \frac{4\pi}{\alpha - \lambda^{\frac{1}{2}} F(\xi)}.$$

The momentum of the particle in the final state must be determined from the energy conservation law:

$$E = (2n_o + 1)\omega_L + k^2_{n_o}/2 = (2n + 1)\omega_L + k_n^2/2,$$

and only transitions with $k^2_n > 0$ are allowed. Scattered waves of equal amplitudes propagate in the positive and negative directions along the z-axis.

Let us assume that the total energy of the particle, E, lies between the mth and (m+1)th Landau levels, that is,

$$E = (2m + 1 + \varepsilon)\,\omega_L ,$$

where $0 < \varepsilon < 1$. Then we find, with the help of an expression for the generalized Riemann zeta function, that

$$F(\xi) \sim -(-\varepsilon)^{-\frac{1}{2}} + O(1),$$

that is, the scattering amplitude vanishes for $\varepsilon \to 0$ and $\varepsilon \to 1$, the leading term of the amplitude being $|F_{n_0 \to n}|^2 \sim \varepsilon$. This conclusion remains true only in the ZRP approximation. If the potential has a finite range, the amplitude will not vanish exactly.

There also exist resonance maxima whose positions are determined by the condition

$$\alpha = \lambda^{\frac{1}{2}} \operatorname{Re} F(\xi).$$

The amplitude in the exact maximum is finite since $\operatorname{Im} F(\xi) \neq 0$. The maximum is narrow if it is at such energy that $\varepsilon$ is close to unity. Evidently, this resonance is due to a bound state created by the external magnetic field. If $n \neq 0$, decay is allowed, that is, the state is quasi-stationary.

## 7.3. WEAKLY BOUND SYSTEMS IN CROSSED ELECTRIC AND MAGNETIC FIELDS

In this section we shall follow Drukarev and Monozon [132] and consider a system with a low binding energy which is under the influence of mutually perpendicular homogeneous electric and magnetic fields. A situation of this kind arises, for instance, in the ionization of highly excited atoms or negative ions moving in a strong magnetic field. In a system of coordinates where the atom is at rest, there will be mutually orthogonal magnetic and electric vectors, the electric field being weaker than the magnetic field. The former can cause detachment of an atomic electron which can tunnel through the potential barrier. This process is usually known as Lorentz ionization.

Another example of the same problem is tunnelling transitions in semiconductors placed in crossed electric and magnetic fields, influenced by weakly bound excitons or impurity states. In this case, the electric field may be stronger than the magnetic field (for references see Drukarev and Monozon (132)).

The Green's function for a particle moving in a combination of an electric field $E$ directed along the y-axis and a magnetic field $H$ directed along the z-axis satisfies the equation

$$\nabla^2 G + 2iHc^{-1}y\frac{\partial G}{\partial x} + (-H^2c^{-2}y^2 + 2Ey - \kappa^2)G = -2\delta(\underline{r}). \quad (7.3.1)$$

In this gauge, only one component of the vector-potential is non-zero, which makes it convenient for calculating the current.

Similarly to the case of Section 7.2, where there was no electric field considered, we shall seek a solution of equation (7.3.1) which depends on one coordinate only.

$$G(0,z) = -2i\,(4\pi i)^{-3/2}L^{-1}\int_0^\infty \frac{dx}{\sin(x/2)\,x^{\frac{1}{2}}}\exp\{i(-\kappa^2L^2x/4$$

$$- s^2x/L^2 + z^2/L^2x + 2^{-1}x^2s^2L^{-2}\cot(x/2))\}, \quad (7.3.2)$$

where $L = (H/2c)^{\frac{1}{2}}$ is the Larmor radius, and $s = c^2E/H^2$ is the separation between the center of the parabolic potential well created by the crossed electric and magnetic fields and the origin of coordinates. The equation for determining the energy thus becomes

$$-\alpha L = -\kappa L - (i\pi)^{-\frac{1}{2}}\int_0^\infty dx\,x^{-3/2}\left(\frac{x}{2\sin(x/2)}\exp\{i(-\kappa^2L^2x/4 -\right.$$

$$\left. - s^2x/L^2 + 2^{-1}x^2s^2L^{-2}\cot(x/2))\} - \exp(-i\kappa^2L^2x/4)\right). \quad (7.3.3)$$

Let us discuss specific features of this problem. For a system bound in the absence of $E$ and $H$, the application of these fields will lead to changes depending on their relative intensities. Outside the well, the motion of the system is governed by $E$ and $H$, the motion along the magnetic field, that is, along the z-axis, remaining free. As a result, the particle outside the well has a continuous energy spectrum with the threshold value $E_b \approx -c^2E^2/2H^2$, provided that the external magnetic field is much smaller than the atomic magnetic field (132). The electric field causes the system to decay. In our particular case, the bound state level will be in the continuum if this electric field is strong enough, $E_o = \alpha^2/2 < |E_b|$. Therefore, the case where the electric field is stronger than the magnetic field requires both the shift of the bound level $E_o$ and its width due to possible ionization of the corresponding quasi-stationary state to be considered. The cross section for ionization was obtained by Kotova et al. (19) in the semiclassical approximation, to within the accuracy of the principal exponential factor, using the method of Feynman integrals generalized to the case of imaginary time.

Another extreme case is when the magnetic field is stronger than the electric field. Then the stabilizing effect of the magnetic field on the system is dominant, and the system cannot decay ($E_o > |E_b|$). The problem is then to find the shift of a stationary state in a magnetic field, caused by a (weak) electric field. Let us introduce the parameter $\gamma$,

$$\gamma = (E_o/|E_b|)^{1/2} = \alpha H/cE, \qquad (7.3.4)$$

which describes possible regimes of the motion of the system. For $\gamma < 1$, disintegration of the system is possible, and it becomes more probable the closer $\gamma$ is to zero. For $\gamma \geq 1$ disintegration of the system cannot take place and the system becomes even more stable for larger $\gamma$.

For $\gamma << 1$, the external electric field is weak in comparison with the atomic field, and it is strong in comparison with the external magnetic field. The integral (7.3.2) can be calculated by the method of steepest descent, and the result is then (132)

$$E = -\alpha^2/2 - E^2/8\alpha^4 + H^2/24\,c^2\alpha^2 + i\Gamma/2\,, \qquad (7.3.5)$$

$$\Gamma(E,H) = 2E_o \frac{E}{2E_o}(1 - \gamma^2/6)\exp\left[-\frac{2}{3}\frac{E_o}{E}(1 + \gamma^2/30)\right],$$

$$E_o = \alpha^3\,. \qquad (7.3.6)$$

This formula shows that the effect of the magnetic field is to narrow the width $\Gamma$ of the level. This stabilization effect agrees with experimental observations. In a qualitative way, this decrease in the decay probability can be understood as being caused by the magnetic field deforming and extending the trajectory through the barrier. The potential barrier is thus effectively increased in strength. The contributions from the electric and magnetic fields to the total shift (7.3.5) of the energy level $E_o$ is additive.

For $\gamma >> 1$, Drukarev and Monozon (132) computed corrections to the results of Section 7.2, where only the magnetic field was considered, due to the introduction of a weak electric field. In this case, the energy level is stationary ($\Gamma = 0$), but the field weakens the binding, leading to compensation of the antidiamagnetic effects discussed in Section 7.2.

From the computational point of view the most difficult case is that of $\gamma \sim 1$, where the effects from both magnetic and electric fields are comparable in magnitude. Drukarev and Monozon considered the special case when the difference $1 - \gamma$ is small and positive. Then the level lies above the continuum boundary and its width can be found in the semiclassical approximation. In reference [132] the authors also assumed that the transition through the barrier takes place mainly along the direction of the electric field, that is, along the y-axis. Then the variables in the Schrödinger equation can be approximately separated in parabolic coordinates. The probability of ionization, W, obtained in this way is given by

$$W = |E_o| \frac{\mathcal{E}}{\mathcal{E}_o} (1 - \gamma^2)^{\frac{1}{2}} \exp\{-\frac{2}{3} \frac{\mathcal{E}_o}{\mathcal{E}} G(\gamma)\}, \tag{7.3.7}$$

where $G(\gamma)$ is an auxiliary function,

$$G(\gamma) = \frac{3}{2\gamma^2}\left[1 + \frac{1 - \gamma^2}{2\gamma} \ln \frac{1 - \gamma}{1 + \gamma}\right]. \tag{7.3.8}$$

## 7.4 A COMBINATION OF ZRPs AND A COULOMB FIELD

Up to now we have been dealing with atomic systems moving in superimposed ZRPs and external homogeneous electric and/or magnetic fields. With respect to the quantum system itself, such fields can be considered as microscopic. If we want to bring into the picture microscopic fields created by the atomic system, it will be necessary to take into account their nonhomogeneity. The most important case is that of the Coulomb field of a charged atomic particle or ion. Here the problem arises of finding the state of the particle in the superimposed field of ZRPs and a Coulomb potential. This problem was considered first by Komarov [133] * and later by several other authors [135 - 140].

We shall solve this problem by considering the analytical expression for the Coulomb Green's function:

$$G(\underline{r},\underline{r}',E) = \frac{\Gamma(1 - n)}{2\pi|\underline{r} - \underline{r}'|}\left[M'_{n,\frac{1}{2}}(\eta/n)\, W_{n,\frac{1}{2}}(\xi/n) - M_{n,\frac{1}{2}}(\eta/n)\, W'_{n,\frac{1}{2}}(\xi/n)\right]; \quad n = (-2E)^{-\frac{1}{2}};$$

$$\xi = r + r' + |\underline{r} - \underline{r}'|, \quad \eta = r + r' - |\underline{r} - \underline{r}'|. \tag{7.4.1}$$

---

* We also mention a paper by Zel'dovich [134], who studied the energy levels in a Coulomb potential distorted, at small distances, by a deep short-range potential.

In (7.4.1) $M_{n,\mu}(z)$ and $W_{n,\mu}(z)$ are Whittaker functions. We shall often omit their subscripts n and $\mu$, as for instance in equation (7.4.2) below, provided that it does not lead to any confusion. In problems involving the Coulomb potential it is sometimes convenient to use the principal quantum number n instead of the energy. In this case, equation (7.1.1) for the determination of the terms of the system takes the form

$$\alpha + \frac{2\Gamma(1-n)}{n}\left[\left(-\frac{1}{4}+\frac{n^2}{2R}\right) M(2R/n)\, W(2R/n) + M'(2R/n)\, W'(2R/n)\right] = 0 . \quad (7.4.2)$$

Following the same arguments as in Section 7.1, we can conclude that only one out of $n^2$ degenerate states corresponding to the same principal quantum number n is influenced by the ZRP. The other $n^2 - 1$ states remain unperturbed. Let us assume that the Coulomb potential describes the field of a positive ion $A^+$, whereas the ZRP does the same for a neutral atom B. Then the unperturbed Coulomb terms will correspond to the system $A^* + B$. On the other hand, the state perturbed by the ZRP will be a linear superposition of hydrogenlike wavefunctions, the coefficients of which are dependent on the internuclear separation R. If $\alpha > 0$, there is also one ionic term which, for large R, corresponds to an electron moving in a ZRP (the system $A^+ + B^-$).

The quantity $r = 2n^2$ is the radius of the nth Bohr orbit, and it determines the turning point of the classical motion of the electron in the Coulomb field. Correspondingly, we can classify the motion of the electron in the neighborhood of the ZRP as "allowed" (if $R < 2n^2 = -1/E$) or as "forbidden" (if $R > 2n^2$). Very simple analytical results can be obtained in the extreme case of low energies ($n \gg 1$). First we shall not make any special assumptions regarding relative values of R and $n^2$. The general asymptotic forms of the Whittaker functions for this case, which are uniform with respect to $x = R/(2n^2)$, were found by Presnyakov [135]

$$M_{n,\frac{1}{2}}(4nx) = (4t/p)^{\frac{1}{2}}\left\{(\pi n)^{\frac{1}{2}}\{\Gamma(1+n)\Gamma(1-n)\}^{-1} u(t) - (\pi n)^{-\frac{1}{2}} \cos(\pi n)\, v(t)\right\} ,$$

$$W_{n,\frac{1}{2}}(4nx) = (4t/p)^{\frac{1}{2}} (\pi n)^{-\frac{1}{2}} \Gamma(1+n)\, v(t) , \quad n \gg 1,$$

$$p(x) = 1 - 1/x , \quad (7.4.3)$$

where u(t) and v(t) are the Airy functions in the notation of Fock which we have already used earlier in Section 7.1. In equation (7.4.3), it is assumed that

$$\frac{2}{3}t^{3/2} = 2n\left(\{x(x-1)\}^{\frac{1}{2}} - \ln\{x^{\frac{1}{2}} + (x-1)^{\frac{1}{2}}\}\right)$$
$$\text{if} \quad x \geq 1 \quad \text{and} \quad t \geq 0,$$

and

$$\frac{2}{3}(-t)^{3/2} = 2n\left(\frac{\pi}{2} - \{x\,(1-x)\}^{\frac{1}{2}}\arctan\left(\{x/(1-x)\}^{\frac{1}{2}}\right)\right.$$
$$\text{if} \quad x \leq 1 \quad \text{and} \quad t \leq 0.$$

Equation (7.4.2) can be approximately written in the form

$$\alpha = \frac{1}{n}(p/t)^{\frac{1}{2}}\left( tu(t)\,v(t) - u'(t)\,v'(t) + \cot(\pi n)\{v'(t)^2 - t\,v(t)^2\}\right), \qquad (7.4.4)$$

which is valid for large n everywhere with the exception of the region $R \ll 1$. In particular, this equation can be used in the vicinity of the turning point. The cases where the motion is well under the barrier and where it is well above the barrier enable further simplifications to be made.

Let us consider first the region under the barrier where the internuclear separation $R \gg 2n^2$. Making use of the asymptotic form of the Airy function, equation (7.4.4) can be written in the form

$$\alpha = \sqrt{\frac{4}{n^2} - \frac{2}{R}} + \frac{e^{-2R/n}}{6n^{n+1}}\left(\frac{2R}{n}\right)^{n-1} e^{-n}\cot\pi n. \qquad (7.4.5)$$

For integer values $n = n_o$ corresponding to the unperturbed Coulomb spectrum, the second term in (7.4.5) has poles and can be written in the vicinity of the poles in the form

$$-2\pi g_{n_o}(E - E_{n_o})^{-1}; \quad E_{n_o} = -1/(2n_o^2),$$

where the function $g_{n_o}(R)$ can be expressed in terms of the Coulomb wavefunctions which belong to the same $n_o$th shell.* Thus

$$g_{n_o}(R) = \sum_{\ell m}|\psi_{n_o\ell m}(R)|^2.$$

The energy of the ionic term is obtained from (7.4.5) if only the first term there is taken into account:

$$E_{ion}(R) = -\alpha^2/2 - 1/R - 1/(8\alpha^4R^4) + O(R^{-5}). \qquad (7.4.6)$$

---

* This formula, as well as subsequent expressions (7.4.6) - (7.4.8), is valid for small n and large R because they can be derived from the asymptotic expansion of the Whittaker function.

The third term on the right-hand side of equation (7.4.6) gives the energy of the polarization interaction between $A^+$ and $B^-$. In order to derive it the next terms in the expansion of the Airy function must be taken into account.

For large distances, the behavior of the terms of the Coulomb series is mainly determined by the second term in equation (7.4.5) thus:

$$E_{n_o}(R) = -\frac{1}{2n_o^2} + \frac{2\pi}{\alpha}\,\frac{g_{n_o}(R)}{\sqrt{2/R - 2E_{n_o}}} . \qquad (7.4.7)$$

Those Coulomb terms are different in their behavior from the ionic term (7.4.6) because the former experience, at large R, only an exponentially small shift.

Within the first-order approximation, the ionic term intersects the Coulomb term at the point

$$R_{n_o} = \{(1/2n_o^2) - \alpha^2/2\}^{-1},$$

so that both terms in expression (7.4.5) must be taken into account in the vicinity of that point. When this is done, the crossing becomes a pseudo-crossing, the least separation between the ionic and Coulomb terms being

$$\Delta E_{n_o} = \left| 8\pi\alpha g_{n_o}(R_{n_o}) \right|^{\frac{1}{2}} . \qquad (7.4.8)$$

The term diagram, for the case of $\alpha > 0$, is shown in Fig. 7.3.

Nonadiabatic Landau-Zener transitions at the points of pseudo-crossing between the ionic and Coulomb terms determine the cross sections for the ion-ion recombination process

$$A^+ + B^- \rightarrow A^* + B .$$

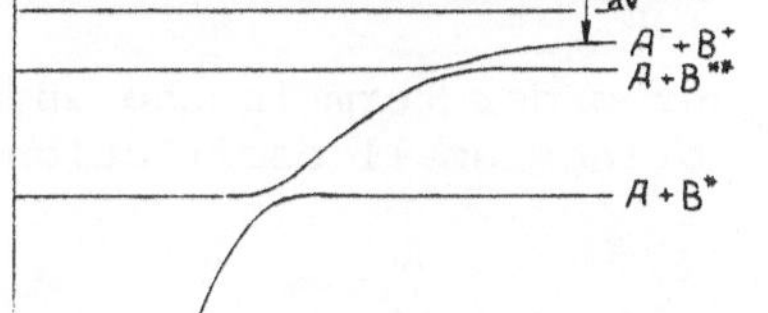

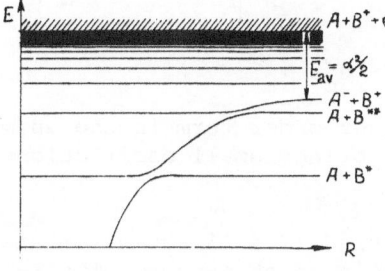

Fig. 7.3. The system of molecular terms for an electron in the field of a ZRP ($\alpha > 0$) and a Coulomb field (a model for the system $A^+ + B^-$).

Komarov et al. [136] computed the cross sections for this process using the term splitting obtained with the help of a ZRP model.

The departure of the field of the positive ion from a Coulomb field near the nucleus, which is especially significant in the s-states, was taken into account by Dalidchik and Ivanov [64] and by Kereselidze and Chibisov [137]. In this case, the degeneracy with respect to the azimuthal quantum number $\ell$ is removed by higher terms of the Coulomb series. An analysis carried out in reference [137] shows that equations (7.4.5) - (7.4.8) retain the same form, provided that $g_{n_o\ell} = |\psi_{n_o\ell 0}(\underline{R})|^2$ (the quantization axis is assumed to be directed along the vector $\underline{R}$). Indeed, these results are actually justified by the use of the pole approximation to the regularized Green's function,

$$G_{reg}(\underline{R},\underline{R},E) = \frac{\sum \psi^2_{n_o\ell m}(\underline{R})}{E - E_{n_o}} + N(\underline{R}), \tag{7.4.9}$$

where summation on the right-hand side takes care of the degeneracy of the terms and $N(\underline{R})$ remains a smooth function as $E \to E_{n_o}$. More detailed analysis of the pole approximation and its use can be found in the works of Dalidchik and Ivanov [64,139,140]. A comparison of the terms obtained from the approximate formulae (7.4.6) - (7.4.8) with those derived numerically from the exact transcendental equation (7.4.2) was carried out in reference [137].

Let us consider now the second special case, $R < 2n^2$ (the region over the barrier). For Coulomb terms lying well above the ionic term ($n \gg \alpha^{-1}$), only the $\cot \pi n$ term has to be considered on the right-hand side of equation (7.4.4). This gives

$$E_{n_o}(R) = -\frac{1}{2n_o^2} + \frac{\alpha}{\pi n_o^4}(p/t)^{\frac{1}{2}}\{v'(t)^2 - tv(t)^2\}, \tag{7.4.10}$$

or, using the asymptotic expression for the Airy functions,

$$E_{n_o}(R) = -\frac{1}{2n_o^2} + \frac{\alpha}{\pi n_o^3}\left[2/R - 1/(n_o^2)\right]^{1/2}, \tag{7.4.11}$$

$$(\alpha/\pi)^2 \ll R^2/2 \ll n_o^2 .$$

Including the next higher order term in the asymptotic expansion made in reference [64] brings small oscillations into formula (7.4.11).

Presnyakov [135] used (7.4.11) to calculate the shifts of highly excited spectral lines in an atmosphere of alkali metal atoms. These approximations give good agreement with experiment because they take into account the formation of negative ions in the ambient gas.

Analytic properties of the terms belonging to the system considered above have been fully discussed so far. One characteristic feature is logarithmic branch points where an infinite number of terms intersect each other. These points occur at zero energy ($n \to \infty$, $\cot \pi n = \pm i$). An approximate position of branch points in the region over the barrier was established in reference [64].

The studies of electron states in a Coulomb field and in the field of a diatomic molecule (a model of two ZRPs) use a Coulomb Green's function with two arguments. Ivanov [141] suggested a convenient quasi-classical approximation for that case.

Apart from the applications considered above, the ZRP approximation was used in many other specific studies (including many-channel problems with ZRPs and a Coulomb potential), particularly in references [64, 138-141].

Superposition of a ZRP and a harmonic oscillator potential was considered by Janev and Maric [142]. Probably it would be also of interest to study a system where the potential is the superposition of a ZRP and a potential of the form $V(r) = -u_o(r^2 + b^2)^{-2}$. The latter has the asymptotic form of a polarization potential because it falls off as $1/r^4$ as $r \to \infty$, and it was used in the past in many semi-empirical calculations of electron scattering. For such a potential, a closed analytic expression for the three-dimensional Green's function is known (see reference [143]), though only in the case of zero energy.

# Chapter 8

# ELECTRON DETACHMENT IN SLOW COLLISIONS BETWEEN A NEGATIVE ION AND AN ATOM

## 8.1 ZRPs IN TIME-DEPENDENT QUANTUM MECHANICAL PROBLEMS

Time-dependent problems with slowly changing parameters of the system, where two discrete energy levels (terms) come close together with a subsequent transition occuring between the two states, have been extensively investigated in quantum mechanics. For calculations of the transition probability the formulae of Landau-Zener or Landau-Teller or Zener-Rosen or some others [26] can be used, depending on the particular conditions. These theories are not applicable, however, to the case where an isolated energy level approaches the continuum boundary. Then the term interacts with an infinite number of energy levels rather than with only one, and the above theories are no longer applicable.

The processes which will be discussed in this chapter are electronic transitions from a bound state to a continuum state (and the reverse transitions) caused by the interaction between two atomic systems. These reactions include:

(i) ionization:

$$A + B \rightarrow A + B^{+} + e,$$

$$A^{+} + B \rightarrow A^{++} + B + e,$$

$$A^{+} + B \rightarrow A^{+} + B^{+} + e,$$

etc.,

(ii) electron detachment from a negative ion:

$$A^{-} + B \rightarrow A + B + e,$$

(iii) associative capture:

$$A + B^- \rightarrow AB + e,$$

(iv) associative ionization:

$$A + B \rightarrow AB^+ + e,$$

where e represents the electron involved in the transition.

Similar problems arise in the theory of thermal ionization (detachment) of electrons in solids [144]. Apart from the direct reactions listed above, the reverse reactions involving electronic transitions from the continuum to bound states such as dissociative recombination:

$$AB^+ + e \rightarrow A + B,$$

as well as electronic transitions between continuum states (the so-called free↔free transitions) resulting in an effective energy exchange between electrons and heavy particles (the atomic nuclei), are also of interest.

The mathematical difficulties encountered in these problems are very substantial. For approximations involving a finite number of states only a few ordinary differential equations need to be solved. However, in the above cases, the reactions involve interactions between an infinite number of states, and they require, even in the simplest approximation, the solution of partial differential equations. In situations where the quantal treatment of the nuclear motion is necessary, the equation to be solved is similar to that of the theory of detachment (see Chapter 11). However, if the energy of the nuclear motion is much higher than that of the electronic transition, it is possible to consider the nuclei as moving along prescribed classical trajectories unperturbed by the change in the electron energy. In this case it is necessary to solve a quantum mechanical time-dependent problem for the electron. In the general case, these problems are of considerable difficulty, and the ZRP approximation turns out to be very useful in producing some very important results.

Let us consider the general equations for a particle moving in a field of a system of ZRPs whose depth varies with time. We assume that there are N such wells situated at points $\underline{R}_j(t)$ and that their depths are determined by parameters $\alpha_j(t)$, $j = 1, 2, \ldots, N$, in the boundary conditions. In the vicinity of each well the wavefunction can be written [145] thus:

$$\psi(\underline{r},t) = c_j(t)\left\{\frac{1}{|\underline{r} - \underline{R}_j(t)|} - \alpha_j(t)\right\} + O(|\underline{r} - \underline{R}_j(t)|). \quad (8.1.1)$$

In order to obtain the boundary conditions for a moving j-th well we have to use a local system of coordinates where the j-th well is at rest at the moment. The wavefunction will then change according to the Galilean transformation [54]:

$$\psi'(\underline{r}, t) = \psi(\underline{r}, t) \exp(i\underline{V}\cdot\underline{r} + iV^2t/2), \tag{8.1.2}$$

where $\psi'$ is the wavefunction in the system of coordinates moving at a speed $\underline{V}$ relative to the laboratory system of coordinates. In the present case, we have to put $\underline{V}_j = d\underline{R}_j/dt$. Then, making use of (8.1.1) and (8.1.2), we obtain the boundary condition in the following form:

$$\psi(\underline{r}, t) = c_j(t)\left(\frac{1}{|\underline{r} - \underline{R}_j(t)|} - \alpha_j(t) + i\frac{\underline{V}_j\underline{r}}{r}\right) + O(|\underline{r} - \underline{R}_j(t)|) . \tag{8.1.3}$$

We see that in the case of a moving ZRP the boundary conditions have to be modified in a way similar to that of an external magnetic field (see Section 3.6). The necessity of this modification, omitted in reference [145], was pointed out to the authors by E. A. Solov'ev.

Let us now write the time-dependent Schrödinger equation for a system of N moving wells:

$$\left(-\frac{1}{2}\nabla^2 - i\frac{\partial}{\partial t}\right)\psi(\underline{r},t) = 2\pi\sum_{j=1}^{N} c_j(t)\,\delta(\underline{r} - \underline{R}_j(t)). \tag{8.1.4}$$

Making use of the Green's function (see, for instance, reference [19]), that is,

$$G(\underline{r}, \underline{r}', t - t') = i\{2\pi i(t - t')\}^{-3/2} \exp\left(i\frac{(\underline{r} - \underline{r}')^2}{2(t - t')}\right) \tag{8.1.5}$$

and

$$\left(-\frac{1}{2}\nabla^2 - i\frac{\partial}{\partial t}\right) G(\underline{r}, \underline{r}', t - t') = \delta(\underline{r} - \underline{r}')\,\delta(t - t'), \tag{8.1.6}$$

we can write a formal solution of the time-dependent Schrödinger equation (8.1.4) in the form

$$\psi(\underline{r}, t) = \frac{1}{\sqrt{2\pi i}} \sum_{j=1}^{N} \int_{-\infty}^{t} c_j(t') \exp\left[ i \frac{(\underline{r} - \underline{R}_j(t'))^2}{2(t - t')} \right]$$

$$\times \frac{dt'}{(t - t')^{3/2}} + \psi_f(\underline{r}, t), \tag{8.1.7}$$

where $\psi_f(\underline{r}, t)$ is a solution of the corresponding homogeneous equation, that is, a free-propagating wave. The function $\psi_f(\underline{r}, t)$ is specified by the initial conditions, and it will be assumed to be known, in the subsequent discussion. For instance, if the particle is in a bound state as $t \to -\infty$, then $\psi_f = 0$. Construction of the unique solution of equation (8.1.4) requires (apart from $\psi_f$) N time-dependent coefficients $c_j(t)$ to be determined.

The equations for the $c_j(t)$ are derived from the boundary conditions for each ZRP. It is convenient to apply the Galilean transformation first in order to consider the corresponding ZRP as being at rest. Then one has to let $\underline{r} \to \underline{R}_j$ in formula (8.1.7), subtract singularities from both sides of the equation, and take into account the boundary conditions in the form (8.1.1). The final form of the resulting system of integro-differential equations for functions $c_j(t)$ is thus:

$$-\alpha_j(t)\, c_j(t) - \psi_f(\underline{R}_j, t) = \frac{1}{\sqrt{2\pi i}} \sum_{m=1}^{N} \int_{-\infty}^{t} \Bigg\{ c_m(t')$$

$$\times \exp\left( i\, \underline{V}_j(t)(\underline{R}_j(t) - \underline{R}_m(t)) + i \frac{|\underline{R}_m(t) - \underline{R}_j(t')|^2}{2(t - t')} \right)$$

$$- c_j(t)\, \delta_{mj} \Bigg\} \frac{dt'}{(t - t')^{3/2}}. \tag{8.1.8}$$

The system of linear algebraic equations, derived earlier in Section 4.5, can be deduced from (8.1.8) if we assume that all ZRPs are at rest and of constant depth, that $\psi_f = \exp(i\, \underline{k}_0 \cdot \underline{r} - ik^2 t)$, and that $c_j(t)$ are harmonic functions of time. Another special case is a single ZRP of constant depth moving at a constant velocity. The solution of equation (8.1.4) can be obtained, in this case, simply by applying the Galilean transformation. It is easy to check that the resulting function is in agreement with equation (8.1.8).

In the general case, the ZRP approximation reduces the partial differential Schrödinger equation (8.1.4) to a system of integro-differential Volterra equations (with a singular kernel) in one variable. This is an essential simplification of the problem.

Now we shall turn to the problem of a particle in a field of N moving separable potentials of variable depth. The time-dependent Schrödinger equation takes the form

$$\left(-\frac{1}{2}\nabla^2 - i\frac{\partial}{\partial t}\right)|\psi(\underline{r},t)\rangle = -\sum_{j=1}^{N} |\phi_j\rangle\, v_j \,\langle\phi_j|\psi(\underline{r},t)\rangle, \quad (8.1.9)$$

where, in the general case, the coefficients $v_j$ and functions $\phi_j$ are time-dependent. Similarly to the case of ZRPs, we shall write the formal solution of equation (8.1.9), with the help of the Green's function $\hat{G}$, as follows:

$$|\psi(\underline{r},t)\rangle = |\psi_f(\underline{r},t)\rangle + \sum_{j=1}^{N}\int_{-\infty}^{t} \hat{G}(t-t')|\phi_j(t')\rangle\, v_j(t')\; c_j(t')dt' ,$$

$$c_j(t) = \langle\phi_j(t)|\psi(\underline{r},t)\rangle . \quad (8.1.10)$$

Projecting both sides of equation (8.1.10) onto the state $|\phi_i\rangle$, we obtain a system of integral equations for N time-dependent coefficients $c_i(t)$:

$$c_i(t) = \langle\phi_i(t)|\psi_f(t)\rangle + \sum_{m=1}^{N}\int_{-\infty}^{t} \langle\phi_i(t)|\,\hat{G}(t-t')|\phi_m(t')\rangle \times v_m(t')\; c_m(t')\; dt' . \quad (8.1.11)$$

According to (8.1.10), the wavefunction of the time-dependent problem (8.1.9) can be obtained as soon as the coefficients $c_j(t)$ have been determined.

In the case of a single separable potential ($N = 1$), with the function $\phi$ being time-independent and $v(t)$ being proportional to t, the solution of the equation can be obtained in a closed form (see Section 9.1 below).

A different class of problems may arise where the depth of the potential wells is constant, but there is a time-dependent external field. The most important physical example of a problem of that kind is photoionization by intense light. The characteristic length associated with changes in such an external electro-magnetic field is much larger than the corresponding atomic scale. Therefore a generalization of the semi-classical approximation to the time-dependent case can be used to describe the motion of electrons in the field of the electro-magnetic wave. These problems will not be considered here, and we refer the reader to the original papers [146, 147] as well as to the monograph [19].

## 8.2 LINEAR APPROXIMATION IN DETACHMENT THEORY

We have seen in Chapter 3 that a term of a negative molecular ion which depends on the internuclear separation R may cross the lower boundary of the continuum. At separation $R = R_c$, say, when the term coincides with the boundary, the bound state disappears and the pole of the S-matrix becomes related either to a virtual or to a quasi-stationary state. Certainly, this takes place, for instance, when a hydrogen atom H and a negative ion of hydrogen $H^-$ are in a close encounter with each other because the negative ion of the united atom, $He^-$, does not exist (with the corresponding electron configuration). The disappearance of the bound state plays a decisive role in the physical picture of the course of the detachment reaction

$$A^- + B \rightarrow A + B + e \qquad (8.2.1)$$

occuring in slow collisions. Indeed, in this time-dependent problem, when the separation R decreases and, eventually, becomes smaller than $R_c$, the wavefunction of the "active" electron takes the form of a spreading wave packet built of continuum states. After the collision, when R increases and becomes greater than $R_c$, the bound state reappears. A part of the wave packet will be trapped in a newly created effective potential well, whereas another part of the packet will continue moving away to infinity.

The questions which this theory is expected to answer are:

(i) What is the probability that the electron remains bound to the nuclei after the collision (electron detachment does not occur)?

(ii) What is the energy spectrum W(E) of the detached electrons?

The physical picture of the process described above is based on considering adiabatic terms of the quasi-molecule $AB^-$, and it is useful only if the velocity of the nuclei is much smaller than 1 a.u. In the opposite case of fast moving nuclei, electron detachment will be caused by the direct scattering of the active electron (the latter can be considered as being free) on atom B. This was confirmed by the Born calculations of Lopantseva and Firsov [148], who also used a ZRP to describe the interaction between the electron and the atom. This case will not be considered in detail here.

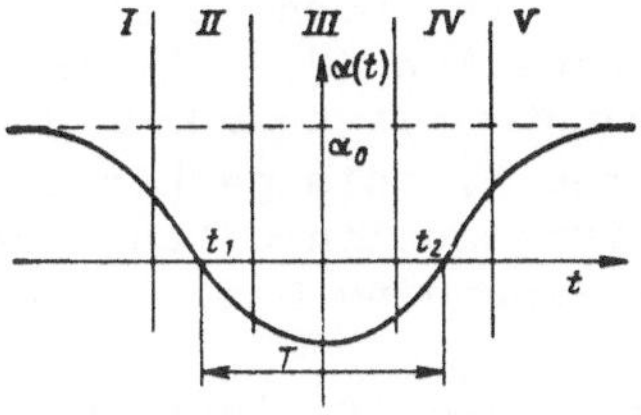


Fig. 8.1. The parameter $\alpha$ in the boundary condition as a function of time in the course of the collision between a negative ion and an atom. Note that $-\alpha(t)$ is the logarithmic derivative of the wavefunction at the edge of the effective potential well.

If a discrete state is initially far from the continuum boundary and the parameters of the system are changing slowly (low velocity of colliding particles), the wavefunction develops adiabatically and transitions occur with an exponentially small probability.* The adiabatic condition is violated only if the discrete level and the continuum are very close to each other. However, in the latter case a single ZRP is a very good approximation to the effective potential well and the time-dependence of the depth of the well can be taken into consideration by a suitable choice of the function $\alpha(t)$.

The function $\alpha(t)$ has to be of the form shown in Fig. 8.1. The bound state exists if $t < t_1$ and $t > t_2$, whereas within the time interval $t_1 < t < t_2$ there is no bound state.

For a single stationary ZRP, which will be considered below in this section, the problem is, in fact, one-dimensional. It will be convenient to use, in this case, the radial function $\Psi(r,t)$, which relates to the three-dimensional $\psi(r,t)$ thus:

$$\Psi(r,t) = r\,\psi(r,t); \qquad c(t) = \Psi(0,t). \tag{8.2.2}$$

The initial condition is imposed in the form

$$\Psi(r,t) = \sqrt{2\alpha_0}\,\exp(-\alpha_0 r + i\alpha_0 t^2/2) \quad \text{as } t\to -\infty\,, \tag{8.2.3}$$

where $\alpha_0$ denotes the limit of the parameter $\alpha(t)$ as $t \to \pm\infty$. For large $t > 0$, the wavefunction is given by

$$\Psi(r,t) = B\sqrt{2\alpha_0}\,\exp(-\alpha_0 r + i\alpha_0 t^2/2) + Q(r,t). \tag{8.2.4}$$

The probability $P_\alpha$ that detachment will not take place, i.e., that

* The probability of such transitions under the barrier into the continuum was obtained by Chaplik (149,150) using adiabatic perturbation theory.

the electron will remain in the bound state, is given by $|B|^2$, where B is the coefficient in equation (8.2.4). The second term in that equation, $Q(r,t)$, is a spreading wave packet, and it goes to zero at any r as t increases. For large t, this packet describes a free particle; therefore the momentum distribution of the detached electrons at $t \to \infty$ can be obtained by expanding Q as a Fourier integral.

Qualitative arguments make it clear that the steeper the function $\alpha(t)$ is, the slower the distribution tail will fall off and the faster the wave packet will spread. It is also clear that the derivative of $\alpha(t)$ at the crossing points $t_1$ and $t_2$, that is, $\alpha'(t_1)$ and $\alpha'(t_2)$, as well as the interval $T = t_2 - t_1$ during which the bound state does not exist, will be important parameters of the theory. The quantity $\alpha'(t_1)$ can be expressed in terms of the relative velocity of the colliding particles or, more precisely, in terms of the quantities $(dE(R)/dt)_{R_c}$ and $(d^2E(R)/dR^2)_{R_c}$, where $E(R)$ is the binding energy of the active electron in the system $(AB^-)$, and $R_c$ is the critical separation at which the linear approximation, considered in this section, is still valid. These quantities are characteristic parameters of the problem.

Let us now write down the general equation (8.1.6) in the specific case of one stationary ZRP of variable depth, which we are considering at the moment. After integrating by parts one obtains

$$-\alpha(t)\,c(t) + \psi_f(0,t) = \frac{2}{\sqrt{2\pi i}} \int_{-\infty}^{t} \frac{\dot{c}(x)}{\sqrt{t-x}}\,dx. \tag{8.2.5}$$

In (8.2.5) we shall put $\psi_f = 0$ because the electron was bound in the initial state. Had this equation contained a known function on its left-hand side, it would have been identical with the well-known integral equation of Abel.

Solving (8.2.5) formally, we shall transform it to a more convenient form, that is,

$$c(t) = -\frac{1}{\sqrt{2\pi}}\, e^{-3\pi i/4} \int_{-\infty}^{t} \frac{\alpha(x)\,c(x)}{\sqrt{t-x}}\,dx\,. \tag{8.2.6}$$

A direct way of obtaining this equation is described by Demkov (18), and we shall follow that paper in the remaining part of this section. Equation (8.2.6) can be formally written as a differential equation using an extension of the concept of n-fold integrals and n-order derivatives to non-integer values of n. In order to achieve this, we shall consider an n-fold integral of a function $\alpha(t)$ and use the following formula:

$$\int_{-\infty}^{t} dt_1 \int_{-\infty}^{t} dt_2 \ldots \int_{-\infty}^{t} dt_n f(t_n) = \frac{1}{\Gamma(n)} \int_{-\infty}^{t} f(x)(t-x)^{n-1} dx.$$

It is readily seen that an integral of the same type as that in equation (8.2.6) can be obtained from the above equation if we put $n = \frac{1}{2}$ on the right-hand side of it. Therefore, a half-fold integral of function f can be defined as

$$\frac{1}{\sqrt{\pi}} \int_{-\infty}^{t} f(x) \frac{dx}{\sqrt{t-x}} .$$

Introducing also the inverse operator, that is, the derivative of order $\frac{1}{2}$, $d^{\frac{1}{2}}/dt^{\frac{1}{2}}$, the integral equation (8.2.6) takes the following form:

$$\left( \frac{d^{\frac{1}{2}}}{dt^{\frac{1}{2}}} - \sqrt{2i}\, \alpha(t) \right) c(t) = 0. \tag{8.2.7}$$

Differential and integral operators of non-integer order are discussed in detail, for instance, by Titchmarsh (151). They have some very useful properties. These operators form a commutative group whose order is an additive parameter of the group. Such rules as integration by parts and many others can be generalized to hold for any operator of this group. Making use of the derivative and integral of order $\frac{1}{2}$, we have transformed the original equation (8.2.6) to a new form (8.2.7) which is more convenient, as we shall see, for further investigations of the problem.

Equation (8.2.7) is a particular case of a linear differential equation and, by expanding the function $\alpha(t)$ in powers of t and retaining only terms of order one, we obtain a linear differential equation with coefficients which are linear functions of the argument t. It is well known that this type of equation of an arbitrary order n can be solved by applying the Laplace transformation to the function c(t). If we employ, in equation (8.2.7); the linear approximation to $\alpha(t)$, then, of course, we cannot obtain a solution of the exact equation. However, we can expect that the approximate solution found in this way will be good enough in the neighborhood of the crossing points $t_1$ and $t_2$ where the term enters the continuum. Let us put

$$\alpha(t) = -\beta t \tag{8.2.8}$$

and seek a solution of equation (8.2.7) in the form of a contour integral:

$$c(t) = \int_C e^{iut} Z(u)\,du \; , \qquad (8.2.9)$$

where the contour C will be defined later. Substituting (8.2.9) for c(t) in the equation (8.2.6) and using (8.2.8), we derive, in the usual way, an equation for Z(u):

$$Z(u) = -\frac{i\beta}{\sqrt{2u}} Z'(u) \; . \qquad (8.2.9^*)$$

Solving this equation, we obtain c(t) in the form of the following contour integral:

$$c(t) = g \int_C \exp\{(i/3\beta)(2u)^{3/2} + iut - \sqrt{2u}\, r\}\, du. \qquad (8.2.10)$$

The path of integration in (8.2.10) has to be chosen in such a way that the exponential integrand would fall off as both ends of the contour go to infinity.

Substituting the solution (8.2.10) for c(t) into equation (8.1.7) for $\Psi(r,t)$ and carrying out the necessary integration, we obtain

$$\Psi(r,t) = g \int_C \exp\{(i/3\beta)(2u)^{3/2} + iut - \sqrt{2u}\, r\}\, du \; ,$$

or, on replacing $\sqrt{2u}$ by a new, single-valued variable v,

$$\Psi(r,t) = g \int_C \exp\{(i/3\beta)\, v^3 - v r + i \frac{v^2}{2} t\}\, v\, dv \; . \qquad (8.2.11)$$

Equation (8.2.11) shows that $\Psi$ is a solution of the Schrödinger equation for a free particle. It is also easy to check, by differentiating (8.2.11) with respect to the parameter $\beta$ and integrating by parts, that $\Psi$ satisfies the boundary condition (8.1.1).

Now we shall determine the contour C. For this it is sufficient to take into account the behavior of $\Psi$ at $r = 0$ and $t \to \pm\infty$. For $t \to -\infty$, we can use the adiabatic approximation and write for $\Psi$ thus:

$$\Psi(r,t) \sim \sqrt{-2\beta t}\, \exp\{\beta t r + i\beta^2 t^3/6\} \; . \qquad (8.2.12)$$

For $t \to +\infty$ , the wavefunction $\Psi$ has to fall off, for any finite r, as a power of t.

For the integral (8.2.11) the saddle points, at $r = 0$, are $v = 0$ and $v = -\beta t$. Figure 8.2 shows the behavior of the modulus of the exponential in (8.2.11), with those areas being hatched where that quantity is larger than unity. The right and left saddle points correspond to the values 0 and $-\beta t$, depending on which one is larger. Calculating this integral by the method of steepest descent, we find that the integration path (contour C) has the form shown in Fig. 8.2. With this choice of contour, the asymptotic form of $c(t)$, as $t \to -\infty$, is given by

$$c(t) \sim \sqrt{2\beta t}\ \exp(i\beta^2 t^3/6), \tag{8.2.13}$$

as it ought to be. For $t \to +\infty$, $c(t)$ falls off according to

$$c(t) \sim -i\sqrt{2/(\beta^3 t^5)}\ .$$

The constant $g$ in (8.2.11) is $(i\beta\pi)^{-\frac{1}{2}}$. Therefore, the final expression for the normalized wavefunction is

$$\Psi(r,t) = \frac{1}{\sqrt{i\beta\pi}} \int_C \exp\left\{\frac{iv^3}{3\beta} + i\frac{v^2}{2}t - vr\right\} v\,dv. \tag{8.2.14}$$

For $r \neq 0$, the right saddle point shifts downwards and to the right, whereas the left one shifts upwards and to the left. The asymptotic form of $\Psi$ is also easy to investigate. As $t \to -\infty$, we obtain the asymptotic form (8.2.12); however, only if $r \ll \beta t^2$. For very large $r$, that is, $r \gg \beta t^2$, $\Psi$ falls off faster than for $r = 0$; thus:

$$\Psi \sim \exp\{-(2/3)\sqrt{-i\beta r^3}\}\ .$$

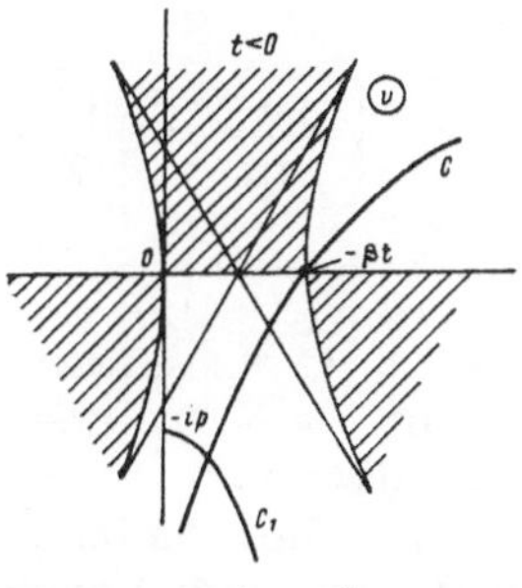


Fig. 8.2. Integration contours in the complex v plane for calculating the integral (8.2.5). Modulus of the exponential integrand is greater than unity in the hatched areas of the plane. At the initial moment, $t \to -\infty$, contour C corresponds to a bound state of the electron, and contour $C_1$ corresponds to the electron being in the continuum, with momentum p.

It is not difficult to understand this result. The changes at the point $r = 0$ cannot instantly be felt in the domain where r is large, so that the adiabatic approximation is not valid in the latter case. At large r, the function $\Psi$ retains information about the values of $\alpha(t)$ at previous moments and, correspondingly, falls off more rapidly.

Investigation of the asymptotic form of $\Psi$ at $r \neq 0$ and $t \to +\infty$ shows that this function does fall off as a power of t, at any finite r. The wavefunction $\Psi(r,t)$ can be expressed in terms of the Airy function (see, for instance, reference [122]). We obtain

$$\Psi(r,t) = -2\beta^{-1/6} e^{i(\beta^2 t^3 + \pi)/12} \times \frac{\partial}{\partial x} \left\{ e^{\beta tr/2} v\{\beta^{1/3} e^{-i\pi/6}(r + i\beta t^2/4)\}\right\} .$$

However, in the actual investigations it is easier to proceed by first obtaining expression (8.2.14) and using the contour integral representation for $\Psi$.

Now we shall derive the momentum representation for $\Psi(r,t)$ as $t \to +\infty$ which will give us the momentum distribution of the detached electrons. Let us multiply (8.2.14) by the normalized wavefunction of a free particle, $(2/\pi)^{1/2} \sin kr$, and integrate the product over r from zero to infinity. We obtain

$$\phi(k,t) = (2/(i\beta\pi^2))^{1/2} \int_0^\infty \sin kr\, dr \int_C \exp\left(\frac{iv^3}{3\beta} + i\frac{v^2}{2}t - vr\right) v\, dv.$$

If we deform the contour C to its new position C' which is entirely in the right half-plane, then the order of integration can be changed with the integration over r carried out first. The result is thus:

$$\phi(k,t) = \frac{1}{\pi}\sqrt{2/i\beta} \int_{C'} \exp\left(\frac{iv^3}{3\beta} + i\frac{v^2}{2}t\right) \frac{kv\, dv}{v^2 + k^2} .$$

If this integral is taken, as $t \to \infty$, along the contour C which passes through the saddle point $v = 0$, the resulting function will tend to zero at any finite r. However, this contour passes between the pair of poles, $v = \pm ik$, whereas the deformed contour C' has both poles on its left. The difference between the two integrals taken along C' and along C is an integral along a closed contour encircling the pole $v = -ik$. Taking the residue of the integrand, we obtain the following expression for $\phi(k,t)$:

$$\phi(k,t) = (2/\beta)^{\frac{1}{2}} k \; e^{-\frac{k^3}{3\beta} + \frac{i k^2}{2} t - \frac{3 i \pi}{4}}$$

$$+ \{2/(i\pi\beta^2)\}^{\frac{1}{2}} \int_C \frac{e^{i \frac{v^3}{3\beta} + i \frac{v^2}{2} t}}{v^2 + k^2} k\, v\, dv \; . \qquad (8.2.15)$$

The first term in (8.2.15) is due to integration over the residue. It does not fall off as $t \to \infty$ and has the form of a wavefunction of a free particle in the momentum space. That is the part of (8.2.15) which gives us the sought momentum distribution of the detached electrons,

$$W(k) = \frac{2}{\beta} k^2 \exp\left(-\frac{2k^3}{3\beta}\right) . \qquad (8.2.16)$$

The distribution W(k) satisfies the normalization condition

$$\int_0^\infty W(k)\, dk = 1.$$

A qualitative prediction made earlier about the spectrum of the detached electrons agrees with the present results. For large $\beta$, that is, for a rapid change in $\alpha(t)$, the spectrum is shifted towards higher energies.

Obviously, we cannot correctly predict the probability of ionization if we use a linear approximation to the function $\alpha(t)$. Indeed, in this case $\alpha(t) \to -\infty$ as $t \to +\infty$ and ionization occurs with the probability $P_\alpha = 1$. This, of course, differs from the real case. The distribution W(k) derived above has a meaning only if the ionization probability obtained with the exact function $\alpha(t)$ is close to unity. Then the system is almost certain to disintegrate soon after the separation R becomes critical, $R = R_c$, and the linear approximation used for $\alpha(t)$ is still valid.

We shall now outline how to estimate the probability that the electron will remain bound after collision. Let us assume that $\Psi_1(r,t)$ and $\Psi_2(r,t)$ are two solutions of the Schrödinger equation such that they satisfy the same boundary condition (8.1.1) but correspond to different initial conditions. Then the overlap integral

$$\int_0^\infty \Psi_2^* \Psi_1 \, dr = I \tag{8.2.17}$$

is time-independent. We choose $\Psi_1$ to satisfy condition (8.2.3) as $t \to -\infty$ and $\Psi_2$ to satisfy the same condition as $t \to +\infty$. The condition (8.2.3) means that the particle is bound. Therefore, $\Psi_1$ corresponds to the usual formulation of the collision problem, and $\Psi_2$ corresponds to the particle being in a bound state after the collision. The required probability is then given by $|I|^2$.

The argument t of $\alpha(t)$ changes in the interval $-\infty < t < +\infty$. Let us divide this interval into five regions shown in Fig. 8.1. During the entire course of the collision, the molecular term merges with the continuum boundary twice, within regions II and IV. For $\Psi_1$, the adiabatic solution in region I goes smoothly into a solution corresponding to a linear $\alpha(t)$ in region II. Similarly for $\Psi_2$, we can construct a solution which goes into the adiabatic solution in region V and into a solution for a free particle in region III. Then the integral (8.2.17) can be calculated at some time t within region III. We see that we can avoid, in this way, to continue $\Psi_1$ into regions IV and V, and $\Psi_2$ into regions II and I.

It is more convenient to carry out computations of (8.2.17) in the momentum space. The function $\phi_1(k,t)$ is given in region III by the first term in (8.2.15). In the same way, it is not difficult to obtain for $\phi_2$ the expression

$$\phi_2(k,t) = (2/\beta)^{\frac{1}{2}} k \exp\left(-\frac{k^3}{3\beta} + \frac{ik^2}{2}(t-T) + \frac{3\pi i}{4}\right),$$

where T is the shift of the origin on the t-axis.

Then we obtain, for the probability $P_\alpha$, that

$$P_\alpha = \left| \int_0^\infty \phi_2^* \phi_1 dk \right|^2 = \left| \frac{2}{\beta} \int_0^\infty \exp\left(-\frac{2k^3}{3\beta} + \frac{ik^2T}{2}\right) k^2 dk \right|^2. \tag{8.2.18}$$

Assuming in (8.2.18) that T is large and neglecting the first term in the argument of the exponential, we finally obtain the following expression for the probability $P_\alpha$:

$$P_\alpha = 2\pi / \beta^2 T^3. \tag{8.2.19}$$

We shall clarify the meaning of the parameter $\beta$ in (8.2.19) in the following way. For $t < 0$ and in the vicinity of the crossing point,

$$E(R) = -\frac{\beta^2 t^2}{2} .$$

Then

$$\beta = \left|\frac{d^2E(R)}{dt^2}\right|^{\frac{1}{2}}_{t=0} = \left|\frac{d^2E(R)}{dR^2}\right|^{\frac{1}{2}}_{R_c} \left(\frac{dR}{dt}\right)_{R_c} .$$

With the help of these expressions, the energy distribution can be written using formulae (8.1.16) - (8.1.18), in arbitrary units, thus:

$$W(E) = \gamma E^{\frac{1}{2}} \exp\left[-\frac{2}{3}\gamma E^{3/2}\right] , \tag{8.2.20}$$

where

$$\gamma = 2^{3/2} \left(\frac{dR}{dt}\right)^{-1}_{R_c} \left(\frac{d^2E}{dR^2}\right)^{-\frac{1}{2}}_{R_c} .$$

Also

$$P_\alpha = \frac{2\pi}{T^3} \left(\frac{d^2E}{dR^2}\right)^{-1}_{R_c} \left(\frac{dR}{dt}\right)^{-2}_{R_c} . \tag{8.2.21}$$

The quantity $(dR/dt)_{R_c}$ in the formulae above is the component of the relative velocity of the nuclei along the internucleus axis, taken at time $t = t_1$. In various estimates, this quantity can be replaced by the relative velocity of the nuclei $V_\infty$, for the infinite separation, that is, as $t \to \infty$.

We shall now establish criteria for the validity of the theory developed above. The distribution (8.2.16) has a maximum at $k = \beta^{1/3}$ which corresponds to the energy $\beta^{2/3}/2$. Obviously, the particle can be treated as free only if $\alpha(t)^2/2$, where $\alpha < 0$ (this quantity is analogous to the binding energy if $\alpha > 0$), is much greater than $\beta^{2/3}/3$, thus:

$$\alpha^2/2 = \beta^2 t^2 \gg \beta^{2/3}/2, \quad \beta^2 t^3 \gg 1 .$$

The linear approximation for $\alpha(t)$ is expected to be adequate for all t up to the time determined by the inequality above. Replacing t by the quantity $\delta R/V$, where $\delta R$ is an interval in the neighborhood of $R_c$ within which the linear approximation is valid, we obtain

$$\left(\frac{d^2E(R)}{dR^2}\right)_{R_c} \frac{(\delta R)^3}{v} \gg 1 . \qquad (8.2.22)$$

This condition is similar to that known as the Massey adiabatic criterion. It follows from the same formula (8.2.22) that the probability w is small in those cases where the theory is applicable, and that the method of steepest descent can be used to calculate the integral (8.2.18). If $\beta^2T^3 \sim 1$, the theory is inapplicable and the probability of ionization can differ substantially from unity.

Another important restriction relates to the replacement of a system of two atoms by a point singularity placed at the origin of coordinates. Evidently such a model is valid only for momenta $k = \beta^{1/3}$ which are associated with de Broglie wavelengths much greater than the size of the colliding systems. In this way, we obtain the inequality

$$\beta^{1/3} = v^{1/3} \left(\frac{d^2E(R)}{dR^2}\right)_{R_c}^{1/6} \ll \frac{2\pi}{R}, \qquad (8.2.23)$$

or

$$v^{1/3} \ll 2\pi R^{-1} \left(\frac{d^2E(R)}{dR^2}\right)_{R_c}^{-1/6}, \qquad (8.2.24)$$

where R is the effective size of the system.

The latter condition turns out to be rather restrictive because the velocity occurs raised to the 1/3 power. Even at very low velocities, the left-hand side of (8.2.23) may be smaller than the right-hand side by only a factor of two or three.

One can expect that even if the inequality (8.2.23) does not hold, the energy distribution of the detached electrons will be given by formula (8.2.19), provided that the energy is low, whereas the position of the maximum and the behavior at high energies will be determined by the finite size of the colliding system. These expectations will be shown to be right in the next section.

## 8.3 ACCOUNT OF THE FINITE SIZE OF THE COLLIDING SYSTEM

In this section we shall take into account the finite size of the system of colliding atoms, thus removing the most severe restriction imposed on the theory. Strictly speaking, this account requires a consideration of two three-dimensional ZRPs of constant depth whose motion along their relative trajectory is a given function of time. A similar one-dimensional model of Breit [152] (see also references [153-155]) has a limited value. Apart from that, the calculations in these early papers have not been carried out analytically to the end and the energy distribution for detached electrons has not been found. In the three-dimensional case, even the simplest model with a straight-line trajectory cannot be solved in a closed form.

We shall consider below a different approach to the problem. As long as the colliding atoms move at a low relative velocity, the process takes place chiefly in the neighborhood of $R = R_c$. Therefore it is possible to assume that the colliding particles are represented by two stationary ZRPs of variable depth separated by distance $R_c$. The rate of change of the depths of the ZRPs will be determined by the condition that the change in the model binding energy must be the same as that in the real collision. Furthermore, because the behavior of the terms is essential only near the continuum boundary, we can use the linear approximation developed in the previous section for $\alpha_1$ and $\alpha_2$ and neglect their properties as $t \to \pm\infty$. This would not affect the energy spectrum of the detached electrons.

The case of two identical wells (Demkov [145]), which will be considered below, is of principal interest and admits to an analytical solution. Let us put

$$\alpha_1 = \alpha_2 = -\beta t ,$$

$$c^{\pm} = c_1 \pm c_2 ,$$

$$\underline{R}_2 - \underline{R}_1 = \underline{R} = \text{const.}$$

Then we obtain, from the system of equations (8.1.8), an equation for $c^{\pm}$ in the form

$$\beta t c^{\pm}(t) = \frac{1}{\sqrt{2\pi i}} \int_{-\infty}^{t} \left\{ c^{\pm}(t') - c^{\pm}(t) \pm c^{\pm}(t') \exp\left[\frac{iR^2}{2(t-t')}\right] \right\} \times (t-t')^{-3/2} \, dt' .$$

The kernel of this integral equation depends on the difference t - t' only, and the coefficient of $c^{\pm}$ on the left-hand side is linear in t. Therefore, we can apply a Laplace transformation to solve this integral equation. We find that

$$c^{\pm} = g \int_C \exp\left\{ \frac{i}{\beta} \left( v^2\beta t/2 + v^3/3 \pm e^{-vR}(1/R^3 + v/R^2)\right)\right\} vdv.$$

Considering the $t \to -\infty$ limit in a similar way to that in Section 8.2, we shall determine the contour of integration C whose ends go to infinity at the angles $\arg v = -\pi/2$ and $\arg v = \pi/4$.

We substitute the integral above for c(t) in equation (8.1.7) and obtain the wavefunction,

$$\psi(\underline{r}, t) = g \int_C \exp\left\{ \frac{i}{\beta}\left[ v^3/3 \pm \frac{e^{-vR}}{R^3}(1 + vR)\right] + iv^2t/2 \right\}$$

$$\times \left\{ \frac{\exp(-v|\underline{r} - \underline{R}_1|)}{|\underline{r} - \underline{R}_1|} \pm \frac{\exp(-v|\underline{r} - \underline{R}_2|)}{|\underline{r} - \underline{R}_2|}\right\} vdv. \quad (8.3.1)$$

It follows in a straightforward way from (8.3.1) that $\psi$ is a solution for a free particle everywhere with the exception of the points $\underline{r} = \underline{R}_1$ and $\underline{r} = \underline{R}_2$. In the momentum representation,

$$\phi(\underline{k}, t) = \frac{2g}{\sqrt{2\pi}} \{ \exp(-i\, \underline{k}\cdot\underline{R}_1) \pm \exp(-i\, \underline{k}\cdot\underline{R}_2) \}$$

$$\times \int_C \exp\left\{\frac{i}{\beta}\left[\frac{v^3}{3} \pm e^{-vR}(1 + vR)/R^3\right] + \frac{iv^2t}{2}\right\} \frac{vdv}{v^2 + k^2}. \quad (8.3.2)$$

Repeating the arguments of Section 8.2, we find that the momentum distribution of the detached electrons is determined, as $t \to +\infty$, by the residue of the integrand in (8.3.2) at the point $v = -ik$. The distribution is given by

$$W(k) = \frac{k^2}{2\pi\beta} \{1 \pm \cos(kR\cos\theta\}$$

$$\times \exp\left\{ -\frac{2}{\beta}\int_0^k (1 \pm (\sin k'R)/k'R)\, k'^2 dk'\right\}, \quad (8.3.3)$$

or, after averaging over the angles, by

$$W(k) = \frac{2k^2}{\beta}\left(1 \pm \frac{\sin kR}{kR}\right) \exp\left\{-\frac{2}{\beta}\int_0^k\left(1 \pm \frac{\sin k'R}{k'R}\right)k'^2 dk'\right\}. \quad (8.3.4)$$

The undetermined factor g in (8.3.2) has been replaced in the last two expressions by the factor obtained from the normalization condition.

Now we shall express the parameter $\beta$ in equation (8.3.4) in terms of the initial physical parameters of the problem, that is, the relative radial velocity of the colliding nuclei, $dR/dt$, and some characteristic related to the behavior of the terms in the vicinity of the merging point with the continuum boundary. It follows from (3.2.4) that at this point we obtain, for the symmetric case,

$$\alpha = 1/R_c \,, \qquad (d\kappa/d\alpha)_{R_c} = -1/2 \,,$$

and

$$(d\kappa/dR)_{R_c} = -\{d^2E(R)/dR^2\}^{\frac{1}{2}}_{R_c} = -\{E''(R_c)\}^{\frac{1}{2}} \,. \quad (8.3.5)$$

Equating the changes in $\kappa$ due to the motion of the nuclei and due to the variation in the depths of the wells, we find that

$$\left(\frac{\partial\kappa}{\partial\alpha}\right)_{R_c}\left(\frac{d\alpha}{dt}\right)_{R_c} = \left(\frac{\partial\kappa}{\partial R}\right)_{R_c}\left|\frac{dR}{dt}\right|_{R_c} \,;$$

hence

$$\beta = 2\left(E''(R_c)\right)^{\frac{1}{2}}\left|\frac{dR}{dt}\right|_{R_c} \,. \quad (8.3.6)$$

This definition of $\beta$ does not contain $R_c$ and differs only by a factor of two from the analogous parameter in Section 8.2.

Assuming that $kR$ is small and expanding $\sin kR$ in (8.3.4) in powers of $kR$, we obtain, retaining the first-order terms in the expression, that

$$W(k) = \frac{4k^2}{\beta}\exp\left(-\frac{4k^3}{3\beta}\right) \,. \quad (8.3.7)$$

This formula is identical with formula (8.2.16) on account of equation (8.3.6). This confirms the assumption made in Section 8.2 that formula (8.2.16) is valid, for k small enough, even when $\bar{k}R_c$ ($\bar{k}$ being the average momentum) is comparable with or larger than unity. On the other hand, if $\beta R_c^3 \ll 1$, the formula (8.3.7) is applicable practically for all $k$, and this condition coincides with the condition (8.2.23).

It ought to be mentioned that the symmetrical bound state disappears, during the course of the collision, at a later stage when the ZRPs move apart from each other ( $\partial\kappa/\partial R < 0$) rather than during the initial stage when they approach each other. The disappearance of the bound state takes place only if $\alpha < 0$, that is, when there is no bound state for a single ZRP (see Section 3.2). In such a case, the approximation of two moving ZRPs of constant depth is not applicable, and the present model of two stationary ZRPs of variable depth is more suitable to describe the detachment process.

For the antisymmetrical case, we obtain from equation (8.2.5) that

$$(\partial E/\partial \alpha)_{R_c} = -1/R_c , \qquad \alpha = -1/R_c ,$$

and

$$\left(\frac{\partial E}{\partial \alpha}\right)_{R_c} \left(\frac{d\alpha}{dt}\right)_{R_c} = \left(\frac{\partial E}{\partial R}\right)_{R_c} \left|\frac{dR}{dt}\right|_{R_c} . \tag{8.3.8}$$

In this case, the parameter $\beta$ has the form

$$\beta = R_c \left|\frac{\partial E}{\partial R}\right|_{R_c} \left|\frac{dR}{dt}\right|_{R_c} = R_c E'(R_c) \left|\frac{dR}{dt}\right|_{R_c} . \tag{8.3.9}$$

For $kR \ll 1$, we obtain the distribution

$$W(k) = \frac{k^4 R_c^2}{3\beta} \exp\left(-\frac{k^5 R_c^2}{15\beta}\right) . \tag{8.3.10}$$

Therefore, the approximation employed in Section 8.2 is not applicable in the antisymmetrical case. The weakly bound antisymmetrical state retains the antisymmetry as $R \to R_c$ and, at large distances from the nuclei, it depends on the angle $\theta$ as $\cos\theta$; i.e., it is a p-function. Instead of having a contact point where the term merges with the continuum boundary, in the antisymmetrical case the term crosses the boundary at a certain angle and passes into the continuum. There it corresponds to a quasi-stationary state produced by the centrifugal barrier which exists for the states with $\ell \neq 0$ (see Chapter 2). This leads to the stabilization of the electronic state and to an increase in the average momentum of the emitted electrons, as can be seen from equation (8.3.10). For $\ell \neq 0$ states, the ZRP approximation is not directly applicable. However, equation (8.3.10) shows that the transition is still possible if $\beta \to 0$

and $R_c \to 0$ simultaneously, provided that we demand that the ratio $R_c^2/\beta$ remain finite.

The dependence of the distribution upon the angle $\theta$ is contained in the pre-exponential factor only. This factor has a typical interference form and originates from a superposition of two spherical outgoing waves $\exp(ikr)/r$ shifted a distance R relative to each other. For the symmetrical case, the waves are in phase, and for the antisymmetrical case, they are out of phase by $\pi$. The angular dependence is symmetrical with respect to the angle $\theta = \pi/2$. At this angle, the momentum distribution has a maximum for the symmetrical case, and is zero for the antisymmetrical case. If $kR < \pi/2$, the distribution is a monotonic function in the intervals $0 < \theta < \pi/2$ and $\pi/2 < \theta < \pi$. However, if $kR > \pi$, a series of interference maxima and minima appear, with the minima tending to zero. In the real collision, this interference picture is expected to be blurred due to factors such as the motion and finite size of the atom and the ion which have been neglected in the present discussion. In addition, averaging over the impact parameter is necessary because the angle $\theta$ is measured with respect to the internuclear line at the moment when $R = R_c$ whose direction does not coincide with the direction of the incident particle. This will result in an additional smoothing of the angular distribution. The last averaging effect can be eliminated if the direction of the heavy particle is measured in coincidence with the direction of the emitted electron. However, this is a very difficult experiment to carry out. We shall therefore consider here only the average cross section.

Let us introduce the dimensionless quantities $s = \beta R_c^3$ and $q = kR_c$, to be used instead of $\beta$ and $k$. In the new variables, the normalized momentum distribution (8.3.4) takes the form

$$w(s,q) = f'(q)\, s^{-1} \exp\{-f(q)/s\}, \qquad (8.3.11)$$

where

$$f(q) = \frac{2}{3} q^3 \pm (\sin q - q\cos q). \qquad (8.3.12)$$

If we now take into account the collision geometry and assume that the ion travels past the atom along a straight line at a constant velocity V, then the quantity $(V_R)_c$ entering in $\beta$ and s is equal to $V\cos\alpha$, where $\alpha$ is the angle between vector $R_c$ and the direction of the projectile motion.

Introducing the parameter $s_c = s/\cos\alpha$ which characterizes the collision as a whole and assuming that the momentum distribution (8.3.11) is correct for all impact parameters $\rho$, we obtain the effective cross section $\sigma(s_c, q)$, after averaging over $\rho$, in the following form:

$$\sigma(s_c,q) = 2\pi \int_0^{\pi/2} w(s_c\cos\alpha, q) R_c \sin\alpha R_c \cos\alpha \, d\alpha \,, \qquad (8.3.13)$$

so that, making use of equations (8.3.11) and (8.3.12), we obtain

$$\frac{\sigma(s_c, q)}{\pi R_c^2} = \frac{2 f'(q)}{s_c} \int_0^1 \exp(-f(q)/s_c x) \, dx$$

$$= \left[2 f'(q)/s_c\right] F(f(q)/s_c) \,, \qquad (8.3.14)$$

where

$$F(y) = e^{-y} + y \, Ei(-y) \,, \qquad (8.3.15)$$

the function $\sigma(s_c, q)/(\pi R_c^2)$ being normalized to unity with respect to $q$.

The cross sections given by (8.3.14), computed for both symmetrical and antisymmetrical cases, are shown in Fig. 8.3 for two numerical values of $s_c$, $s_c = 1$ and $s_c = 10$. The broken curves in Fig. 8.3 are those for $s_c = 10$ using the approximate expressions (8.3.7) and (8.3.10), where we have used the approximate values $f^+(q) = 4q^3/3$ and $f^-(q) = q^5/15$. Although this assumption is valid, strictly speaking, only for $s_c \ll 1$, the graphs show that even for $s_c = 10$ it leads to good results everywhere, with the exception of large q, where the cross section is small anyway. For $s_c = 1$, within the scale of the drawing, the exact and approximate curves are indistinguishable from each other. It can be seen that the ZRP approximation turns out to be unexpectedly good even for rather large average velocities of the emitted electrons, when their wavelength is comparable with the characteristic size of the system itself ($q \sim 1$). It can be assumed that even a more accurate account of the finite size of the atoms will change cross sections appreciably only for large q.

In a real collision, the symmetrical and antisymmetrical states enter with equal weights so that the spectrum of the emitted electrons is a superposition of two different distributions. One can expect that a curve with two maxima will be obtained, in particular, at a certain ratio of the parameters $E_c'$, $E_c''$, $R_c^{(+)}$, and $R_c^{(-)}$. Unfortunately, the parameters $E_c'$ and $E_c''$ are unknown, and the formulae developed above cannot be directly used in the actual calculations of various ion-atom pairs.

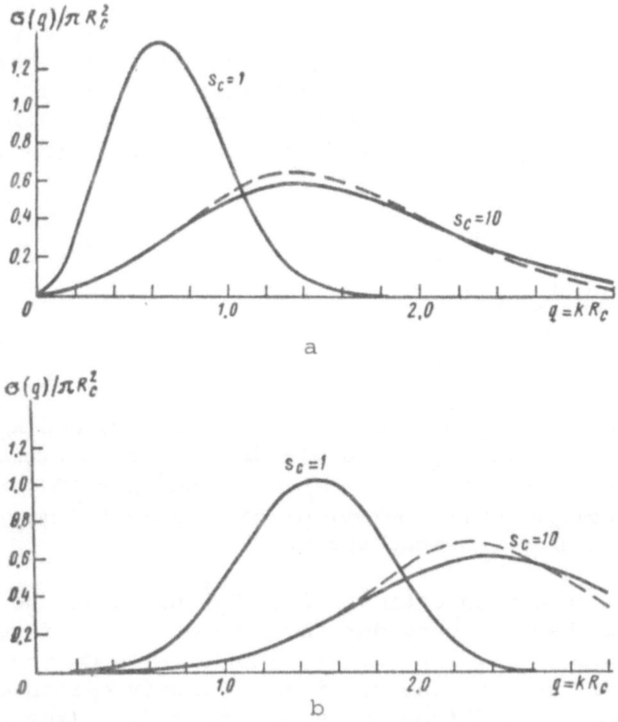


Fig. 8.3. The total cross section for electron detachment with momentum k (in dimensionless units) as a function of the parameter $s_c$ (reference [145]). (a) for the symmetric case; (b) for the antisymmetric case. The broken line corresponds to a calculation where the finite size of the system of colliding atoms has been neglected.

The velocity criterion analogous to that of Massey's adiabatic criteria can be derived as in Section 8.2 and is

$$\left.\begin{aligned} &E''(R_c)\,(\delta R)^3/V \gg 1, \quad \text{symmetrical case,} \\ \text{and} \\ &E'(R_c)\,(\delta R)^2/V \gg 1, \quad \text{antisymmetrical case,} \end{aligned}\right\} \tag{8.3.16}$$

where $\delta R$ corresponds to the interval within which $\alpha(t)$ is a linear function. Obviously, $\delta R$ must be small enough in comparison with $R_c$, that is, $\delta R \ll R_c$. It is possible that the formulae are valid in a somewhat wider interval so that both conditions can be combined together and we can write, in place of (8.3.16),

$$\left.\begin{aligned} &V \lesssim E''(R_c)\,R_c^3, \quad \text{symmetrical case,} \\ \text{and} \\ &V \lesssim E'(R_c)\,R_c^2, \quad \text{antisymmetrical case.} \end{aligned}\right\} \tag{8.3.17}$$

Even a straightforward calculation of the critical separation $R_c$ for a particular quasi-molecule $AB^-$ presents considerable difficulties. A calculation or even an estimate of $E_c'$ and $E_c''$ is still more difficult. If we use the ZRP approximation of two identical wells approaching each other, assuming that their depths do not vary, then we shall find

$$
\begin{aligned}
E_c' &= R_c^{-3} = (2E_{aff})^{3/2}, \\
q &= k/k_o = k/(2E_{aff})^{1/2}, \\
s_c &= VR_c = V/(2E_{aff})^{1/2},
\end{aligned}
\tag{8.3.18}
$$

where $E_{aff}$ is the energy of the electron affinity, and $k_o$ is the average momentum of a weakly bound electron of the system. Putting $E_{aff} = 1$ eV, we find that $s_c = 1$ and $s_c = 10$ correspond to relative velocities of nuclear motion of $V = 0.6 \times 10^8$ cm/sec, and $V = 6 \times 10^8$ cm/sec, respectively.

In order to obtain an estimate for $E_c''$ based on the ZRP approximation, we have to consider the case of two different wells (as shown in Section 3.2, for $\alpha_1 > 0$ and $\alpha_2 > 0$, the contact point between the term and the continuum boundary exists only if $\alpha_1 \neq \alpha_2$ ). Then we find from the equation (3.2.1) (see reference [156]) that

$$
E_c'' = \left(\frac{d\kappa}{dR}\right)^2, \quad \left(\frac{d\kappa}{dR}\right)_{R_c} = \frac{2(\alpha_1\alpha_2)^{3/2}}{(\alpha_1^{1/2} - \alpha_2^{1/2})^2} = \alpha_1\alpha_2\delta\,,
\tag{8.3.19}
$$

where the parameter $\delta$ is defined by formula (3.3.8). When $\alpha_1$ and $\alpha_2$ are not very different from each other, the numerical value of $E_c''$ obtained from (8.3.19) is large. This means that the region $\delta R$, where the linear approximation for $\kappa(R)$ (i.e., quadratic approximation for $E(R)$) is valid, becomes very narrow. The same can be deduced from Fig. 8.2. A more accurate determination of $\delta R$ is possible from the condition that the coefficients of the linear and quadratic terms in the expansion of $\kappa$ in a power series of $R - R_c$ are of the same order of magnitude. It gives

$$
\left(\frac{\partial^2\kappa}{\partial R^2}\right)_{R_c} = \frac{8(\alpha_1\alpha_2)^3}{(\alpha_1^{1/2} - \alpha_2^{1/2})^6}, \quad \delta R = \frac{(\alpha_1^{1/2} - \alpha_2^{1/2})^2}{8(\alpha_1\alpha_2)^{3/2}}.
\tag{8.3.20}
$$

The first condition in (8.3.16) in this case takes the following form:

$$\frac{1}{V}\,(2E_1/m)^{\frac{1}{2}} \gtrsim \frac{2^9 g^{3/4}}{(1-g^{\frac{1}{4}})^8}\,, \tag{8.3.21}$$

where

$$0 < g = E_1/E_2 < 1,$$

$$E_1 = \alpha_1^2/2; \quad E_2 = \alpha_2^2/2\,.$$

This condition leads to very low velocities V even if $g = 0.1$. Thus in the model considered here, the change in energy of the bound state proceeds at a rate which is too high for all realistic values of the ratio $E_1/E_2$.

It is possible, however, that the results of the model discussed above are largely due to the ZRP approximation employed there and that, in reality, smaller values for $E_c''$ and greater values for $\delta R$ are correct. Apart from that, it can also happen that even with a function $E(R)$ which is not quadratic in the vicinity of $R_c$, the momentum distribution for the emitted electrons will still remain the same. Bearing these comments in mind and taking into account that any reasonably accurate theoretical determination of $E_c'$ and $E_c''$ requires rather cumbersome computations, it is expected that these quantities will be first accurately estimated with the help of experimental data.

## 8.4 PRODUCTION OF NEGATIVE IONS IN THREE-BODY COLLISIONS

One of the important sources of negative ions in plasma is three-body collisions of the type

$$e + A + B \rightarrow A^- + B\,. \tag{8.4.1}$$

When the plasma density is not too low, this reaction dominates over radiative capture of an electron by an atom, which also leads to the creation of negative ions. Because of the great disparity between the masses, the direct exchange of energy between the electron and the atom during the course of reaction (8.4.1) is not important. However, when the system $AB^-$ has a quasi-stationary term, the energy exchange can be realized within the adiabatic approximation via a two-stage mechanism where a continuum state is "captured" into the quasi-stationary state with a subsequent decay of the quasi-stationary state when the system is at a different nuclear separation, i.e., at a different energy. Below we shall discuss the simplest example of such a reaction where an approximation of one ZRP and the linear approximation to $\alpha(t)$ can be used. This problem was solved by Vitlina and Chaplik [157] and by Borodin [158]. More general cases will be discussed later in Chapters 9, 10, and 11 of the present monograph.

We shall seek the wavefunction in the form

$$\Psi(r, t) = \int_{C_1} e^{iv^2t/2 - vr} Z(v)\, v dv + (2/\pi)^{1/2} e^{-ip^2t/2} \sin pr , \qquad (8.4.2)$$

where p is the initial momentum of the electron in the continuum state. The second term in (8.4.2) has the form of a free wave, and it represents the wavefunction as $t \to \pm\infty$, that is, where as shown in Section 8.2, the existence of a ZRP does not influence the wavefunction. We take the contour $C_1$ in the complex plane v that starts at $v = -ip$. Substituting (8.4.2) into the boundary condition

$$\left.\frac{d\Psi}{dr}\right|_{r=0} = \beta t\, \Psi(0, t) \qquad (8.4.3)$$

and integrating by parts, we obtain an equation for Z(v), namely,

$$i v^2 Z(v) = \beta Z'(v),$$

which coincides (with a suitable change of the variable) with equation (8.2.9*). The sum of the term outside the integral, obtained by integration by parts, and the second term in (8.4.2) must add up to zero so that we obtain the condition

$$- i \beta Z(-ip) + (2/\pi)^{1/2} p = 0, \qquad (8.4.4)$$

which can be used to normalize the wavefunction Z(v). The final form of the wavefunction is

$$\Psi(r, t) = (2/\pi)^{1/2} \frac{ip}{\beta} \int_{C_1} \exp\{ iv^2t/2 - vr + \frac{1}{3\beta}(iv^3 + p^3)\}\, v dv + (2/\pi)^{1/2} \exp(-ip^2t/2) \sin pr . \qquad (8.4.5)$$

In order to ensure that the probability vanishes so that the system is in a bound state as $t \to -\infty$, the contour $C_1$ has to be specified in the way shown in Fig. 8.2.

Let us consider now the final state of the system ($t \to +\infty$). In this case, the contour $C_1$ has to be shifted to the lower left quadrant of the v plane. Going over to the momentum representation (similarly to the procedure described in Section 8.2), we shall see that this transformation of $C_1$ gives an additional term which arises from the residue at $k = -iv$, and we obtain the scattering matrix $S(p, k) = \delta(k - p) + F(p, k)$, F being the amplitude of the momentum distribution, in the form

$$S(p, k) = \begin{cases} \delta(k-p) - (2kp/\beta)\exp\{(p^3 - k^3)/3\beta\}, & k > p, \\ 0, & k < p. \end{cases} \tag{8.4.6}$$

Therefore the spreading of the initial "monochromatic" wave of the electron takes place with the energy of the electron always increasing.

In order to consider capture of the electron from the continuum state into a bound state, it is necessary to investigate the case where $\beta < 0$. This can be done by analogy with the previous derivation. The result is

$$S(p, k) = \begin{cases} 0, & k > p, \\ \delta(k-p) - (2kp/\beta)\exp\{(p^3 - k^3)/3\beta\}, & k < p. \end{cases} \tag{8.4.7}$$

The probability that the electron will be captured into a bound state is given by

$$P_c = \frac{2}{\beta} p^2 \exp(-2p^3/3\beta) . \tag{8.4.8}$$

During the course of a three-body collision (8.4.1) the nuclei first come closer together ($\beta > 0$) and then move away from each other ($\beta < 0$). The final distribution can be obtained by treating these two processes as independent (for details, see reference [157]*).

---

* In this reference, however, there are misprints in the expressions for the transition amplitudes.

## Chapter 9

# TIME-DEPENDENT QUANTUM MECHANICAL PROBLEMS SOLVABLE BY CONTOUR INTEGRATION

## 9.1 GENERAL TIME-DEPENDENT PROBLEMS SOLVABLE BY CONTOUR INTEGRATION

Only a few quantum mechanical problems are known with a time-dependent Hamiltonian $H(t)$ whose solutions can be expressed in a closed analytical form, and an extension of the list of such problems is of great interest to many particular branches of physics. In this chapter, we shall show that a rather wide class of time-dependent quantum mechanical problems can be solved in a closed form if the method of contour integration is applied to construct the solution.

A system of parallel terms (energy eigenvalues with time-independent separations) which is crossed, in the zero-order approximation, by a term of a different nature is a case naturally arising in many real quantum mechanical problems. Such a system of terms in the zero-order approximation is often called diabatic. When the interactions existing in the atomic system are taken into account, term crossings become pseudo-crossings and the diabatic picture is transformed to the exact (adiabatic) picture of terms, as Fig. 9.1 shows.

If the term splitting at each pseudo-crossing is small in comparison with the term separation, the calculation may neglect the existence of all other terms. Each encircled region in Fig. 9.1 is then considered one at a time, with the help of the Landau-Zener formula, and the total transition probability is obtained by multiplying the individual probabilities. In the opposite case, this method is, generally speaking, not applicable.

Under certain conditions, however, these limitations can be lifted and the Landau-Zener method can be successfully applied to study the structure of the S-matrix in problems far beyond the formal limits of the validity set for this theory.

We shall follow Demkov and Osherov [144, 159-162] and consider a time-dependent Hamiltonian H(t) such that $H(t) = H_o + V(t)$, where $H_o$ is time-independent and the perturbation V(t) is linear in t and is a projection operator onto a certain state $|\phi>$. In other words, we assume that V(t) is a separable potential. Let us seek the solution of the corresponding Schrödinger equation

$$(H_o + |\phi>\beta t<\phi|)\,|\psi> = i\frac{\partial}{\partial t}|\psi> \quad (\beta > 0) \qquad (9.1.1)$$

in the form of a contour integral,

$$|\psi> = \int_C G(E)\,|\phi>\,Z(E)\,e^{-iEt}\,dE\,, \qquad (9.1.2)$$

where $G(E) = (H_o - E)^{-1}$ is the resolvent operator whose kernel is the Green's function of the operator $H_o$. Substituting equation (9.1.2) into equation (9.1.1), integrating by parts, assuming that the term outside the integral vanishes, and equating the integrands, we obtain an equation for Z(E):

$$Z(E) = i\beta\frac{d}{dE}\left(Z(E)<\phi|G(E)|\phi>\right). \qquad (9.1.3)$$

Thus, the solution of equation (9.1.1) has the form

$$|\psi> = N\int_C \frac{G(E)\,|\phi>}{<\phi|G(E)|\phi>}\exp\left(-\frac{i}{\beta}\int^E\frac{dE'}{<\phi|G(E')|\phi>} - iEt\right)dE. \qquad (9.1.4)$$

Therefore, formula (9.1.4) gives the contour integral representation of the solution to an entire class of time-dependent problems of the form (9.1.1) with an arbitrary function $|\phi>$ provided that the Green's function for $H_o$ (that is, the general solution of the corresponding stationary problem) is known. No assumption has been made here with regard to the number of degrees of freedom or about the character of the spectrum of the operator $H_o$.

The condition to be imposed on the contour C in (9.1.4) follows from the requirement that the outside term, after integrating by parts, must vanish, thus:

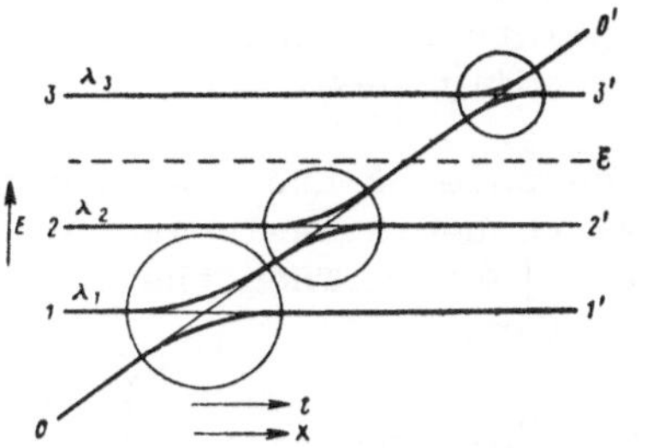


Fig. 9.1. A typical system of molecular terms of time-dependent and stationary problems solvable by the method of contour integration (only four terms are shown in the diagram).

$$\exp\left[ -\frac{i}{\beta} \int^{E} \frac{dE'}{\langle \phi | G(E') | \phi \rangle} - i\,Et \right] \Bigg|_{C} = 0. \tag{9.1.5}$$

Let us now show that the saddle points of the exponential function in (9.1.4) and (9.1.5) are the instantaneous eigenvalues of the energy operator H(t),

$$H(t) = H_o + | \phi \rangle \beta t \langle \phi | . \tag{9.1.6}$$

Indeed, the condition which determines the saddle points is

$$\langle \phi | G(E) | \phi \rangle = -(\beta t)^{-1} . \tag{9.1.7}$$

Multiplying the equation for eigenfunctions $| \psi \rangle$,

$$\left( (H_o - E) + | \phi \rangle \beta t \langle \phi | \right) | \psi \rangle = 0, \tag{9.1.8}$$

by $\langle \phi | (H_o - E)^{-1}$ from the left we obtain equation (9.1.7) also.

It should be noted that if the operator $H_o$ has a continuous spectrum, the operator G(E) will have a cut along a path bordering the continuum. Then equation (9.1.7) may have solutions on the nonphysical sheet which correspond to quasi-stationary or virtual states of the system, the roots being allowed to move from the physical sheet onto the nonphysical sheet as the variable t changes (see Chapter 2).

As $t \to \pm \infty$, the eigenvalues of H converge to the limiting values determined by the equation

$$\langle \phi | G(E) | \phi \rangle = 0 . \tag{9.1.9}$$

For t = 0, the saddle points coincide with the eigenvalues of the operator $H_0$ as can be readily seen.

We now consider the case of the operator $H_0$ having only discrete eigenvalues. We choose to use a basis whose zeroth row and column correspond to the state $|\phi\rangle$. Then the matrix representation of H will take the following form:

$$H = \begin{pmatrix} h_{00} + \beta t & h_{01} & h_{02} & \cdots \\ h_{10} & h_{11} & h_{12} & \cdots \\ \cdots & \cdots & \cdots & \cdots \end{pmatrix} . \qquad (9.1.10)$$

Selecting the origin of time so that $h_{00} = 0$ and diagonalizing the submatrix h,

$$h = \begin{pmatrix} h_{11} & h_{12} & \cdots \\ h_{21} & h_{22} & \cdots \\ \cdots & \cdots & \cdots \end{pmatrix} ,$$

we obtain the representation which is most suitable for the problem in question,

$$H = \begin{pmatrix} \beta t & h_1 & h_2 & \cdots \\ h_1 & \lambda_1 & 0 & \cdots \\ h_2 & 0 & \lambda_2 & \cdots \\ \cdots & \cdots & \cdots & \cdots \end{pmatrix} . \qquad (9.1.11)$$

The constants $h_i$ in (9.1.11) can always be made real and positive by a suitable choice of the phase factors. It is easy to see that the eigenvalues of H are close to $\beta t$ and $\lambda_i$, provided that $\beta t$ is large in comparison with $h_i$ and $\lambda_i$ so that the roots $\lambda_i$ are the asymptotic eigenvalues of the operator H and the solutions of equation (9.1.9). We shall note here that $\lambda_i$ and eigenvalues of $H_0$ are zeros and poles, respectively, of the function $\langle \phi | G(E) | \phi \rangle$. Thus, they never coincide and will alternate with each other.

Figure 9.1 gives a general picture of the terms in the case where there exist only four states. If $h_i$ is small in comparison with the difference $|\lambda_{i+1} - \lambda_i|$, the term splitting in the neighborhood of the pseudo-crossing $\beta t = \lambda_i$ will be small and equal to $2h_i$. In the two-state approximation, the Landau-Zener formula for the probability $p_i$ that a nonadiabatic transition occurs (that is, the probability that the system will move along the unperturbed term) will be $p_i = \exp(-2\pi h_i^2/\beta)$. The probability $q_i$ that the system will move along the adiabatic term is $q_i = 1 - p_i$.

We shall now write the function (9.1.4) in the representation (9.1.11). The equation to determine the eigenvalues of the operator $H_o$ is

$$|H_o - E| = -\prod_i (\lambda_i - E)\Big( E + \sum_j h_j^2 (\lambda_j - E)^{-1}\Big) = 0. \qquad (9.1.12)$$

It is easy to calculate the inverse matrix $(H_o - E)^{-1}$ and the quantities $\langle \phi | G(E) | \phi \rangle$ and $G | \phi \rangle$. We obtain

$$\langle \phi | \psi \rangle = \langle 0 | \psi \rangle = N \int_C L(E)\, dE,$$

$$\langle n | \psi \rangle = N h_n \int_C (\lambda_n - E)^{-1} L(E)\, dE,$$

where

$$L(E) = \prod_m (\lambda_m - E)^{-ih_m^2/\beta} \exp(iE^2/2\beta - iEt). \qquad (9.1.13)$$

We see that the asymptotic eigenvalues $\lambda_m$ are poles and branch points of the integrand. For large t, the saddle points are close to $\lambda_i$, and they move from $\lambda_i$ to $\lambda_{i+1}$ as t changes from $-\infty$ to $+\infty$. The first saddle point is near to $\beta t$ as $t \to -\infty$, and to $\lambda_1$ as $t \to +\infty$. The last saddle point (if it exists) is close to the largest $\lambda_N$ as $t \to -\infty$, and to $\beta t$ as $t \to +\infty$.

Now we have to specify the integration path. If the system is in the state $|\phi\rangle$ as $t \to -\infty$, the contour has to pass only through the saddle point $E \approx \beta t$. The ends of this contour must go to infinity in such a way as to ensure that the exponential in the integrand vanishes; that is, they must be in the first and third quadrants. Figure 9.2(a) shows an example of such a contour. If $|\beta t| \gg |\lambda_m|$, then the moduli of the factors

$$(\lambda_m - E)^{-ih_m^2/\beta}$$

in (9.1.13) are close to unity, for $E \approx \beta t$, and, calculating the integral by the method of steepest descent, we obtain

$$\left| \int L(E)\, dE \right| \to (2\pi\beta)^{1/2}, \qquad (9.1.14)$$

so that the normalization constant N in equation (9.1.13) is $(2\pi\beta)^{-1/2}$.

It is clear that all other components vanish as $t \to -\infty$, on account of the factors $(\lambda_m - E)^{-1}$.

For $t \to -\infty$, the factor $e^{-iEt}$ falls off rapidly in the lower half-plane and the saddle point $E \approx \beta t$ is to the right of the branch points $\lambda_n$. In order to determine the asymptotic behavior the contour C has to be shaped in the form of a sum of loops $\omega_n$ going around the branch points $\lambda_n$ to form the contour C' in Fig. 9.2 (a). Only the loop $\omega_n$ yields a finite contribution to the component $\langle n | \psi \rangle$ as $t_n \to +\infty$, and only that part of the contour is essential where the factors

$$\left| (\lambda_m - E)^{-ih_m^2/\beta} \right|^2$$

are close to unity for $m > n$, and close to $p_m$ for $m < n$.

Thus for the transition probabilities we obtain

$$\begin{aligned} |S_{On}|^2 &= \lim_{t \to \infty} |\langle n | \psi \rangle|^2 \\ &= h_n^2 (2\pi\beta)^{-1} p_1 p_2 \cdots p_n \times \left| \int_{C'} (\lambda_n - E)^{-1-ih_n^2/\beta} e^{iEt} dE \right|^2 \\ &= h_n^2 (2\pi\beta)^{-1} p_1 p_2 \cdots p_{n-1} p_n^{\frac{1}{2}} \left| 2\pi / \Gamma(1 + ih_n^2/\beta) \right|^2 \\ &= p_1 p_2 \cdots\cdots p_{n-1} q_n , \end{aligned} \qquad (9.1.15)$$

and

$$|S_{OO}|^2 = \lim_{t \to \infty} |\langle O | \psi \rangle|^2 = p_1 p_2 \cdots p_N . \qquad (9.1.16)$$

If the initial state is $|n\rangle$ $(n \neq 0)$, as $t \to -\infty$, the corresponding contour C will pass only through the n-th saddle point, as shown in Fig. 9.2 (b). In the vicinity of the saddle point $\lambda_n$, the contour follows the path of steepest descent for the function

$$(\lambda_n - E)^{-1-ih_n^2/\beta}$$

and, thus, is close to the logarithmic spiral. Making use of the formula

$$\int_\sigma e^t t^{-z} dt = -2\pi i \{\Gamma(z)(1 - e^{2\pi iz})\}^{-1}, \quad \text{Im } z > 0,$$

where the contour $\sigma$ has the form of a spiral in the vicinity of the

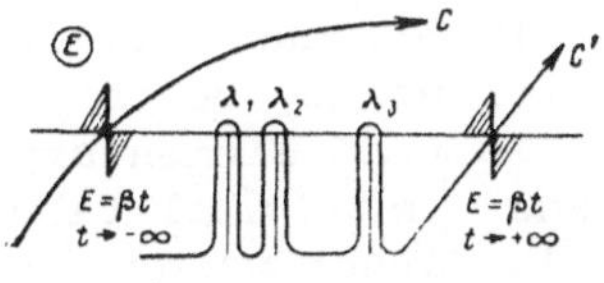


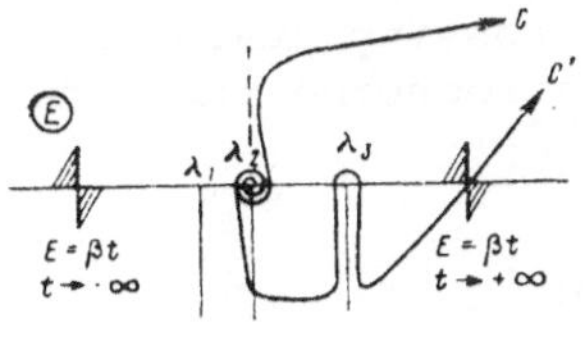


Fig. 9.2. Integration contours for calculating the transition probabilities. (a) System is in the state $|\phi\rangle$ as $t \to -\infty$; (b) system is in the state $|n\rangle$, $n = 2$, as $t \to -\infty$.

the origin and runs to infinity along the upper rim of the cut in the left half-plane, it is easy to obtain the normalization factor

$$N = (2\pi\beta\, p_1 p_2 \cdots p_{n-1})^{-\frac{1}{2}} (1 - p_n)^{\frac{1}{2}} . \tag{9.1.17}$$

For $t \to +\infty$, the contour C has to be deformed to a contour C', as is shown in Fig. 9.2 (b).

For the transition probabilities we find

$$\begin{aligned} |S_{nm}|^2 &= 0, \quad 0 < m < n, \\ |S_{nn}|^2 &= p_n , \\ |S_{nm}|^2 &= (1 - p_n) p_{n+1} \cdots p_{m-1} (1 - p_m), \quad m > n, \\ |S_{n0}|^2 &= (1 - p_n) p_{n+1} \cdots p_N . \end{aligned} \tag{9.1.18}$$

We see that the transition probability is zero for $m < n$ because the contour does not pass through the corresponding saddle points. This means that while $t$ changes from $t \to -\infty$ to $t \to +\infty$ the energy of the system cannot decrease. This property may be called the triangularity of the S-matrix, though this term is somewhat imprecise because the matrix elements $S_{n0}$ differ from zero. The operator $\hat{V}(t)$ increases monotonically so that the average energy of any state must also increase with time. The statement obtained here is much stronger. The principle of detailed balance and time reversal invariance are violated by the zeroth state $|\phi\rangle$ because it has a different energy as $t \to -\infty$ and as $t \to +\infty$.

Formulae (9.1.15), (9.1.16), and (9.1.18) also show that the probabilities of all transitions may be obtained as successive products of the probabilities that elementary transitions $p_n$, $q_n$ between the states $|0>$ and $|n>$ will take place at each pseudo-crossing (see Fig. 9.1). In other words, the S-matrix may be decomposed into a product of simple factors. Each of these couples only one pair of pseudo-crossing states in accordance with the Landau-Zener formula. Introducing the notation $S_n$ for each of these matrices, we obtain

$$S = S_1 S_2 S_3 \ldots S_N . \tag{9.1.19}$$

The triangularity property follows automatically from (9.1.19).

If the quantities $h_n$ are of the same order of magnitude as the term separations $|\lambda_n - \lambda_{n+1}|$ and $|\lambda_n - \lambda_{n-1}|$, they cannot be determined directly from the term splitting at the points of pseudo-crossing. It is seen, from equation (9.1.11), that

$$h_n = < n | H | 0 > . \tag{9.1.20}$$

One of the conditions ensuring the validity of the method is that the quantities $h_n$ can be considered as time-independent for a sufficiently long interval of time.

Formulae (9.1.15) and (9.1.16) show that the transition probability $w(E)$ that the system which is initially in a state $|\phi>$ goes over to a state with energy larger than E takes an especially simple form:

$$w(E) = \prod_{\lambda_n < E} p_n = \exp\left\{ - 2\pi\beta^{-1} \sum_{\lambda_n < E} h_n^2 \right\} . \tag{9.1.21}$$

Now it is not difficult to discuss the case where the operator $H_o$ has a spectrum, part of which is continuous. If we extend, for instance, the domain of space where the particle moves, the eigenvalues $\lambda_n$ in a certain energy interval will become closer to each other, whereas the corresponding matrix elements $h_n$ will converge to zero. In the limit, the function $< \phi | G(E) | \phi >^{-1}$ will have a cut along the real axis, with the discontinuity $\Delta(E)$ of the imaginary part of the function across the cut equal to

$$\Delta(E) = 2\pi \left( \frac{dn}{dE} h_n^2 \right)_{\lambda_n = E} . \tag{9.1.22}$$

Therefore, for an operator $H_o$ having a continuous spectrum only, the probability $w(E)$ is given by

$$w(E) = \exp\left(-\beta^{-1}\int_{E_L}^{E}\Delta(E')\,dE'\right), \qquad (9.1.23)$$

where $E_L$ is the lower limit of the spectrum.

If the initial state $|n>$ lies in the continuum, the passage to the limit requires changing the normalization of the initial state (without doing that the function originally normalized to unity will converge to zero everywhere). We shall use the $\delta$-function normalization in the energy:

$$\psi_E = (dn/dE)^{\frac{1}{2}}\,\psi_n .$$

On account of the convergence of $h_n^2$ and, therefore, $q_n$ to zero we obtain the transition probability $w(E_o, E)$ from a state with energy $E_o$ to a state with a larger energy E, given by

$$w(E_o, E)$$

$$= \lim_{\Delta\lambda_n \to 0}\left(\frac{dn}{dE}\,\frac{2\pi}{\beta}\,h_n^2\right)_{\lambda_n = E_o} \times \exp\left(-\frac{1}{\beta}\int_{E_o}^{E}\frac{dn}{dE'}\,h_n^2\,dE'\right)$$

$$= \frac{1}{\beta}\,\Delta(E_o) \times \exp\left(-\frac{1}{\beta}\int_{E_o}^{E}\Delta(E')\,dE'\right). \qquad (9.1.24)$$

The probability of a transition to any state with $E < E_o$ is zero.

The simplest example of the class of problems considered above is the Landau-Zener model itself. In the latter case we have a two-component wavefunction $|\psi>$, and the energy operator H has the form

$$H = \begin{pmatrix} \beta t & h \\ h & \lambda \end{pmatrix}. \qquad (9.1.25)$$

The solution of this problem is well known, of course. Another example is the case of one degree of freedom x, with $|\phi> = \delta(x - X)$. Then $<\phi|G(E)|\phi> = G(X,X,E)$, and the solution of equation (9.1.1) is

$$|\psi> = N\int_C \frac{G(x,X,E)}{G(X,X,E)}\exp\left(-\frac{i}{\beta}\int^{E}\frac{dE'}{G(X,X,E)} - iEt\right)dE. \qquad (9.1.26)$$

It is convenient to use this example for a more detailed discussion of the asymptotic states $|n\rangle$ and energies $\lambda_n$. Indeed, if $H_o$ is the energy operator for a particle moving in a potential well, then

$$G(X,X,E) = \psi_1(X,E)\,\psi_2(X,E)\,,$$

where $\psi_1$ and $\psi_2$ are the solutions of the Schrödinger equation which satisfy the boundary conditions as $x \to -\infty$ and $x \to +\infty$, respectively. Then it follows that the roots of the equation $G(X,X,E) = 0$ are the energy levels in the right and left parts of the potential well divided by an impenetrable partition at the point X. Therefore, in this case we have two independent systems of energy levels corresponding to the two parts of the well, and an additional energy level localized in the vicinity of X. As t changes from $-\infty$ to 0, the partition vanishes gradually and all levels mix. With a further change of t from 0 to $+\infty$, the partition re-appears but without a state localized at X. The theory developed above enables the transition probabilities between the states of such a problem to be easily determined.

If we put $H_o = -(1/2)(\partial^2/\partial x^2)$, assuming that $H_o$ is the operator of a free particle, we find immediately that

$$\left.\begin{aligned} G(x,X,E) &= (-2E)^{\frac{1}{2}} \exp(-\sqrt{-2E}\,|x - X|)\,, \\ G(X,X,E) &= (-2E)^{\frac{1}{2}}, \end{aligned}\right\} \qquad (9.1.27)$$

and

$$\Delta(E) \sim E^{\frac{1}{2}}\,, \qquad -\ln w(E) \sim E^{3/2}\,. \qquad (9.1.28)$$

The latter result has already been obtained in Section 8.1 and corresponds to the energy spectrum of the emitted particles when a bound state has been "driven out" from the ZRP into the continuum.

Another application of the method can be obtained if we consider a three-dimensional ZRP of variable depth. This problem can be solved exactly if we impose, at the point $\underline{R}$ where the well is positioned, the time-dependent boundary condition (8.1.1), assuming there that $\alpha(t) = -\beta t$. Unlike the previous case of a one-dimensional ZRP, in the present problem the well vanishes as $t \to +\infty$ and $H \to H_o$. For $t \to -\infty$, we can also assume that $H \to H_o$ because in the latter case the well becomes very deep, thus causing the binding energy in the well, $E = -\beta^2 t^2/2$, to be very large. Consequently, the existence of the bound state in the well cannot influence other states of the system. The partitioning of the well at $t \to \pm\infty$ does not take place in the latter example and, in this respect, the three-dimensional problem appears to be even more simple than that of the one-dimensional.

The solution of the time-dependent Schrödinger equation

$$\left(-\frac{1}{2}\nabla^2 + V(\underline{r})\right) |\psi> = i\frac{\partial}{\partial t} |\psi> \qquad (9.1.29)$$

with the boundary conditions (8.1.1) can be written as follows:

$$|\psi> = N\int_C G(\underline{r},\underline{R},E)\exp\left(\frac{2\pi i}{\beta}\int^E G_r(\underline{R},\underline{R},E')dE' - iEt\right)dE, \qquad (9.1.30)$$

where $G_r$ is the regularized Green's function from which the singular part has been subtracted. This function is defined by equation (3.1.8).

We can see that the problem has been reduced to the determination of the residues of $G_r(E)$ at the poles, that is, at the points corresponding to discrete eigenvalues of $H_o$ as well as to the determination of the discontinuity in $G_r(E)$ on the cut on the complex E plane.

If $V(\underline{r}) = 0$, then $G(\underline{r},0,E) = (1/2\pi r)\exp(-\sqrt{-2E}\,r)$, and $G_r(E) = -(1/2\pi)\sqrt{-2E}$, and it is easy to obtain the same result (9.1.28) as that obtained in the case of the one-dimensional problem. If we assume that $V(\underline{r}) = 0$ for a particle moving in a half-space with the conditions imposed on the boundary plane, being either

$|\psi> = 0$, antisymmetrical case,

or

$(\underline{n}\cdot\underline{\nabla})|\psi> = 0$, symmetrical case,

then the Green's function can be easily obtained by the method of images, and we shall arrive at the model already used* in Section 8.2.

We note that those quantum mechanical problems which can be solved in a closed analytical form usually have only a few (one or two) nontrivial parameters. This generally limits the scope of application of the solutions. In the class of problems considered above there may be an infinite number of parameters (such as $h_n$, $\lambda_n$, for instance), and it is possible to consider the Hamiltonian operators which have discrete, continuous, mixed (discrete and continuous), and zonal spectra. Therefore the mathematical approach formulated above is expected to be applicable to many physical situations.

---

*Further applications to the theory of electron detachment can be found in reference [163].

## 9.2 ADIABATIC APPROXIMATION AND TRAJECTORIES OF THE POLES OF THE S-MATRIX

In the adiabatic case, the Hamiltonian of a general time-dependent quantal problem varies slowly with time so that there always exists a small parameter (the rate of change) or an equivalent large parameter (characteristic time of change) which is related to the Hamiltonian. The adiabatic approximation employs power expansions in terms of this small parameter.

Let us consider the instantaneous Hamiltonian $H(t)$ whose eigenfunctions $\phi_n(t)$ and eigenvalues $E_n(t)$ depend upon time as a parameter:

$$H(t)\phi_n(t) = E_n(t)\,\phi_n(t)\,. \tag{9.2.1}$$

In the adiabatic case, the solution of a time-dependent problem corresponds to the development of the system in terms of the adiabatic eigenvalues $E_n(t)$. In other words, it is assumed that the coefficients $c_n(t)$ in the expansion of the exact solution in terms of the complete adiabatic basis $\{\phi_n(t)\}$,

$$\psi(t) = \sum_n c_n(t)\,\phi_n(t)\,\exp\left(-i\int^t E_n(t')\,dt'\right), \tag{9.2.2}$$

are slowly varying functions of time. Transitions between the adiabatic states $\phi_n(t)$ occur chiefly in the regions where adiabatic terms come closer to each other (pseudo-crossing). If the term is an analytic function of time (for more details see Section 3.3), then there is a relation between the pseudo-crossing of the two levels and the existence of a branch point of the function $E(t)$. In this case, the transition probability is exponentially small [54]:

$$w = \left|\exp\left(-i\int_C E(t)\,dt\right)\right|^2. \tag{9.2.3}$$

Both ends of the contour C in equation (9.2.3) lie on the real axis, and the contour loops around the branch points.

For a Hamiltonian whose spectrum also includes the continuum, we encounter a problem of calculating, in the adiabatic approximation, the probability of the transition from the bound state to a continuous state. As we have shown in Section 3.3 (see also reference [62]), in this case there exists a branch point which corresponds to the coincidence of S-matrix poles on the imaginary axis k (see Chapter 2).

The solution of the time-dependent quantal problem considered in Section 9.1 can be used to ascertain the general expression for the probability of a transition to the continuum. Following Demkov and Ostrovskii (62), we shall write the probability amplitude b(E) in the equation $W(E) = |b(E)^2|$ in the form

$$b(\omega) = |2 \text{ Im } t(\omega)|^{\frac{1}{2}} \exp \left( \int_{(C)E_i}^{\infty} t(\omega')\, d\omega' \right), \tag{9.2.4}$$

where $\omega$ denotes the energy of the particle in the continuum state, and t(E) is the inverse function of E(t). For the function t(E), the continuum boundary is a branch point. The plane E has a right-hand branch cut and the integration path C runs along the upper rim of the cut. The contour C starts at $E_i$ which is the value of the energy term as $t \to -\infty$. Thus we should have put $E_i = -\infty$ in the model (9.1.1). However, we note that a particular choice of $E_i$ affects only the phase of the transition amplitude b which is determined in equation (9.2.4) only to within an additive term independent of $\omega$. The validity of equation (9.2.4) can be tested by comparing it with equation (9.1.23), allowing for equation (9.1.7) and the definition of $\Delta(E)$ as the discontinuity of the function $(\langle\phi|H_o - E|^{-1}\phi\rangle)^{-1}$ across the cut.

After integrating by parts in equation (9.2.4), the amplitude $b(\omega)$ takes the form

$$b(\omega) = |2 \text{ Im } t(\omega)|^{\frac{1}{2}} \exp \{i\omega t(\omega) - i\int_{-\infty}^{t(\omega)} E(t')dt'\}, \tag{9.2.5}$$

where the complex time $t(\omega)$ corresponds to the time in the complex plane t when the term energy is $\omega$.

It is easy to check that the probability that a transition between discrete states has taken place is given by (9.2.3). Thus, the solution of the problem considered in Section 9.1 is entirely determined by the analytic properties of E(t) or, in other words, by the trajectories of the poles of the S-matrix.

An approximation applied above to model actual collision problems can be called the approximation of the standard equation. According to this, we describe the decay of a bound state in terms of the trajectories of the poles of the S-matrix. A given standard equation is applicable to a particular real problem if the trajectories of the poles, in both cases, correspond to each other in the specific domain of the complex energy plane, that is, the functions E(t) for the actual problem and for the standard equation must be close to each other in the considered region of t and E. The general problem that is being discussed in this chapter may be considered as the standard problem for the case where the quasi-stationary term runs in the continuum at a certain angle to the continuum boundary. For instance, the particular case solved in Section 8.2 corresponds to the passage of the S-matrix pole along the imaginary axis on the complex k plane. For the problem studied in Section 8.3, two poles may fuse together with subsequent production of quasi-stationary states (see Chapters 2 and 3). The general Hamiltonian (9.1.1) admits a rather wide class of trajectories of the poles of the S-matrix. As shown in reference [62], under certain conditions, the Hamiltonian $H_o$ and vector $|\phi\rangle$ can be reconstructed, given the function E(t).

If a term, after entering the continuum, corresponds to a quasi-stationary state, and the trajectory of the pole of the S-matrix lies close to the real axis E, that is,

$$\frac{1}{2}\Gamma(t) = \mathrm{Im}\,E(t) \ll \mathrm{Re}\,E(t), \tag{9.2.6}$$

then, to good accuracy, the equation

$$\mathrm{Im}\left(\int_{E_i}^{\omega} t(\omega')\,d\omega'\right) \approx \mathrm{Im}\left(\int_{-\infty}^{\tilde{t}} E(t')\,dt'\right) \tag{9.2.7}$$

holds, where the real $\tilde{t}$ is determined by the condition $\mathrm{Re}\,E(\tilde{t}) = \omega$. Therefore we obtain the approximate expression

$$\int_0^{\infty} |b(\omega')|^2\,d\omega' = 1 - \exp\left(-\int_{-\infty}^{\tilde{t}} \Gamma(t')\,dt'\right). \tag{9.2.8}$$

Equation (9.2.8) has an obvious meaning based on the natural interpretation of the second term on the right-hand side there, namely, as the probability that the system will "survive," remaining on the quasi-stationary term with width $\Gamma(t)$. The same type of formulae was used by Demkov et al. [69] in calculations of the energy spectrum of the electrons emitted as a result of collisions between $H^-$ and H as well as in calculations of the total probability of electron detachment. A similar approach was used by Demkov and Kuchinskii [113] to estimate the population of the vibrational states of the $H_2$ molecule in the same reaction. The widths used in these works were obtained in the ZRP approximation.

If condition (9.2.6) does not hold, then formula (9.2.8) is not applicable and the process cannot be adequately modeled by quasi-stationary states decaying exponentially. For instance, if an s-state term enters the continuum (see Section 8.2) we have, for low energies, $E(t) = -\beta^2 t^2/2$ and $\Gamma = 0$. Nevertheless formula (9.2.4) or (9.2.5) reproduces the exact result derived in Section 8.2.

We note that the approximate formula

$$\text{Im } t(E) \approx \frac{1}{2}\,(dt/dE)\;\Gamma(\tilde{t}) \qquad (9.2.9)$$

can be used to calculate the pre-exponential factor in equations (9.2.4) and (9.2.5), provided that condition (9.2.6) holds.

Equations (9.2.4) and (9.2.5) correspond to the case when the poles of the S-matrix describe the trajectories only once while moving in the complex plane in one direction. Had we introduced a quadratic term for the temporal dependence of the Hamiltonian H(t) (see Chapter 10 for further discussion) the poles would have described each trajectory twice: firstly, for the nuclei coming closer and, secondly, for the nuclei moving apart from each other. A qualitative description using a quasi-stationary term suggests that a given state in the continuum is populated at two different moments in time when the real part of the energy term coincides with the energy of the continuum state. In this case, the function t(E) has, apart from the continuum boundary, additional branch points related to the maxima of the term E(t). Therefore, there exist several contours of integration, $C_j$, which connect the initial energy of the terms with the energy of the continuum state in question. Correspondingly, the amplitude b(ω) is given by the sum of several terms of the type (9.2.4). If conditions (9.2.6) and (9.2.9) hold and if the model Hamiltonian is used, then it can be shown (164) that

$$b(\omega) = \sum_j \left| -\frac{\pi}{2}\frac{dt}{d\omega} \right|^{\frac{1}{2}} \exp\left( i \int_{\substack{E(-\infty)\\(C)}}^{\omega} t(\omega')\, d\omega' \right) . \qquad (9.2.10)$$

The presence of several terms in equation (9.2.10) leads to the interference oscillations in $|b(\omega)|^2$.

Finally we point out that the probability amplitude for a free-free transition from a continuum state with energy ω' to a continuum state with energy ω can also be written with the help of the adiabatic term. Thus:

$$b(\omega, \omega') = \left[2 \text{ Im } t(\omega)\right]^{\frac{1}{2}} \left[2 \text{ Im } t(\omega')\right]^{\frac{1}{2}}$$

$$\times \exp\left( i \int_{\omega'}^{\omega} t(E)\, dE \right) . \qquad (9.2.11)$$

## 9.3 IONIZATION IN SLOW ATOMIC COLLISIONS

The ionization process

$$A + B \rightarrow A + B^{+} + e \tag{9.3.1}$$

is more difficult to describe theoretically than electron detachment (8.2.1) because, at large distances, the emitted electron moves in the effective Coulomb field of the system $AB^+$. Apart from the continuum, in this case, there will be an infinite number of bound Rydberg states with arbitrary small binding energies. The corresponding energy terms of the system AB will be nearly parallel to the ground state term of the system $AB^+$ lying below the latter and converging to the continuum limit. A theory of reaction (9.3.1) must take into account the existence of all these terms and, therefore, consider, apart from ionization, excitation of high-lying bound states of B. According to the von Neumann-Wigner theorem, the initial term cannot cross the Rydberg terms on account of the interaction. Hence the general picture of the adiabatic terms in the neighborhood of the point $R_c$ will be similar to that shown in Fig. 9.3.

It will become clear that this problem can be solved with the help of the theory developed in Section 9.1. Following Demkov and Komarov (165), we shall first choose a time-independent operator $H_o$ to describe the system of parallel Rydberg terms and the continuum of AB. In our particular case, the simplest choice of $H_o$ is the energy operator for an electron moving in a Coulomb field of the system $AB^+$. The state corresponding to the initial term will be denoted $\phi$. This state can be assumed time-independent, provided that during reaction (9.3.1) R remains close to the point $R_c$ where the term enters the continuum.

It follows from Section 9.1 that the determination of the probabilities reduces to the investigation of the properties of the function

$$| < \phi | (H_o - E)^{-1} | \phi > |^{-1} .$$

More precisely, we have to determine the residues at the poles of this function as well as the discontinuity $\Delta(E)$ of its imaginary part across the cut for $E > 0$ (we have chosen the energy of $AB^+$ to be zero). The analytical form of the Green's function for a particle moving in the Coulomb field is known (see Section 7.4). Therefore for the case of a pure Coulomb potential in the operator $H_o$, the problem can be solved using quadratures. However, such a solution is not of any special interest to us because the Coulomb field is only the asymptotic limit of the real field at large distances, and both spectra coincide only for small energies, that is, for large n.

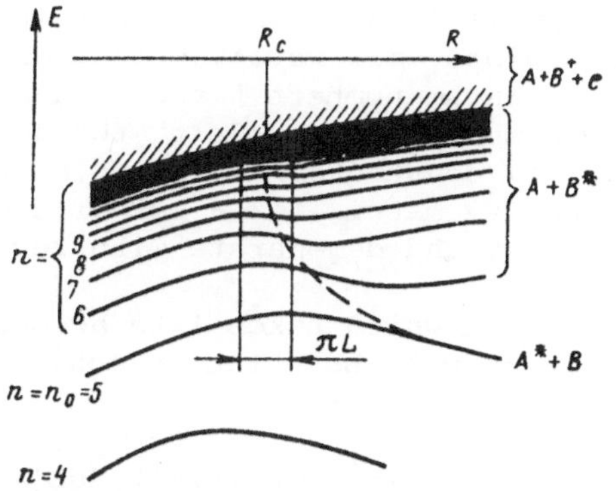


Fig. 9.3. The neighborhood of the point where a molecular term enters the continuum (the case of the Coulomb continuum limit).

In order to achieve consistency between the accuracy of the solutions and that of the initial approximations, we shall use the asymptotic form of the Green's function for large n which can be derived with the help of the asymptotic formulae for the Whittaker functions with large first indices. Then we obtain from equation (7.4.1) that

$$G \sim \frac{1}{x-y}\left(\frac{\partial}{\partial x}-\frac{\partial}{\partial y}\right)(xy)^{1/2}\; J_1(2\sqrt{y})$$
$$\times \left|\,\cot \pi n\, J_1(2\sqrt{x}) + N_1(2\sqrt{x})\,\right| , \qquad (9.3.2)$$

where $J_1$ and $N_1$ are the Bessel and von Neumann functions, respectively. The term with $N_1$ above does not depend upon n and diverges if $|\underline{r} - \underline{r}'| \to 0$. We shall regularize the function G by ignoring that term in (9.3.2). Then it follows from (9.3.2) that the energy- and coordinate-dependent terms separate. For large n, the matrix element $<\phi|G|\phi>$ has the form

$$<\phi\,|G|\,\phi> \;\approx\; -\;D\cot \pi n\,, \quad D > 0\,, \qquad (9.3.3)$$

so that the total information about the system is contained, in this model, in the constant D. Formula (9.3.3) is valid for not too large values of $\underline{r}$ and $\underline{r}'$, when the points $\underline{r}$ and $\underline{r}'$ lie far from the boundary enclosing the domain classically accessible at a given energy E.

Formula (9.3.2) is not applicable if $n < 0$ because the true Green's function has no poles if n is negative (however, it has a logarithmic cut in this case). On the right half-plane where $n > 0$ (the physical sheet of E) including the imaginary axis (i.e., $E > 0$), formula (9.3.2) holds.

The asymptotic formula (9.3.2) infers that in the region far away from the turning point, all wavefunctions of highly excited states with identical quantum numbers $\ell$ and $m$ and differing only in the principal quantum number $n$, are close to each other and differ only in the normalization constants. A factor D in (9.3.3) is proportional to the density matrix for a set of the Coulomb wavefunctions with zero energy and degenerate with respect to $\ell$ and m.

Thus, the wavefunction which takes into account the contribution from highly excited states will have the following form:

$$|\psi\rangle = N \int_C G|\phi\rangle \exp\left(-iB\int^E \cot\pi(-2E')^{-\frac{1}{2}} dE' - iEt\right) dE, \quad (9.3.4)$$

where the constant B incorporates both the constant D of equation (9.3.3) and the constant $\beta$ from equation (9.1.1). We are reminded that the latter specifies the rate of change in the term corresponding to the initial function $\phi$.

The residues of the $\cot \pi n$ function are

$$\tilde{h}_n = \frac{1}{\pi}\frac{dE}{dn} = \frac{1}{\pi n^3}, \quad (9.3.5)$$

and the discontinuity of the imaginary part across the cut is

$$\tilde{\Delta} = 2 \coth \pi\nu, \quad \nu = (2E)^{-\frac{1}{2}}. \quad (9.3.6)$$

For small E (the only case when this theory is, in fact, applicable), the discontinuity term is $\tilde{\Delta} \approx 2$.

Making use of the general formulae of Section 9.1, we obtain the following simple result for the total probability $w(E)$ that this system makes a transition to a state whose energy is higher than the initial energy E. If $E = E_n = -(2n^2)^{-1} < 0$, then

$$w(E_n) = w_o \exp\left(-2\pi B \sum_{j=n_o}^{n} \tilde{h}_j\right)$$

$$= w_o \exp\left(-2B \sum_{j=n_o}^{n} j^{-3}\right). \quad (9.3.7)$$

If the initial energy $E > 0$ (the continuous spectrum), the corresponding expression for the total probability is

$$w(E) = w_o \exp\left(-2\pi B \sum_{j=n_o}^{\infty} \tilde{h}_j - B\int_0^E \tilde{\Delta}(E')\,dE'\right)$$

$$= w_o \exp\left[-2B\left(E + \sum_{j=n_o}^{\infty} j^{-3}\right)\right]. \qquad (9.3.8)$$

In equations (9.3.7) and (9.3.8), summation over j starts from a certain value of $n = n_o$ which is large enough to ensure that application of the asymptotic expression is meaningful. The quantity $w_o$ is the probability that the system, with R decreasing during the course of the collision, reaches the term $E_{n_o}(R)$. This probability depends on the detailed behavior of the energy terms at small n and it cannot be obtained in a general form.

If $R_c$ is large enough, the system is in the region under the barrier, provided that n is small. In this case, the residues are exponentially small and can be ignored. Thus $n_o$ is determined by the equation $E_o = (2n_o^2)^{-1} \approx 1/R_c$. The same can be said if the energy of the electron detachment, before the collision, is low (for instance, if atom A or atom B is in an excited state). Then the term of the initial state does not approach the terms corresponding to small n, and the probability that the system makes a transition to one of these states is small. Thus, the order of magnitude of $E_o$ is determined by this detachment energy. Therefore, under favorable conditions for ionization, $w_o$ can be only slightly less than unity and the energy $E_o$ can be rather low, 1 eV or even less (for $n_o \sim 4 - 5$).

Let us introduce the function F(E) defined by

$$F(E) = \begin{cases} E, & \text{if } E > 0, \\ -\sum_{j \geq (-2E)^{-1/2}}^{\infty} j^{-3}, & \text{if } E < 0. \end{cases} \qquad (9.3.9)$$

The graph of this function is shown in Fig. 9.4. We can combine formulae (9.3.7) and (9.3.8) together, and write

$$w(E) = w_o \exp\left[-2B(E_o + F(E))\right]. \qquad (9.3.10)$$

The graph shows that the definition of F for E < 0, that is, F(E) = E, can, in fact, be extended to a region of small negative E where the step-wise character of F(E) is hardly noticeable. Then we obtain an approximate expression, namely:

$$w(E) = w_o \exp\left[-2B(E_o + E)\right], \tag{9.3.11}$$

which is sufficiently accurate for our approximation in both the discrete and continuous spectra.

We shall now investigate the constant B in equation (9.3.11). Using the fact that the saddle points in equation (9.3.4) coincide with the eigenvalues of the instantaneous Hamiltonian (see Section 9.1), we obtain from equation (9.3.4) that

$$B \cot \pi n + t = 0, \tag{9.3.12}$$

and

$$n = -\frac{1}{\pi} \arctan \frac{t}{B} = -\frac{1}{\pi} \arctan \frac{R_c - R}{vB}, \tag{9.3.12}$$

where v is the relative velocity of the approaching nuclei at $R = R_c$.

Therefore, if the variable n is used in place of E, then all curves n(R) for highly excited states have a similar pattern of behavior in the vicinity of $R_c$. This is clearly seen in Fig. 9.5. We introduce a quantity L with the dimension of length defined by $vB = L$. The meaning of this constant is illustrated by Fig. 9.5. The change in R by 2L in the vicinity of $R_c$ leads to a change in the principal quantum number n of the highly excited states by ½. The constant L can also be expressed in terms of the squared modulus of the projection of $\phi$ onto the subspace of the zero-energy wavefunctions. Namely, L is the derivative of that quantity with respect to R, taken at $R = R_c$. If this calculation is too difficult to carry out, L remains the only undetermined constant of the theory, which specifies the relative magnitude of the probability for excitation of the system to the total probability for ionization with the emitted electron having all possible energies.

Going over to arbitrary units, we obtain, for the probability w(E), the following expression:

$$w(E) = w_o \exp\left[-\frac{2L}{\hbar v}(E_o + E)\right]. \tag{9.3.13}$$

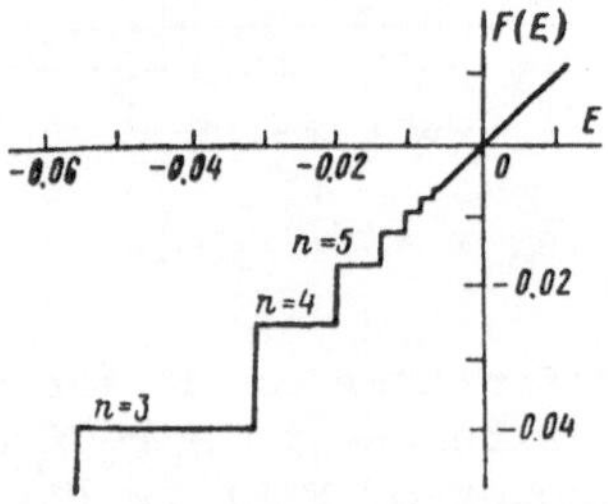


Fig. 9.4. Function F(E) determined by equation (9.3.9).

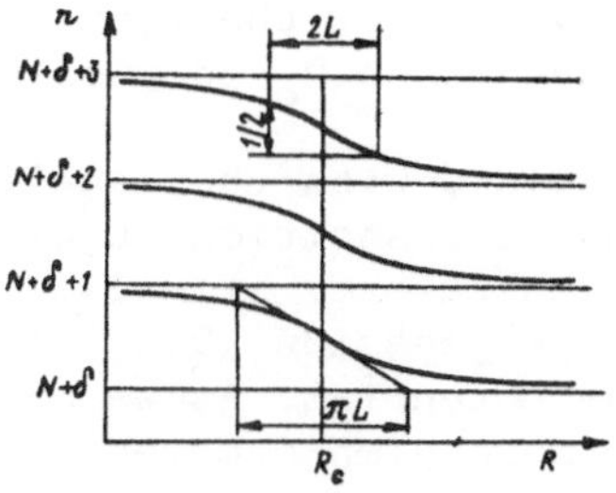


Fig. 9.5. Change of the effective principal quantum number n in the neighborhood of the point where a molecular term enters the continuum.

Equation (9.3.13) holds in the interval $-E_0 \lesssim E \lesssim E_0$. The parameter L describes the behavior of the system in the vicinity of $R = R_c$, $E = 0$. Other parameters describe the behavior of the system far away from the continuum boundary and at $R > R_c$. The total probability of ionization is

$$w(0) = w_0 \exp\left[ - 2LE_0 / \hbar v \right] . \quad (9.3.14)$$

The results obtained above are valid only if the probabilities of all individual transitions, $\exp(-2L/n^3 v)$, are near to unity. In the opposite case, that is, when n and v are small enough, the second passage of the system through the point $R_c$ (when the atoms move away from each other) must be taken into account. For the discrete spectrum, this correction does not essentially alter the population of the excited levels and it will not be considered here. For the continuum, the result depends on the spreading of the wave packet formed at $R = R_c$, during the time T when the separation between the atoms remains to be $R < R_c$. The spreading may be considered as a development of the incoherence of the wave packet components. Using the same technique as in Section 8.2, we can obtain the following expression for the probability $P_\alpha$ of the recapture of the emitted electron:

$$\begin{aligned} P_\alpha &= \left| \frac{2L}{v} \int_0^\infty \exp\left[ - (2L/v + iT) E \right] dE \right|^2 \\ &= \left( 1 + v^2 T^2 / 4L^2 \right)^{-1} . \end{aligned} \quad (9.3.15)$$

Because the magnitude of $vT$ is comparable with that of $R_c$, the criterion for the validity of the theory, $P_\alpha \ll 1$, can be expressed in a natural way by requiring that $L \ll R_c$. Then the change in n by ½, say, will require only a small change in R, in comparison with $R_c$. It appears, however, that the results obtained above will be qualitatively correct even when $R_c$ only slightly exceeds L.

Formula (9.3.13) coincides with the usual formulae for the probability that nonadiabatic transitions take place between discrete states of the system, and therefore, it may appear to be self evident. However, the simplicity of expression (9.3.13) is essentially related to the Coulomb form of the potential. As we have seen in Chapter 8, in some other, apparently more simple cases, the expressions obtained are more complex. Indeed, the energy distribution for electrons emitted in low-energy collisions $A^- + B$ or $A^- + A$ was proportional to $E^{1/2}$ and $E^{3/2}$. In the present case, the distribution is finite at the threshold $E = 0$ itself and falls off monotonically as E increases. The reason is that at $E = 0$, the density of states is increased due to the Coulomb accumulation of the levels, leading to a smoother transition from the discrete spectrum to the continuum. (This is reflected in the form of the function F in Fig. 9.4.) Similar results are known for other problems where the Coulomb field is essential, for instance, in the case of ionization of an atom or ion by electron impact or for ion excitation.

Another merit of equation (9.3.13) is that the characteristic length in the adiabatic criterion is the same for all highly excited states ($E < 0$) as well as for the continuum ($E > 0$), and it is related in a simple manner to the behavior of the energy terms in the neighborhood of $R = R_c$ and $E = 0$.

The presence of the Coulomb field, despite a formal complication that it introduces, in fact, simplifies the treatment of the problem and gives formula (9.3.13). Another advantage of the theory is that the changes which have to be made if the initial state possesses higher symmetry ($\ell \neq 0$, etc.) are not significant. The higher symmetry does not affect the Coulomb accumulation point, and a new expression will be obtained only for the coordinate part of the Green's function (9.3.2). In other words, the change affects only the determination of the constant B. The influence of the Coulomb field on the system is so profound that the final result is only modified slightly if symmetry considerations imply that some values of the quantum numbers $\ell$ and m are to be excluded.

Only one linear combination of the states with differing $\ell$ and m but having the same n, corresponding to the projection of $\phi$ onto the subspace, will be excited. All other combinations can be considered as orthogonal to $\phi$ and, in the present approximation, they will not be excited (compare with Section 7.4).

If the effective potential in $H_0$ differs from the Coulomb potential at small distances, this will mainly affect the form of the Green's function at large energies $|E|$, that is, at small $|n|$. For large n, this influence can be taken into account by the quantum defect $\delta(n)$. The latter is a measure of the departure of the effective quantum number $n^*$ from the integral quantum number n, and it goes into a constant $\delta$ as $n \to \infty$ (the case of highly excited energy levels). Then the matrix element $< \phi | G | \phi >$, for large n, takes the form

$$< \phi | G | \phi > \sim B \cot \pi (n - \delta), \tag{9.3.16}$$

and all our previous results are still applicable, provided that n has been replaced by $n^* = n - \delta$.

Departures from the Coulomb field and from the spherical symmetry, at small distances, will remove the degeneracy of the energy levels with respect to $\ell$ and m. This will not change the main formulae and, as far as the function F is concerned, will increase the number of steps in F for $E < 0$, hence making the linear approximation to F in that region of E even more justified. Of course, the height of some steps may change considerably, from level to level, resulting in significant changes in the population of some states (some of them can be nearly orthogonal to $\phi$). However, the averaged formula (9.3.13) will still be valid.

The problem becomes especially simple if $R_c = 0$ and $\phi$ is a delta-function, i.e., if apart from the Coulomb field, there exists a ZRP of variable depth placed at the origin of the coordinates. Then the regularized Green's function at $r = r' = 0$ can be easily obtained from formula (7.4.1), if expressions for the Whittaker functions at small values of the argument are used. We obtain

$$\begin{aligned} G_r(0,0,E) &= \frac{1}{\pi} \left[ \ln n - \Psi(1 - n) - 1/2n \right] \\ &= \frac{1}{\pi} \left[ \ln n - \Psi(n) - \pi \cot \pi n - 1/2n \right], \end{aligned} \tag{9.3.17}$$

where $\Psi$ is the logarithmic derivative of the gamma-function. The residues of $G_r$ in the E plane at integral n are given exactly by

$$h_n = (\pi n^3)^{-1}$$

and the discontinuity of the imaginary part across the cut, $\Delta$, is given by

$$\Delta = 2 / (1 - e^{-2\pi\nu}), \tag{9.3.18}$$

where $\nu = (2E)^{-\frac{1}{2}}$. The same quantities can be obtained directly from the normalized wavefunctions for the discrete and continuous spectra. In this particular case,

$$h_n = |\psi_{n,0,0}(0)|^2 , \quad \Delta = \frac{1}{\pi}|\psi_{E,0,0}(0)|^2 .$$

For large n, we can make use of the asymptotic formula for $\Psi$. Thus

$$G_r = -\cot \pi n + O(1/n^2) . \tag{9.3.19}$$

The representation (9.3.19) for $G_r$ is valid everywhere except for a narrow sector cutting out the (negative) real half-axis. Thus, for the discrete spectrum, we can use, as before, formula (9.3.7). However, in that case $n_0 = 1$, and the results are valid for all (including small) n. For the continuum, the distribution is found to be

$$w(E) = \exp\left\{-2B\left[\zeta(3) + \int_0^E \left(1 - \exp\left[\frac{-2\pi}{\sqrt{2E'}}\right]\right)^{-1} dE'\right]\right\} , \tag{9.3.20}$$

where

$$\zeta(3) = \sum_{n=1}^{\infty} n^{-3} \approx 1.202.$$

For small energies, expression (9.3.20) coincides with the one found before. For larger energies, $n \ll 1$, we have $\Delta \approx (2\pi)^{-1}\sqrt{2E}$, that is, the same discontinuity as that for a free particle. This result can be expected because, for large energies, the Coulomb accumulation limit ceases to play a significant role and the spectrum must become the same as that for a short-range potential with one discrete state vanishing as it fuses together with the continuum (Section 8.2).

This last result, however, has little of practical importance because the real potential well cannot be considered as being short-ranged even for $n \lesssim 1$, and the distribution function will have, in this region, a different form. Nevertheless, the general statement that the distribution function will fall off, at large n, more rapidly than formula (9.3.13) predicts remains correct.

Finally, we note that the possibility for the system AB to form a quasi-stationary state at $R < R_c$ has not been taken into account above. If this is done, the probability that the wave packet decays when $R < R_c$ may appreciably decrease, whereas the probability of recapturing the particle into a bound state may appreciably increase. This means that the constant L in the formulae above may change and become very small. As a result, the theory will be correct for small E as before; however, the probability that the energy of the electron after collision is small and lies within the interval $-E_0 < E < E_0$ will become small. The emitted electrons will mainly have high energies and their distribution functions can only be obtained from a more detailed study of the process. In this case, the low-energy electrons will dominate the distribution only in very slow collisions. Obviously, the same consideration shows that the theory is not accurate for high energies. The theory is quasi-adiabatic. It is based on the wavefunction of the instantaneous energy operator of the time-dependent problem. Therefore, it ceases to be valid if the relative velocity of the nuclear motion becomes of the same order of magnitude as the velocity of the atomic electron in the initial (but not in the final) state.

## Chapter 10

# NONLINEAR APPROXIMATIONS IN THE THEORY OF ELECTRON DETACHMENT

## 10.1 NONLINEAR PROBLEMS SOLVABLE BY CONTOUR INTEGRATION. SUDDEN APPROXIMATION

In Section 9.1 we considered a Hamiltonian with a separable potential multiplied by a coefficient linearly dependent on time, and showed that the solution of the corresponding time-dependent equation could be obtained with the help of contour integration. An equation where the coefficient of the separable potential is a linear function of time can be solved in a similar way. In the particular case when the coefficient of the separable potential is inversely proportional to time, the equation

$$\left[ H_o + |\phi>\gamma t^{-1} <\phi| \right] |\psi> = i \frac{\partial}{\partial t}|\psi> \qquad (10.1.1)$$

has a solution of the form

$$|\psi> = N \int_C G(E)\,|\phi> \exp\left\{ i \int^{E} <\phi|G(E')|\phi> dE' - i\,Et \right\} dE, \qquad (10.1.2)$$

which appears to be even simpler than the form of equation (9.1.4). Unfortunately, the perturbation term in (10.1.2) is singular at $t = 0$, which makes it difficult to apply this equation to real systems. The general behavior of the terms for a simplified problem which has only three discrete energy levels is illustrated in Fig. 10.1. In certain cases when the terms which go to infinity may be ignored, this model can still be used. In the particular case of the time-dependent equation

$$\left( H_o e^{at} + | \phi > b < \phi | \right) | \psi > = i \frac{\partial}{\partial t} | \psi > , \qquad (10.1.3)$$

there is a system of exponentially diverging terms which interact with a single horizontal term. Making the substitution exp(at) = s, we transform equation (10.1.3) to the form (10.1.1); hence, the exact solution of equation (10.1.3) can be obtained making use of equation (10.1.2). A two-state system of this type was considered by Nikitin and also by Demkov (for references see [26]), and it was applied to problems of non-resonant charge exchange and to the calculation of the fine-structure transitions in alkali metals caused by impact.

The time-dependent problem

$$( H_o + | \phi > e^{at} < \phi | ) | \psi > = i \frac{\partial}{\partial t} | \psi > \qquad (10.1.4)$$

leads to a functional equation for Z(E) in equation (9.1.2) which connects values of Z(E) and Z(E + a). On certain occasions this equation can be either investigated qualitatively or solved exactly. In the two-level approximation, elementary transformations reduce equation (10.1.4) to equation (10.1.3).

A disadvantage of these models as well as the model solved in Section 9.1 is that they cannot take into account the time symmetry which exists usually in physical processes. The quadratic dependence on time is one of the simplest forms of a time-dependent perturbation which would ensure such symmetry. However, in the latter case, the solution of the equation with a separable potential cannot be obtained in a closed form. Yet the similar problems with a ZRP can still be analyzed and solved exactly (see Sections 10.2 and 10.3 below).

Our discussion in Chapter 9 mentioned that the model considered there could be regarded as a generalization of the adiabatic approximation. It is also of interest to consider the opposite case of the

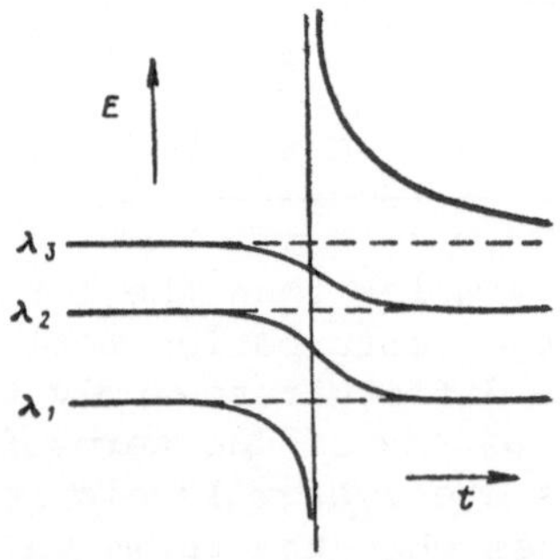


Fig. 10.1. A typical system of molecular terms for the problem defined by equation (10.1.1) with the potential energy inversely proportional to time t.

perturbation being introduced by sudden switching. Within the single ZRP approximation the latter model was studied by Bronfin and Ermolaev [166]. The parameter $\alpha$ in the boundary condition was taken in the form of a step-wise function of time, thus:

$$\alpha(t) = \begin{cases} \alpha_o, & \text{if } t < t_1, \\ -f_o, & \text{if } t_1 < t < t_2, \\ \alpha_o, & \text{if } t > t_2. \end{cases} \qquad (10.1.5)$$

The solution of the problem is constructed by matching wavefunctions at $t_1$ and $t_2$, that is, at the moments of the sudden change in the depth of the well.

Assuming that the particle is in a bound state at $t < t_1$, we shall write for the wavefunction $\Psi(r,t)$, thus:

$$\Psi(r,t) = \sqrt{2\alpha_o} \exp(-\alpha_o r + i\alpha_o^2 t/2), \quad t < t_1. \qquad (10.1.6)$$

In the intermediate region $t_1 < t < t_2$ the solution $\Psi$ is sought in the form of an expansion in terms of the eigenfunctions of the continuum. We have

$$\Psi(r,t) = \sqrt{2/\pi} \int_0^\infty a(k) \sin(kr+\delta) e^{-ik^2t/2} dk, \quad t_1 < t < t_2, \qquad (10.1.7)$$

where

$$k \cot\delta(k) = f_o .$$

Finally, for $t > t_2$, the spectrum of the Hamiltonian includes both a discrete state and the continuum so that the corresponding expansion of $\Psi$ takes the following form:

$$\Psi(r,t) = B\sqrt{2\alpha_o} \exp(-\alpha_o r + i\alpha_o^2 t/2)$$
$$+ \sqrt{2/\pi} \int_0^\infty b(k) \sin(kr+\tilde{\delta}) e^{-ik^2t/2} dk, \quad t > t_2, \qquad (10.1.8)$$

where

$$k \cot\tilde{\delta}(k) = -\alpha_o .$$

The coefficients $a(k)$ and $b(k)$ are found from the continuity condition for the solution $\Psi$ imposed at $t = t_1$ and $t = t_2$ and using the completeness of the ortho-normalized system of eigenfunctions. This leads to the following expressions for the probabilities:

$$W(k) = |b|^2 = \frac{4\alpha_o}{\pi}\frac{(f_o+\alpha_o)^2 k^2}{k^2+\alpha_o^2}\left|\frac{k^2 - f_o\alpha_o}{(k^2+\alpha_o^2)(k^2+f_o^2)}\right.$$

$$\left. - \frac{2}{\pi}(f_o+\alpha_o)\, e^{ik^2T/2}\int_0^\infty e^{-k'^2 iT/2}\frac{k'^2\,dk'}{(k'^2+\alpha_o^2)(k'^2+f_o^2)(k'^2-k^2)}\right|^2, \tag{10.1.9}$$

and

$$P_\alpha = |B|^2 = \frac{16\alpha_o^2}{\pi^2}(f_o + \alpha_o)^2$$

$$\times \left|\int_0^\infty e^{-ik^2T/2}\frac{k^2\,dk}{(k^2+\alpha_o^2)^2(k^2+f_o^2)}\right|^2, \tag{10.1.10}$$

where $T = t_2 - t_1$. The integrals in equations (10.1.9) and (10.1.10) can be expressed in terms of the Fresnel integrals. For $k = k'$, the principal value of the integral (10.1.9) is to be used.

Let us consider some special cases of these expressions. For $T \to \infty$, we obtain

$$W(k) \to \frac{\alpha_o}{\pi}\frac{4(\alpha_o + f_o)^2 k^2}{(k^2+\alpha_o^2)(k^2+f_o^2)} = |a(k)|^2.$$

A simple physical interpretation can be given to this result. If a bound state in the well disappears at $t_1$ and does not exist for a very long time (i.e., T is large), then the electron moves far away from the well and the reappearance of the bound states in the well at $t_2$ cannot influence the spectrum of the emitted electrons. For large T, the probability that the electron remains bound in the well is given by

$$P_\alpha = \frac{(\alpha_o + f_o)^2}{2\pi\,\alpha_o^6 f_o^4}\frac{1}{T^3} + O(T^{-5}).$$

The latter result coincides with that found in Section 8.2. In the limit of $T \to 0$ when the perturbation vanishes, the natural results $P_\alpha \to 1$ and $W(k) \to 0$ follow from the two general expressions above. For small k, similarly to the earlier result obtained in Section 8.2, we now have $W(k) \sim k^3$ as $k \to 0$. However, for large k, the present distribution falls off only as $1/k^4$, whereas in the adiabatic case the distribution falls off exponentially. This difference appears to be quite natural from the physical point of view (see Section 8.3).

Bronfin and Ermolaev [166] also applied this general theory to the particular case of electron detachment in the reaction

$$H^- + H \rightarrow H + H + e .$$

The most serious difficulty in the application of the sudden approximation is the absence of a clear physical guidance as to how to choose the parameter $f_o$ in the model. The overall good agreement between the theoretical and experimental total cross sections for detachment, for a wide range of relative velocities ($V \lesssim 1.2$ a.u.), can be achieved if one puts $f_o = 0.76$. Nevertheless, the physical background of this approximation requires further investigation.

## 10.2 QUADRATIC APPROXIMATION IN THE THEORY OF ELECTRON DETACHMENT

Within the model of a single ZRP, the quadratic approximation in time requires the following form for the coefficient $\alpha$ in the boundary condition (8.1.1) (see references [167], [168]):

$$\alpha(t) = a t^2 - b , \qquad a > 0 . \tag{10.2.1}$$

For $b > 0$, the term enters the continuum (see Fig. 10.2). If the positive b is large, the probability that the electron remains bound can be calculated using the same method as that in Section 8.2. In the opposite case, when $b < 0$, the term never reaches the continuum boundary and always corresponds to the bound state. The probability that detachment will take place is expo-

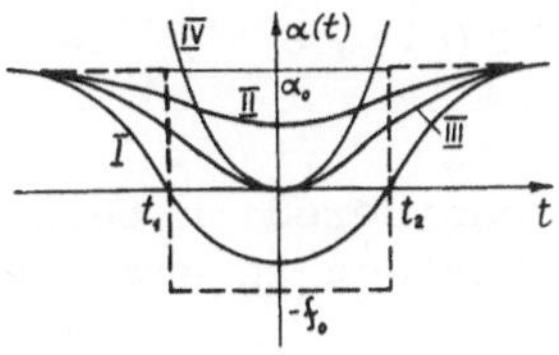


Fig. 10.2. Different cases of the parameter $\alpha(t)$ in the boundary condition. I - a term entering the continuum; II - a term corresponding to a bound state for all times; III - an intermediate case between I and II, $\alpha(t)$ is tangent to the t-axis at $t = 0$; IV - a quadratic approximation to III. Broken line - sudden approximation to $\alpha(t)$.

nentially small and can be obtained in the adiabatic approximation. For such under-barrier transitions the theory was developed by Chaplik [149, 150].

The general solution of the problem in the quadratic approximation in time, for arbitrary coefficients a and b in equation (10.2.1), would allow the two extreme regions pointed out above to be connected. The simplest case to be treated theoretically is that where b = 0, where there is a contact point between the term and the continuum boundary. This case will be taken as the basis for subsequent study of the general problem.

Applying the Laplace transformation to the function, we shall seek the solution in the form

$$\Psi(r,t) = \int_{L_1} \exp\{ iu^2t/2 - ur \}\, Z(u)\, u^{3/2}\, du . \qquad (10.2.2)$$

Substituting this integral into the boundary condition

$$\left.\frac{\partial\Psi}{\partial r}\right|_{r=0} = -\alpha(t)\,\Psi(0,t)$$

and making use of equation (10.2.1), we obtain, after integrating twice by parts, the following equation for Z(u):

$$\frac{d^2Z}{du^2} + \left( u^3/a + bu^2/a - 3/4u^2 \right) Z = 0 , \qquad (10.2.3)$$

together with the conditions for the integration contour $L_1$ given by

$$i\,a\,t\, e^{iu^2t/2}\, u^{\frac{1}{2}}\, Z(u) \Big|_{L_1} = 0, \qquad (10.2.4)$$

and

$$-\,a\, e^{iu^2t/2}\, (1/u)\, d\{u^{\frac{1}{2}} Z(u)\}/du \Big|_{L_1} = 0. \qquad (10.2.5)$$

For b = 0 (the case being considered in this section), the solution of equation (10.2.3) was obtained in a closed form by Devdariani and Demkov [167]:

$$Z(u) = u^{\frac{1}{2}}\, \Xi_{2/5}\{2\,u^{5/2}/5\sqrt{a}\} , \qquad (10.2.6)$$

where $\Xi_{2/5}$ is a cylindrical function. Bearing in mind that we have to satisfy the contour conditions (10.2.4) and (10.2.5), we shall use, for $\Xi_{2/5}$, the Hankel functions $H^{(1)}_{2/5}$ and $H^{(2)}_{2/5}$ because they are the only cylindrical functions that fall exponentially in the required region of the complex argument, ensuring the conditions (10.2.4) and (10.2.5).

Figure 10.3 shows schematically the surface generated by the function

$$|u^2 H^{(1)}_{2/5}(2u^{5/2}/5a^{1/2})| , \qquad -\pi < \arg u < \pi .$$

For this function, there exists only one sector of damping, namely, $0 < \arg u < 2\pi/5$. The antigradient directions are shown, in the diagram, by arrows where also the positions of the zeros are marked.

The similar surface generated by using $H^{(2)}_{2/5}$ in place of $H^{(1)}_{2/5}$ is obtained by rotating the figure about the origin through an angle of $-2\pi/5$ and is shown in the same diagram.

The conditions (10.2.4) and (10.2.5) will be satisfied as $t \to -\infty$ if we put $Z = u^2 H^{(1)}_{2/5}$ and choose $L_1$ as a contour whose upper end is in the sector of damping and with the lower end going to infinity along the ray $\arg u = -2\pi/5$. The required conditions (10.2.4) and (10.2.5) will be satisfied at the upper end of $L_1$ due to the exponential decrease of the function $H^{(1)}_{2/5}$ which behaves as $\exp(i\,2u^{5/2}/5a^{1/2})$ for $|u| \to \infty$. At the lower end of $L_1$, the decrease of $\exp(iu^2 t/2)$ for $t < 0$ ensures the same conditions. We shall note that the conditions (10.2.4) and (10.2.5) can also be satisfied by a contour with one of its ends going to infinity along

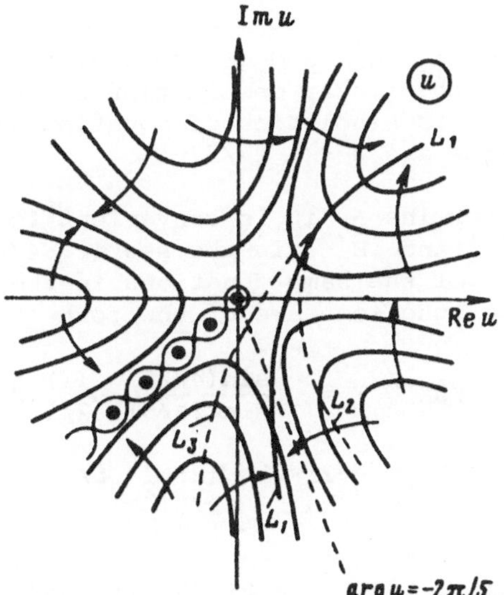


Fig. 10.3. Integration contours in the case of quadratic approximation for $\alpha(t)$ (reference [168]).

the ray arg $u = 4\pi/5$. However, analysis similar to that carried out above shows that in this case we obtain a solution which is exponentially growing in r as $t \to -\infty$. The use of $u^2 H^{(2)}_{2/5}$ as the integrand leads to a wavefunction $\Psi(r,t)$ which satisfies different initial conditions.

The function $H^{(1)}_{2/5}$ oscillates along the ray arg $u = -2\pi/5$, and it is convenient, for the purpose of further discussions, to divide the sector which includes this ray into two parts: one to the left and another to the right of the line arg $u = -2\pi/5$. Then we shall represent $H^{(1)}_{2/5}$ as the sum of two functions, each of them falling off, at large $|u|$, in the corresponding subsector. At the lower end, the contour $L_1$ must be split into two different paths of integration, $L_2$ and $L_3$, for these two functions, as shown in Fig. 10.3 above. In the right subsector, we shall use $H^{(2)}_{2/5}$, whereas in the left subsector, we shall use a linear combination of functions $H^{(1)}_{2/5}$ and $H^{(2)}_{2/5}$. Thus, the solution (10.2.2) finally takes the form

$$\Psi(r,t) = g_1 \int_{L_1} e^{iu^2t/2 - ur} u^2 H^{(1)}_{2/5} du + g_2 \int_{L_2} e^{iu^2t/2 - ur} u^2 H^{(2)}_{2/5} du + g_3 \int_{L_3} e^{iu^2t/2 - ur} u^2\{H^{(1)}_{2/5} + A H^{(2)}_{2/5}\} du. \qquad (10.2.7)$$

In equation (10.2.7), $g_1$, $g_2$, and $g_3$ are constants which will be specified later, and the linear combination of the Hankel functions in the third integral is a solution of equation (10.2.3) falling off in the sector $-4\pi/5 < \arg u < -2\pi/5$.

In order to determine A, it is convenient to express the values of the functions $H^{(1)}_{2/5}$ and $H^{(2)}_{2/5}$ in the sector $-4\pi/5 < \arg u < 0$ in terms of the values of the same functions in the sector $-2\pi/5 <$ arg $u < 2\pi/5$. These functions are of the form

$$H^{(1)}_{2/5}(2\, e^{-i\pi} u^{5/2}/5a^{1/2}) = \frac{\sin(4\pi/5)}{\sin(2\pi/5)} H^{(1)}_{2/5}(2\, u^{5/2}/5a^{1/2}) + e^{-2i\pi/5} H^{(2)}_{2/5}(2\, u^{5/2}/5a^{1/2}) ,$$

and

$$H^{(2)}_{2/5}(2\, e^{-i\pi} u^{5/2}/5a^{1/2}) = -\, e^{2i\pi/5} H^{(1)}_{2/5}(2u^{5/2}/5a^{1/2}). \qquad (10.2.8)$$

The constant A in equation (10.2.7) is then obtained in the form

$$A = e^{-2i\pi/5} \frac{\sin(\pi/5)}{\sin(2\pi/5)}, \qquad (10.2.9)$$

using equation (10.2.8) and the condition that the solution must tend to zero in the sector $-4\pi/5 < \arg u < 2\pi/5$, as $|u| \to \infty$. Hence the required solution has the form

$$H^{(1)}_{2/5}(2\, e^{-i\pi}\, u^{5/2}/5a^{1/2}) + A\, H^{(2)}_{2/5}(2\, e^{-i\pi}\, u^{5/2}/5a^{1/2})$$

$$= e^{-2i\pi/5}\, H^{(2)}_{2/5}(2\, u^{5/2}/5a^{1/2}). \qquad (10.2.10)$$

From the conditions (10.2.4) and (10.2.5) at the point where the contours in Fig. 10.3 intersect each other, we find that

$$g_3 = g_1, \quad g_2 = -A\, g_1 . \qquad (10.2.11)$$

The value of $g_1$ is obtained by comparing, as $t \to -\infty$, the solution constructed here with the adiabatic wavefunction which has the form

$$\Psi(r,t) \sim t\sqrt{2a}\, \exp\{-a t^2 r + i(a^2/10)\, t^5\} . \qquad (10.2.12)$$

Considering the surface determined, at $r = 0$, by the integrand of the first integral in (10.2.7), one finds that it has a saddle point, at large t, situated on the real positive axis. Therefore we can use the simple asymptotic form of the Hankel function for $|u| >> a^{1/5}$. The saddle point is at $u = a t^2$. Finally we find, to a non-essential phase factor, that the first integral is

$$I_1 \sim g_1\, 2\sqrt{5}\, a t\, \exp(i a^2 t^5/10). \qquad (10.2.13)$$

The remaining two integrals in equation (10.2.7) have no saddle points, and they will fall off exponentially as $t \to -\infty$. Consequently, $\Psi(0,t) \approx I_1$. By comparing, at $r = 0$, equation (10.2.13) with the adiabatic solution (10.2.12), one finds that $g_1 = (10a)^{-1/2}$.

If $r \neq 0$, the turning point is shifted downwards and to the right, and the asymptotic form of $\Psi$ can still be investigated. At $a|t|^3 >> 2r$, we obtain, as expected, the formula (10.2.12). However, for large r when $at^3 << 2r$, the function decreases at a faster rate:

$$\Psi(r,t) \sim \exp\{-\tfrac{3}{5}(-ar^5)^{1/3}\, e^{i\pi/3}\} . \qquad (10.2.14)$$

This violation of the adiabatic approximation, at large r, is in agreement with the similar result of Section 8.2 and is due to retardation: the wavefunction at large r cannot respond instantly to changes occuring at r = 0.

Let us consider the behavior of the solution as $t \to \infty$. Now the first integral has no saddle point. The second integral can be calculated in the same way as the first integral as $t \to -\infty$. On account of equation (10.2.11) it gives, at r = 0,

$$I_2 \sim \frac{\sin(\pi/5)}{\sin(2\pi/5)} t (2a)^{1/2} e^{ia^2 t^5/10}, \tag{10.2.15}$$

that is, it describes a bound state as $t \to \infty$. The third integral, which will be considered later, represents the wave packet of the emitted electron. Therefore, we obtain for the probability that the recapture of the emitted electron takes place,

$$P_\alpha = \frac{\sin^2(\pi/5)}{\sin^2(2\pi/5)} = 0.38. \tag{10.2.16}$$

In the present theory, this probability does not depend upon any parameter.

Let us now find the momentum distribution of the emitted electrons. We have to multiply equation (10.2.7) by a normalized wavefunction of the free particle, $\sqrt{2/\pi} \sin kr$, and integrate the product over r from zero to infinity. We then have to calculate the limit as $t \to \infty$.

The result of integration of the first term in (10.2.7) with the sine function goes to zero exponentially as $t \to \infty$, whereas the second integral gives the usual momentum distribution for the wavefunction of the bound state (see equation (10.2.12) above).

For the third integral, we obtain

$$\phi_3(k,t) = g_3(2/\pi)^{1/2} \int_0^\infty \sin kr\, dr \int_{L_3} e^{iu^2t/2 - ur} \times \{H_{2/5}^{(1)} + AH_{2/5}^{(2)}\} u^2\, du. \tag{10.2.17}$$

If we deform the contour $L_3$ so that it lies entirely in the right half-plane, the order of integration in equation (10.2.17) can be changed and the integration over r performed first. The result is

$$\phi_3(k,t) = g_3(2/\pi)^{1/2} \int_{L_3} e^{iu^2t/2} \{H_{2/5}^{(1)} + A\, H_{2/5}^{(2)}\} \frac{ku^2 du}{k^2 + u^2}. \tag{10.2.18}$$

This integral is equal to the residue of the integrand at $u = -ik$ plus the integral along the contour $L_3$ which passes to the left of the point $-ik$, and goes to zero exponentially as $t \to \infty$ since both the $H^{(1)}_{2/5} + A\,H^{(2)}_{2/5}$ and $\exp(iu^2t/2)$ fall off. After taking the residue, we obtain the result

$$\phi_3(k,t) = F(k)\, e^{-ik^2t/2},$$

where

$$F(k) = g_3(2\pi)^{\frac{1}{2}}\, k^2\, \{H^{(1)}_{2/5} + A\, H^{(2)}_{2/5}\}\Big|_{u=-ik}. \tag{10.2.19}$$

Finally, using equations (10.2.10) and (10.2.11), we derive the following expression for the momentum distribution $W(k) = |F(k)|^2$ of the emitted electrons,

$$W(k) = \frac{\pi}{5a}\, k^4\, |\, H^{(2)}_{2/5}(2\,k^{5/2}\, e^{-i\pi/4}/5a^{1/2})\,|^2. \tag{10.2.20}$$

Making use of the expansions of the Hankel functions for small and for large values of the argument, respectively, we obtain the two limiting cases of the distribution $W(k)$:

$$W(k) = \frac{\pi}{5^{1/5}\,\sin^2(2\pi/5)\,\Gamma^2(3/5)}\, \frac{k^2}{a^{3/5}} + O\left[\frac{k^4}{a^{6/5}}\right], \qquad k \ll 1$$

and

$$W(k) = \frac{k^{3/2}}{a^{1/2}}\, \exp\left(-\frac{2}{5}\sqrt{2/a}\, k^{5/2}\right) \left\{1 + O\left[\frac{a^{1/2}}{k^{5/2}}\right]\right\}. \qquad k \gg 1 \tag{10.2.21}$$

The maximum of the distribution, $W_{max} = 0.55\, a^{-1/5}$, lies approximately at $k_{max} = a^{1/5}$.

Let us now find the probability that the energy of the emitted electron lies between zero and E, i.e.,

$$w(E) = \int_0^E W(\sqrt{2E'})\, \frac{dE'}{\sqrt{2E'}}. \tag{10.2.22}$$

Changing from the variable E' to a new variable z, according to $2(2E')^{5/4}/5\,a^{1/2} = z$, and using

$$\left(H^{(2)}_{2/5}(z\, e^{-i\pi/4})\right)^* = H^{(1)}_{2/5}(z\, e^{i\pi/4}), \tag{10.2.23}$$

we reduce equation (10.2.22) to a table (indefinite) integral whose integrand is a weighted product of Hankel functions. The result of integration can be written in the following form:

$$w(E) = \frac{i\pi}{4} z \left| D(z) \right|_{z = z_0}^{z = z_1} , \tag{10.2.24}$$

where

$$D(z) = \begin{vmatrix} \frac{d}{dz} H_{2/5}^{(1)}(z\,e^{i\pi/4}) & H_{2/5}^{(1)}(z\,e^{i\pi/4}) \\ \frac{d}{dz} H_{2/5}^{(2)}(z\,e^{-i\pi/4}) & H_{2/5}^{(2)}(z\,e^{-i\pi/4}) \end{vmatrix} , \tag{10.2.25}$$

and

$$z_0 = 0, \quad z_1 = 2\,(2E)^{5/4} / 5a^{1/2} .$$

For small E, the simplest way to obtain an approximate formula for w(E) is to use directly equation (10.2.21):

$$w(E) \approx - \frac{\pi}{3\cdot 5^{1/5} \sin^2(2\pi/5)\, \Gamma^2(3/5)} \, \frac{(2E)^{3/2}}{a^{3/5}} . \tag{10.2.26}$$

For $\sqrt{2E} < a^{1/5}$, it is more convenient to use equation (10.2.24). Expanding the Hankel functions at the vicinity of zero, we find that

$$D(0) = \frac{4i}{5} \frac{\sin(\pi/5)}{\sin(2\pi/5)} .$$

With the help of the asymptotic formulae for the Hankel functions for large arguments, we find that

$$w(E) = \frac{\sin(\pi/5)}{\sin(2\pi/5)} - \frac{1}{\sqrt{2}} \exp\left( - \frac{2}{5} (2/a)^{1/2} (2E)^{5/4} \right). \tag{10.2.27}$$

Finally, for the total probability of detachment, we obtain

$$P_i = \frac{\sin(\pi/5)}{\sin(2\pi/5)} = 0.62. \tag{10.2.28}$$

We note that, in this problem, the normalization condition $P_\alpha + P_i = 1$ is satisfied. Another interesting feature is that the specific relation $P_\alpha = P_i^2$ also holds.

Devdariani and Demkov (167) also found the probabilities for continuum-continuum and continuum-bound state transitions. As in Section 8.4, the integration is carried out along a contour which starts at a finite point $u = -ip$, where p is the initial momentum of the particle. Here we shall give only the final results of those calculations. In accordance with the principle of detailed balance we obtain formula (10.2.20) for the transition probability to a bound state where k is replaced by p. The S-matrix and the transition amplitude F to a continuum state are

$$S(p,k) = \delta(p-k) + F(p,k), \qquad (10.2.29)$$

where

$$\begin{aligned} F(p,k) = & -\frac{\pi}{5\alpha} p^2 k^2 \Big\{ H^{(2)}_{2/5}(2p^{5/2} e^{-i\pi/4}/5a^{1/2}) \\ & \times H^{(1)}_{2/5}(2k^{5/2} e^{-i\pi/4}/5a^{1/2}) \times \theta(p-k) \\ & + H^{(2)}_{2/5}(2k^{5/2} e^{-i\pi/4}/5a^{1/2}) \times H^{(1)}_{2/5}(2p^{5/2} e^{-i\pi/4}/5a^{1/2}) \\ & \times \theta(k-p) \Big\}, \end{aligned}$$

and

$$\theta(x) = \begin{matrix} 1, & \text{if } x > 0, \\ 0, & \text{if } x < 0. \end{matrix}$$

The following approximate expressions can be derived for the momentum distribution of the emitted electrons, $W(p,k) = |F(p,k)|^2$, in several special cases:

$$W(p,k) = \frac{\pi^2}{5^{2/5} \sin^4(2\pi/5)\, \Gamma^4(3/5)} \frac{p^2 k^2}{a^{6/5}}, \qquad (10.2.30)$$

$$k < p \ll a^{1/5};$$

$$W(p,k) = \frac{\pi}{5^{1/5} \sin^2(2\pi/5)\, \Gamma^2(3/2)} \frac{p^{3/2} k^2}{a^{11/10}} e^{-\frac{2}{5}(2/a)^{\frac{1}{2}} p^{5/2}},$$

$$k \ll a^{1/5}, \quad p \gg a^{1/5}; \qquad (10.2.31)$$

$$W(p,k) = \frac{p^{3/2} k^{3/2}}{a} e^{-\frac{2}{5}(2/a)^{\frac{1}{2}}(p^{5/2} - k^{5/2})},$$

$$k \gg a^{1/5}, \quad p \gg a^{1/5}. \qquad (10.2.32)$$

In the above formulae, $p > k$. The case when $p < k$ can be obtained from equations (10.2.30) - (10.2.32) by interchanging p and k. These formulae show that the distribution maximum lies at $k = p$.
For small p,

$$W_{max} = \pi^2 p^4 / \{5^{2/5} a^{6/5} \sin^4(2\pi/5)\Gamma^4(3/5)\},$$

and for $p \gg a^{1/5}$,

$$W_{max} = p^3/a .$$

The first derivative of the momentum distribution W has a discontinuity $(4p^3/a)\,\mathrm{Im}\,F(p,p)$, at $k = p$. This result differs from that obtained in Section 8.4 for the case of $\alpha(t)$ linear in time t. There the distribution itself had a discontinuity point due to the slow decrease of perturbation at large $|t|$. In the quadratic case the perturbation is falling off at a faster rate than it does in the linear case because the instantaneous Hamiltonian approaches its free-particle limit faster as $|t| \to \infty$. As a result, the diagonal elements of the S-matrix have a weaker singularity in the quadratic case.

The principle of detailed balance was referred to above in connection with transitions into bound states. In the case of free-free transitions, this principle is ensured by the equation $S(p,k) = S(k,p)$. It is also possible to check the unitarity of the S-matrix, namely, that the equation

$$\int_0^\infty F(p,k)\, F^*(p',k)\, dk + S_{po} S^*_{p'o} = \delta(p-p') \qquad (10.2.33)$$

is satisfied. In (10.2.23), $S_{po}$ is an element of the S-matrix corresponding to the transitions from the continuum into a bound state, $W(k) = |S_{ko}|^2$.

Ostrovskii (79) considered the quadratic problem using the spectral properties of the S-matrix. It has been shown in Chapter 4 that the representation formed by eigenfunctions of the S-matrix is most natural for scattering problems. This approach can be extended to include the time-dependent case, with many expressions of the original stationary theory retaining their form in the extended version (79). The quadratic problem is one of few known cases where the solution can be obtained in closed form. Another important feature of this problem is that it is time-symmetrical unlike, for instance, the linear problems of Chapters 8 and 9. Therefore, it is of particular interest to investigate the spectrum of the corresponding S-matrix (in the original work (79), some other time-dependent problems were also considered). It turns out that the eigenfunctions $\chi_\lambda$ of the S-matrix, which satisfy the equation

$$S\,\chi_\lambda = e^{2i\delta_\lambda}\,\chi_\lambda\,, \qquad (10.2.34)$$

have, in the case of the quadratic problem with $a < 0$ and $b = 0$ (see equation (10.2.1) for the definition of the parameters a, b), the following form in the momentum representation:

$$\chi_\lambda(k) = k^2 \,\mathrm{Re}\left\{\left[1 - (\cot\delta_\lambda)^{2/5}\, e^{-3\pi i/5}\right]\right.$$

$$\left.\times\; H^{(1)}_{2/5}\left(\frac{2}{5}\{\cot\delta_\lambda/|a|\}^{1/2}\, k^{5/2}\right)\right\} . \qquad (10.2.35)$$

The eigenphases $\delta_\lambda$ satisfy $0 < \delta_\lambda < \pi/2$, and hence the spectrum of the S-matrix is continuous and it occupies the internal region of the upper semi-circle of unit radius.

## 10.3 QUADRATIC APPROXIMATION (GENERAL CASE)

In the general case where the parameter $\alpha(t)$ defined by equation (10.2.1) is quadratic in t, with $b \neq 0$, the solution of the equation (10.2.3) cannot be obtained in closed form. Nevertheless, the most important qualitative features of the general problem are the same as those in the special case where $b = 0$ (the contact case). As a result, we can formulate the general problem in a form suitable for numerical integration, and we can obtain approximate analytic expressions which are simple enough and convenient for applications. We shall follow below Devdariani (168).

For large $|u|$, we can use the quasi-classical approximation to obtain any solution of equation (10.2.3) in an appropriate form as a linear combination of the following two functions:

$$Y_{1,2}(u) = \left[\frac{u^3}{a} + \frac{b}{a}u^2\right]^{-1/4} \exp\left\{\pm\int_{u_0}^{u}\left[\frac{u'^3}{a} + \frac{bu'^2}{a}\right]^{1/2} du'\right\} . \qquad (10.3.1)$$

We shall use two independent solutions $Z_{1,2}$ of equation (10.2.3) such that, for large $u > 0$, these solutions go into $Y_1$ and $Y_2$, respectively. The functions $Z_1$ and $Z_2$ are a generalization of the contact solutions obtained in Section 10.2 in terms of the Hankel functions of the first and second kinds. The functions $Y_1$ and $Y_2$ are good approximations to $Z_1$ and $Z_2$, correspondingly, not only on the real axis, but also in the sectors with boundaries determined by the equation

$$\mathrm{Im}\int_{u_0}^{u}\left[\frac{u'^3}{a} + \frac{bu'^2}{a}\right]^{1/2} du' = 0 . \qquad (10.3.2)$$

These boundaries go asymptotically into pairs of rays: $\arg u = -2\pi/5$ and $\arg u = 4\pi/5$, for $Y_1$, and $\arg u = 2\pi/5$ and $\arg u = -4\pi/5$, for $Y_2$. In the remaining part of the complex u plane (which includes the region outside the two sectors), the solutions $Z_1$ and $Z_2$ can be written, at large $|u|$, as linear combinations of $Y_1$ and $Y_2$. The coefficients of these sums are determined by the Stoke's phenomenon for second-order differential equations. Qualitatively, the surface $|Z_1|$ remains the same as that in the contact case (see Fig. 10.3).

It follows from Section 10.2 that the choice of the integration contours L satisfying the required boundary conditions is determined by the asymptotic properties of the solution. Hence in the general case, the contours remain qualitatively the same as in the contact case. However, now the coefficient A in formula (10.2.7) cannot be obtained in closed form. We remember that A is chosen from the condition that the sum $Z_1 + A Z_2$ falls off in the sector $-4\pi/5 < \arg u < -2\pi/5$, as $|u| \to \infty$. In the general case, A is Stoke's constant for the differential equation (10.2.3). This follows from equation (10.3.2), which determines a Stoke's line whose asymptote is the ray $\arg u = -2\pi/5$.

We conclude that, in the general case, the wavefunction is given by the same formula (10.2.7) where, however, the term $u^{\frac{1}{2}} H^{(1,2)}_{2/5}(u)$ must be replaced by solutions $Z_{1,2}$ of equation (10.2.2) for the general case. Subsequent calculations are carried out in a similar way to those of Section 10.2. In the general case, the probability that the electron remains in a bound state is

$$P_\alpha = |A|^2, \tag{10.3.3}$$

and the momentum distribution of the emitted electrons is given by the amplitude F(k) thus:

$$\begin{aligned} F(k) &= a^{-1/2} k^{3/2} \{Z_1(u) + A Z_2(u)\} \Big|_{u = -ik} \\ &\equiv a^{-1/2} k^{3/2} G(k) . \end{aligned} \tag{10.3.4}$$

Hence the problem has been reduced to equation (10.2.2) on the imaginary negative semi-axis, that is, to the solution of

$$\frac{d^2G}{dk^2} + \left\{ - \frac{ik^3}{a} + \frac{bk^2}{a} - \frac{3}{4k^2} \right\} G = 0, \tag{10.3.5}$$

where we have to construct a solution whose asymptotic behavior is determined by the expression for $Y_1(u)$ at $u = -ik$. Equation (10.3.5) can be solved numerically. Various limiting cases of the solution of equation (10.3.5) can be derived approximately in analytical form.

We shall give below some approximate solutions obtained by Devdariani (168). Let us introduce a parameter $\lambda = b/a^{1/5}$. The case where the total probability of detachment is close to unity corresponds to large $\lambda > 0$. Then the main term in equation (10.2.3) is proportional to b. Substituting $k = a^{1/5} x^{1/2}$ and $G = x^{-1/4} V(x)$, we transform equation (10.3.5) to the new form

$$\frac{d^2V}{dx^2} + \left[\frac{\lambda}{4} - \frac{i}{4}\sqrt{x}\right] V = 0 . \qquad (10.3.6)$$

An approximate solution of (10.3.6) can be obtained by the method of the standard equation. Introduce the equation

$$\frac{d^2U}{dz^2} + \frac{\lambda}{4}U = 0, \qquad U = \exp(-i\sqrt{\lambda}\, z/2), \qquad (10.3.7)$$

and use U(x) to write the solution of equation (10.3.6) in the form

$$V(x) = \sqrt{dx/dz}\, U(z(x)) . \qquad (10.3.8)$$

Substituting equation (10.3.8) into equation (10.3.6) and using equation (10.3.7) gives the nonlinear equation for the function z(x):

$$\frac{\lambda}{4} z'^2 = \frac{\lambda}{4} - \frac{i}{4} x^{1/2} + \frac{3}{4}\frac{z''^2}{z'^2} - \frac{1}{2}\frac{z'''}{z'} , \qquad (10.3.9)$$

where z' denotes, in the usual way, the derivative of z with respect to the argument x. The solution of equation (10.3.9) is sought in the form of an expansion in terms of $1/\lambda$:

$$z(x,\lambda) = \sum_{n=0}^{\infty} z_n(x)\, \lambda^{-n} , \qquad (10.3.10)$$

where the additional condition $z(0,\lambda) = 0$ is imposed in order to ensure that both equations have coincident singular points. With the first two terms of expansion (10.3.10) taken into account, the amplitude F(k) is obtained in the following form:

$$F(k) = \frac{k}{(ab)^{1/4}\left(1 - ik/2\, a^{1/5}\lambda\right)^{1/2}}$$

$$\times \exp\left\{-\frac{ik^2}{2}\sqrt{b/a} - \frac{1}{6}\frac{k^3}{\sqrt{ab}} + \frac{i\pi}{4}\right\}\left[1 + O(\lambda^{-3/2})\right]. \qquad (10.3.11)$$

The corresponding expression for the distribution $W(k) = |F(k)|^2$ reduces, for $\lambda \to \infty$, to formula (8.2.16) if we put $\beta = 2\sqrt{ab}$. An additional phase factor appears in equation (10.3.11) due to a time shift of $T/2 = \sqrt{b/a}$ in comparison with the resulting distribution W(k) of Section 8.2.

Another extreme case is transitions through the barrier when the term is deep below the continuum boundary ($\lambda < 0$ and $|\lambda|$ is large). Calculations similar to those given above are in accordance with results obtained earlier by Chaplik [143]. They show that

$$F(k) = \frac{k}{(a|b|)^{1/4}\left(1 + ik/2\, a^{1/5}|\lambda|\right)^{1/2}}$$

$$\times \exp\left\{-\frac{4}{15}\frac{|b|^{5/2}}{a^{1/2}} - \frac{|b|^{1/2}}{2a^{1/2}}k^2 - \frac{i}{6}\frac{k^3}{a^{3/2}|\lambda|^{1/2}}\right\}$$

$$\times \left(1 + O(|\lambda|^{-3/2})\right). \qquad (10.3.12)$$

For small $\lambda$, the situation is similar to that arising if there exists a contact point ($b = 0$) between the term and the continuum boundary. In equation (10.3.5), the main term is proportional to $-i/a$. Making the substitutions $k = (a/b)^{1/4}\, x^{2/5}$ and $G = x^{-3/10}V(x)$, we obtain the equation

$$\frac{d^2V}{dx^2} + \left\{-\frac{4i}{25}\frac{1}{\lambda^{5/4}} + \frac{4}{25}\frac{1}{x^{2/5}} + \frac{9}{100\,x^2}\right\}V = 0. \qquad (10.3.13)$$

A standard equation corresponding to equation (10.3.13) and its solution U are

$$\frac{d^2U}{dz^2} + \left[-\frac{4i}{25}\frac{1}{\lambda^{5/4}} + \frac{9}{100z^2}\right]U = 0, \qquad (10.3.14)$$

$$U(z) = z^{1/2}\, H^{(2)}_{2/5}\left(\frac{2z}{5}\, e^{-i\pi/4} / \lambda^{5/8}\right).$$

Retaining, as before, the first two terms in expansion (10.3.10) we derive the following expression for the amplitude $F(k)$:

$$F(k) = (\pi/5a)^{1/2}\, \frac{k^2 \exp(-i\pi/5)}{(1 + ib/2k)^{1/2}}$$

$$\times H^{(2)}_{2/5}\left(2k^{5/2}\exp(-i\pi/4)\,/5a^{1/2} + \frac{k^{3/2}}{3}\exp(i\pi/4)\,(b/a^{1/2})\right)$$

$$\times \left[1 + O(\lambda^{5/2})\right]. \qquad (10.3.15)$$

For $b = 0$, this formula is identical with the result obtained in Section 10.2 above.

The total probability of detachment is obtained by integrating the momentum distribution. Thus, for large $\lambda$,

$$P_i \approx 1 . \tag{10.3.16}$$

For large $|\lambda|$, $\lambda < 0$,

$$P_i = (\pi^{1/2}/4) \frac{1}{|\lambda|^{5/4}} \exp(-8|\lambda|^{5/2}/15) . \tag{10.3.17}$$

Also, for small $\lambda$,

$$P_i = \frac{\sin(\pi/5)}{\sin(2\pi/5)} + \lambda \frac{dP_i}{d\lambda}\bigg|_{\lambda=0} , \tag{10.3.18}$$

where

$$\frac{dP_i}{d\lambda}\bigg|_{\lambda=0} \approx \frac{5^{3/5}}{3.2^{9/10}} \Gamma(8/5) \approx 0.42.$$

An analysis carried out in reference [168] shows that the method discussed here is not sufficient to produce correct expressions for terms of the next order in $1/\lambda$ in the asymptotic expansion. In particular, the extension of the expansion beyond the first two terms is required to confirm equation (8.2.19), which now takes the form

$$P_\alpha = \frac{\pi}{16\lambda^{5/2}} . \tag{10.3.19}$$

The mathematical difficulties arising here are substantial and have not yet been resolved. Therefore comparison of the approximate expressions (10.3.11), (10.3.12), (10.3.15-19) with the results of numerical calculations acquire particular significance. Figure 10.4 presents the total probability of electron detachment $P_i$ and the probability $P_\alpha$ that the electron remains in the bound state. Results of numerical calculations are given by solid curves, whereas those obtained from various approximate expressions are shown, in the regions of $\lambda$ where they are applicable, by broken curves. Figure 10.5 presents the momentum distributions of the emitted electrons computed for several values of $\lambda$. The solid curves were obtained by numerical integration. The dotted curves were obtained from an asymptotic expression for $F(k)$, that is,

$$F(k) = \frac{k}{\{a(-ik+b)\}^{1/4}} \exp\left\{ i\left[ 2(-ik+b)^{5/2}/5a^{1/2} - \frac{2}{3}(b/a^{1/2})(-ik+b)^{3/2} + \pi/4 \right] \right\} . \tag{10.3.20}$$

This formula has been derived using the expression (10.3.1) for

$Y_1(-ik)$ for all k. The expression (10.3.20) gives an approximate distribution which differs from the exact distribution by 10 percent or less, except the case of $\lambda \sim 0$, where the agreement is somewhat worse. The probability of detachment and that of recapture into the bound state computed from equation (10.3.20) are shown in Fig. 10.4 by dotted curves. The maximum error is again at $\lambda \sim 0$ and reaches some 13 percent for $P_i$ (a more detailed discussion is given in reference [168]).

Devdariani [156] discussed some questions related to the application of the quadratic approximation to some experimentally observed reactions where electron detachment takes place. Firstly, we have to integrate the theoretical probabilities over all values of the impact parameter. In this way, we find that the theoretical total cross section for detachment of the electron, measured in units of $\pi R_c^2$, is a universal function depending on a single parameter of the problem, that is, $\lambda_c = (E_c'')^{2/5} R_c^{6/5} 2^{-4/5} v^{-2/5}$. For this function, convenient analytical approximations have been suggested and a numerical calculation has been carried out.

Calculations of the total cross sections for detachment using quasi-stationary terms of the quasi-molecule were performed by Demkov et al. [69]. The computed cross section decreases, with increasing collision velocity, which is a natural result in view of

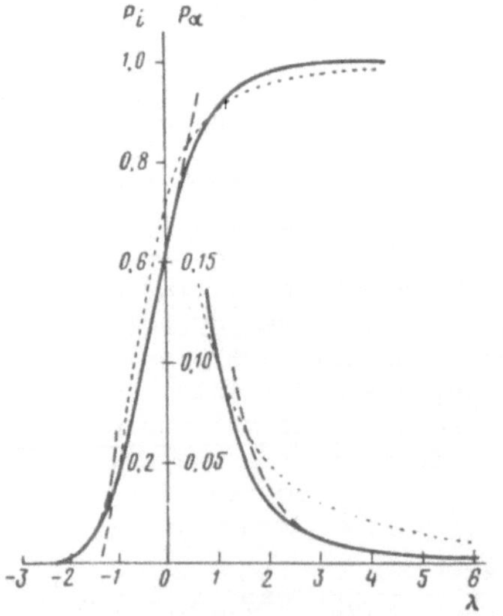


Fig. 10.4. Probability of detachment, $P_i$, and probability of remaining in the bound state, $P_\alpha$, computed in the quadratic approximation for $\alpha(t)$ (see reference [168]). Solid line: numerical integration; broken line: approximate formulae (10.3.17-19); dotted line: approximate expression (10.3.20).

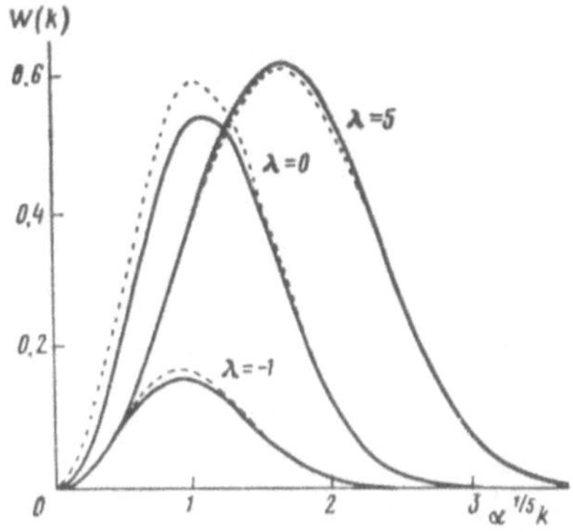


Fig. 10.5. The momentum distribution W(k) of the ejected electrons for several values of the parameter λ [168]. Solid line: numerical integration; dotted line: approximate formula (10.3.20).

the assumed mechanism of detachment. In the quadratic approximation, creation of a quasi-stationary state is not taken into account whereas underbarrier detachment, which may occur before the term enters the continuum, is well described. In this theory, the total cross section for detachment is an increasing function of the relative velocity of colliding particles. This behavior of the cross section is related, in part, to the fact [156] that, in the intermediate (contact) case, the probability of detachment happens to be higher than 0.50 (it is, in fact, $P_i = 0.62$).

In real processes, both effects described above play a certain role, and the resulting behavior of the cross sections depends on the exact balance between these competing phenomena. It appears that one or another factor could dominate, depending on the specific choice of the pair of interacting particles. Indeed, calculations [69] for the pair $H^- + H$ are in good agreement with experiment, which shows the cross section decreasing as velocity increases. On the other hand, the quadratic approximation gives a very satisfactory description of the increase in the experimentally observed cross sections for detachment for the pairs $H^- + He$ and $H^- + Na$. In the latter case, the parameters of the model were selected by comparison with experiment. For instance, in the case of the $H^- + He$ pair, the values $R_c = 1.44$ and $\kappa'(R_c) = 0.2$ were chosen. The first value is in good agreement with $R_c = 1.5$ found from analysis of the molecular terms [156]. It is of considerable interest to perform, in the future, a sufficiently full and rigorous calculation of detachment, taking into account both underbarrier transitions as well as the formation of a quasi-stationary state.

# Chapter 11

# TIME-INDEPENDENT QUANTUM MECHANICAL PROBLEMS

## 11.1 ACCOUNT OF THE QUANTAL MOTION OF THE NUCLEI IN DETACHMENT THEORY

The Hamiltonian of the problems considered in Chapters 8-10 depended explicitly upon time. In other words, there was an external parameter of the system whose dependence upon time was fixed and did not alter during the course of the collision. In reactions where colliding systems are ions and atoms, the nuclear separation R is such a parameter. If the energy of the nuclear motion is low and comparable with that of the electronic transitions, R must be included in the list of the dynamical variables of the system, i.e., the nuclear motion must be treated quantally. This approach is definitely required if, for instance, we deal with reactions near thresholds.

Thus we must consider, under certain conditions, time-independent problems with more dynamical variables, i.e., problems with three or more particles or, after separating the center of mass, problems with two or more particles with reduced masses. A problem of two interacting particles moving in an external field is a special case of the same class of problems. The motion of such general systems has been discussed in the monograph by Baz et al. (19) and also more recently, by Drukarev (107). Therefore, we shall not consider these general problems further here.

Instead, we shall concentrate on the few cases where the solution can be found giving an expression in closed form. It will be possible to view these cases as a natural generalization of the time-dependent problems already considered in Chapters 8 and 9.

We shall start with the detachment problem solved in Section 8.2 where we considered the linear approximation in t to the parameter $\alpha(t)$ in the boundary conditions. Results obtained there had limited application because of the slow dependence of the energy of the emitted electrons on the relative velocity of the colliding particles, $E \sim V^{1/3}$. Indeed, the validity criterion for the ZRP approximation requires the average de Broglie wave of the emitted electron to be appreciably larger than the characteristic size of the colliding particles. This limits the kinetic energy of nuclear motion to a maximum of a few tens of eV. However, these energies are so low that the quantal treatment of the nuclear motion becomes essential.

In the present theory, the nuclear motion will be described, as before, by one effective coordinate, the internuclear separation R. The kinetic energy of the nuclei, $T_n$, will have the form

$$T_n = -\frac{1}{2M}\frac{\partial^2}{\partial R^2},$$

where M is the reduced mass of the nuclei. Introduction of only one relative coordinate does not mean that we shall consider head-on collisions only. It is more accurate to say that we will assume that the region $\delta R$ where we are considering the motion of the nuclear wave packet must be small in comparison with the critical nuclear separation, $R_c$. Then, on account of spherical symmetry, we may assume, to a sufficient accuracy, that the other two coordinates describing the motion of the system in a plane normal to the internuclear axis correspond to free motion so that the components of the momentum conjugate to the two coordinates are conserved. Within this model, for instance, elastic scattering of the nucleus by a sphere of radius $R = R_c$ will be through the mirror angle. The energy loss by the nuclei due to the electron detachment will lead to refraction or reflection through different angles, with the energy required for the detachment being drawn only from the kinetic energy of the radial motion of the nuclei. For the peripheral collision (the wave packet incident on the sphere of radius $R_c$ at a small angle), the energy of the radial motion will not be sufficient to produce electron detachment, and the simple elastic reflection will take place. The curvature of the surface $R = R_c$ must be allowed for, in this model, only if the de Broglie wavelength associated with the nuclear motion is comparable with the distance $R_c$. However, even in the latter case, spherical symmetry can be used, and this problem again reduces to a one-dimensional one by considering motion with a fixed value of the angular momentum $\ell$.

We shall assume that the wavefunction of a weakly bound electron is spherically symmetrical, and we shall introduce a ZRP to approximate the effective potential for the electron. The depth of the ZRP will be a function of R rather than a function of t, as was explicitly assumed in earlier chapters.

The Schrödinger equation takes the form

$$\left\{-\frac{1}{2M}\frac{\partial^2}{\partial R^2} - \frac{1}{2}\frac{\partial^2}{\partial r^2} - E_{AB}(R)\right\}\Psi(r,R) = E\Psi(r,R) \qquad (11.1.1)$$

with the boundary condition

$$\left.\frac{\partial\Psi}{\partial r}\right|_{r=0} = -\alpha(R)\,\Psi(0,R). \qquad (11.1.2)$$

In equation (11.1.1), $E_{AB}(R)$ is the molecular term which determines the continuum boundary for the corresponding molecular ion. The molecular ion term has the form

$$E_{AB^-}(R) = E_{AB}(R) - \alpha^2(R)/2 .$$

As before, the region of R where $\alpha(R) > 0$ is associated with a bound state of the electron and the system AB, whose binding energy is $\alpha^2/2$. The case of $\alpha(R) < 0$ corresponds to a shallow potential well where no bound state exists. We shall consider the motion of the system in the vicinity of $R = R_c$, where $R_c$ is the critical value of the separation, at which $\alpha(R_c) = 0$.

In this region, we shall use the linear approximation to $\alpha$ by writing $\alpha(R) = -\tilde{\beta}(R - R_c)$. For $\tilde{\beta} > 0$, we obtain the usual picture: for $R >> R_c$, the bound state exists, and for $R << R_c$ it disappears. In the vicinity of the critical point $R_c$, it would have been natural to use a similar linear approximation also to the molecular term $E_{AB}(R)$ itself. However, it would lead to considerable mathematical difficulties, and we shall not pursue this course. Instead, we shall follow Devdariani and Demkov [169] and consider the horizontal boundary of the continuum spectrum. In the latter case, we can assume that $E_{AB}(R) = 0$ (some qualitative comments regarding the more general case may be found at the end of this section). In view of this, the present formulation of the problem is applicable to collisions where the influence of acceleration of the atoms on the detachment process can be ignored while the system moves along the molecular term. The passage of the term into the continuum usually takes place at large distances R where the internuclear forces are still small, and this condition justifies the approach described above.

Let us introduce the new coordinates, x and y, as follows:

$$x = \sqrt{M}\,(R_c - R) ,$$

$$y = r .$$

Due to the large factor $\sqrt{M} >> 1$, we can assume that x varies within the infinite limits, $-\infty < x < +\infty$; the case $x \to -\infty$ corresponds to large separations, whereas that of $x \to +\infty$ corresponds to small separations R after detachment of the electron has taken place.* The domain for y is $0 < y < +\infty$.

Introducing the notation $P^2 = 2E$, we obtain a two-dimensional diffraction problem on the half-plane $y > 0$ thus:

$$\left\{\frac{\partial^2}{\partial x^2} + \frac{\partial^2}{\partial y^2} + P^2\right\} \Psi(x,y) = 0 , \tag{11.1.3}$$

with a mixed boundary condition linear in x:

$$\left(\frac{1}{\Psi}\frac{\partial\Psi}{\partial y}\right)_{y=0} = \beta x , \tag{11.1.4}$$

where $\beta = \tilde{\beta} / \sqrt{M} > 0$.

Let us consider, instead of equation (11.1.4), the condition

$$\left(\frac{1}{\Psi}\frac{\partial\Psi}{\partial y}\right)_{y=0} = -\alpha_0, \quad \alpha_0 > 0 . \tag{11.1.5}$$

Then, for $P^2 > -\alpha_0^2$, the solution of equation (11.1.3) can be written as follows:

$$\Psi_0(x,y) = \exp\left(-\alpha_0 y \pm i\sqrt{P^2 + \alpha_0^2}\; x\right) . \tag{11.1.6}$$

We can see that the solution $\Psi_0$ has the form of a surface wave propagating along the x-axis in either direction. The amplitude of this wave falls off exponentially with y increasing, i.e., when moving away from the surface $y = 0$.

In view of the properties of the solution (11.1.6) corresponding to the boundary condition (11.1.5), we may expect that for $x \to -\infty$ and small y, the solution of equation (11.1.3) with the original boundary condition (11.1.4) will be close to a surface wave decreasing as $\exp(\beta xy)$ with y increasing. In order to obtain a complete solution, this decreasing factor must be multiplied by a semi-classical wave propagating in either direction along the x-axis and having its wavenumber specified by energy conservation.

---

*For real systems, R cannot be negative. The case $x \to +\infty$ corresponds to the situation where the two nuclei are still approaching each other after the electron has, with certainty, been lost by the system. This type of motion occurs in the presence of a simple potential field and is beyond our immediate interest.

For $x \to +\infty$, we obtain

$$\Psi(x,y) = \frac{(2\beta x)^{1/2}}{(P^2 + \beta^2 x^2)^{1/4}} \left\{ \exp\left( \beta xy + i \int_0^x \sqrt{P^2 + \beta^2 x'^2}\, dx' \right) + a_{oo'} \exp\left( \beta xy - i \int_0^x \sqrt{P^2 + \beta^2 x'^2}\, dx' \right) \right\} . \quad (11.1.7)$$

In equation (11.1.7), the numerator of the pre-exponential fraction is the normalization constant for the wavefunction of the bound state of the electron, $\exp(-|\beta x|\, y)$, whereas the denominator of the fraction is the normalization constant of a plane wave. The first term in the asymptotic formula (11.1.7) is the incident wave which describes the nuclei being on a collision course with each other and having the electron in a bound state. The second term is the reflected wave. The latter term is at variance with the classical description of the nuclear motion because in the present theory there exists a nonvanishing probability $|a_{oo'}|^2$ that, at the final stage of collision when nuclei move away from each other, the electron will remain in the bound state. This corresponds to an elastic reflection from the sphere $R = R_c$. The schematic position of the terms and the continuum boundary is shown in Fig. 11.1. This problem has one discrete channel (a bound state of the electron) and an infinite number of continuum channels corresponding to electron detachment with an arbitrary distribution of the energy between the nuclei and the electron. The nuclei after detachment are allowed to move along the term either to the right, when they are

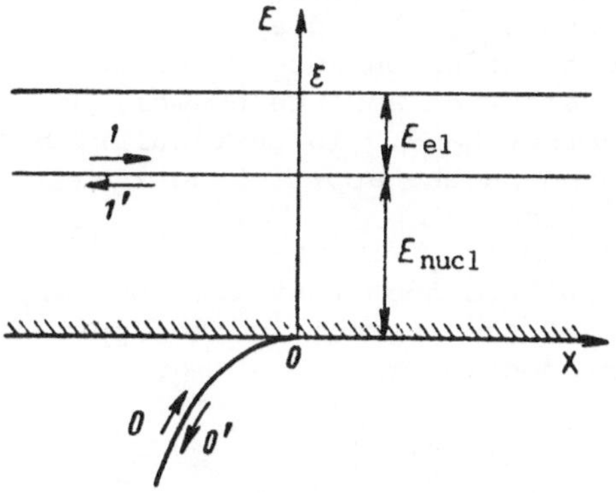


Fig. 11.1. A system of molecular terms for the simplest stationary problem with a horizontal boundary between the discrete and continuous spectra. Arrows indicate the direction of the nuclear motion. Numbers denote the reaction channels: 0 and 0' - motion along the initial term of the negative molecular ion; 1 and 1' - detachment of electron with energy $E_{el}$ and motion of the nuclei along the horizontal molecular term with energy $E_{nucl}$.

still approaching each other, or to the left, having undergone a reflection from the sphere of critical radius $R = R_c$. Thus, in the case where both x and y are large, the principal term in the asymptotic form of the wavefunction is that corresponding to electron detachment. It can be written as follows:

$$\psi(x,y) \approx \int_0^P \sqrt{2/\pi}\, \frac{\sin ky\, dk}{(P^2 - k^2)^{1/4}}$$

$$\times \left\{ a_{01'}(k)\, e^{-i(P^2 - k^2)^{1/2}x} + a_{02'}(k)\, e^{ix(P^2 - k^2)^{1/2}} \right\}. \quad (11.1.8)$$

In equation (11.1.8), the probability amplitudes $a_{01'}(k)$ and $a_{02'}(k)$ correspond to electron transitions to the continuum with final momentum k, where the nuclei after detachment move to the right (i.e., approaching each other) or to the left (i.e., departing from each other), respectively.

Another form of the asymptotic solution (11.1.8) can be obtained by introducing the polar coordinates $\rho$, $\theta$ in the (x,y) plane, that is, putting $x = \rho\cos\theta$ and $y = \rho\sin\theta$. Then we obtain the following asymptotic expression from (11.1.8), for $\rho \to \infty$,

$$\Psi(\rho,\theta) \sim f(\theta)\, \frac{\exp(iP\rho)}{\sqrt{P\rho}}, \quad 0 < \theta < \pi - \varepsilon, \quad \varepsilon > 0. \quad (11.1.9)$$

In equation (11.1.9), the quantity $f(\theta)$ is the scattering amplitude for the two-dimensional diffraction problem, which can be expressed in terms of the amplitudes $a_{01'}(k)$ and $a_{02'}(k)$. In this polar representation, there exists a correspondence between the distribution of energy shared by the electron and the nuclei, on the one hand, and the angle $\theta$, on the other hand. In particular, $\theta < \pi/2$ corresponds to the nuclei remaining on the approaching course, whereas $\theta > \pi/2$ corresponds to reflection.

It is not difficult to show that the boundary condition (11.1.4) ensures conservation of the current in the upper half-plane. Indeed, at y = 0, the current density is

$$j_y = -i \left( \frac{\partial\Psi}{\partial y}\Psi^* - \Psi\frac{\partial\Psi^*}{\partial y} \right)_{y=0} = 0 .$$

This result follows because $\beta$ or, in the general case, $\alpha(R)$ are real. Therefore the current flows, near the y-axis, parallel to this axis. Within the framework of diffraction theory, this problem describes the detachment of a surface wave from the boundary when the boundary condition changes smoothly.

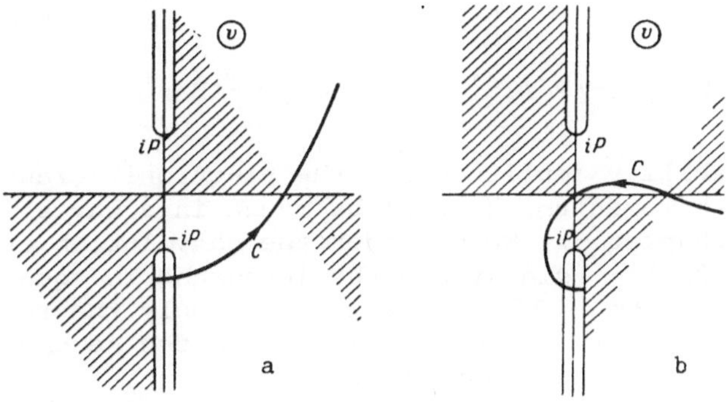


Fig. 11.2. Integration contour for equation (11.1.10) on the two-sheet complex plane v [169]. (a) - first sheet; (b) - second sheet; The cuts are shown. The modulus of the exponential integrand is greater than unity in the hatched areas of the plane.

Let us write the general solution of the problem (11.1.3) with the boundary condition (11.1.4) in the form of a contour integral,

$$\Psi(x,y) = \int_C \exp\left(-vy + i\sqrt{P^2 + v^2}\,x\right) Z(v)\,dv\,. \qquad (11.1.10)$$

Substituting equation (11.1.10) into the boundary condition (11.1.4) and integrating by parts, we find that the boundary condition is satisfied if

$$\frac{\beta}{iv}\sqrt{P^2 + v^2}\;Z(v)\exp\left(i\sqrt{P^2 + v^2}\,x\right)\Big|_{\substack{C\\v}} = 0\,, \qquad (11.1.11)$$

where

$$Z(v) = g\,\frac{v}{\sqrt{P^2 + v^2}}\,\exp\left(\frac{i}{\beta}\int_{v_0}\frac{v'^2\,dv'}{\sqrt{P^2 + v'^2}}\right). \qquad (11.1.12)$$

In equation (11.1.12), g is a normalization constant. The integrand in equation (11.1.10) has a saddle point on each of the sheets of the complex plane v. Calculations of the contribution at the point on the first sheet ( $\sqrt{v^2} = v$ ) gives the incident wave and that on the second sheet gives the reflected wave. Therefore, the asymptotic condition (11.1.7) can be satisfied by a contour situated on both sheets as is shown in Fig. 11.2. The cuts shown there run from $\pm\, iP$ to infinity, and the hatched areas show where the function Z(v) of the integrand grows exponentially so that condition (11.1.11) is fulfilled.

We now calculate the integral (11.1.10), as $x \to \infty$, along the contour C shown above using the method of steepest descent. Then we obtain equation (11.1.7), where g must be taken as $(\pi\beta)^{-\frac{1}{2}}\exp(-i\pi/4)$

and, finally,

$$a_{00'} = -i\exp\{-\pi P^2/2\beta\} = -i\exp\{-\pi E/\beta\}. \qquad (11.1.13)$$

The probability amplitude for the electronic transitions to a free state with momentum $k$ will be found in a way similar to that described in Chapter 3. We consider the convolution of the continuous function $(2/\pi)^{1/2}\sin ky$ and the integrand in equation (11.1.10), assuming that $x \to +\infty$. If we let $x \to -\infty$, only the residues at the poles of the imaginary $v < 0$ semi-axis on the second sheet will contribute, and we shall find that

$$a_{01'}(k) = \sqrt{2/\beta}\,\frac{k}{(P^2-k^2)^{1/4}} \times \exp\left\{\frac{P^2}{\beta}\left[-\frac{\pi}{2} + \int_0^{k/P}\frac{x^2\,dx}{\sqrt{1-x^2}}\right] + i\frac{\pi}{4}\right\}, \qquad (11.1.14)$$

where the integral in the exponent is of a standard type.

In order to obtain the second amplitude, $a_{02'}(k)$, we have to let $x \to +\infty$. Then only the first sheet will give a contribution and we obtain

$$a_{02'}(k) = \sqrt{2/\beta}\,\frac{k}{(P^2-k^2)^{1/4}} \times \exp\left\{-\frac{P^2}{\beta}\int_0^{k/P}\frac{x^2\,dx}{\sqrt{1-x^2}} + i\frac{\pi}{4}\right\}. \qquad (11.1.15)$$

The probabilities of free-free and free-bound state transitions were also found in reference [169]. However, these probabilities can be obtained from a more general theory which will be developed in the next section and, therefore, they are not considered here.

We now introduce arbitrary units in order to determine what the characteristic dimensionless parameters of the problem are. The binding energy of the electron as a function of x, or R, is (for $x < 0$) thus:

$$E(R) = \frac{\tilde{\beta}^2 x^2}{2} = \frac{M\tilde{\beta}^2 (R-R_c)^2}{2}.$$

Differentiating this equation twice with respect to $R_c$, we obtain

$$E''(R_c) = \tilde{\beta}^2 M, \quad \tilde{\beta} = E''(R_c)^{1/2} M^{1/2}.$$

The total energy of the colliding system, measured relative to the merging point of the term with the continuum boundary (that is, with respect to the threshold of the detachment reaction), is $E = MV_o^2/2 = P^2/2$, where $V_o$ is the relative velocity of the nuclei at $R_c$. Therefore, an effective dimensionless parameter entering into this result is

$$\varepsilon = P^2/\tilde{\beta} = \frac{2E\,M^{1/2}}{E''(R_c)^{1/2}} = \frac{M^{3/2}\,V_o^2}{E''(R_c)^{1/2}},$$

and the characteristic energy of the colliding particles which corresponds to $\varepsilon = 1$ is $E_c = E''(R_c)^{1/2}/2M^{1/2}$.

We proceed further by introducing a dimensionless parameter $\tau$, $0 < \tau < 1$, which measures, in relative units, the energy transmitted from the nuclei to the electron, during the course of the reaction, that is,

$$\tau = E_{el}/E_{nucl} = k^2/P^2 .$$

Writing the total probability that the electron goes to a continuum state with energy $k^2/2$, in terms of these dimensionless variables, gives

$$W(\tau,\varepsilon) = \left\{ |a_{01'}|^2 + |a_{02'}|^2 \right\} \frac{dk}{d\tau}$$

$$= \frac{2\varepsilon\sqrt{\tau}\,\exp(-\pi\varepsilon/2)}{\sqrt{1-\tau}} \cosh\left[\varepsilon \arccos\sqrt{\tau} + \varepsilon\sqrt{\tau(1-\tau)}\right]. \quad (11.1.16)$$

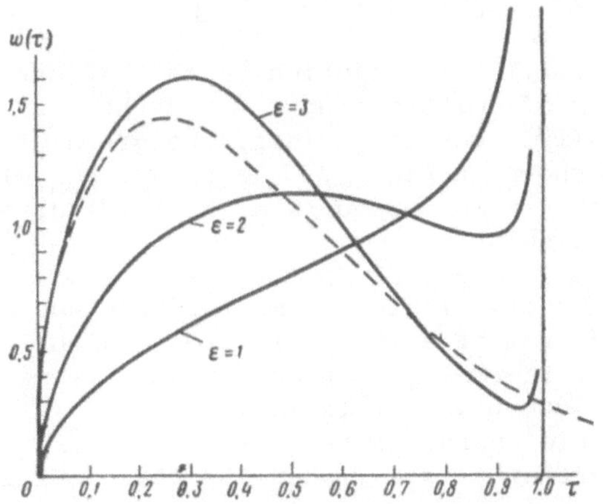


Fig. 11.3. The energy distribution of electrons, $W(\tau,\varepsilon)$, expressed in dimensionless units [169]. Solid lines: quantal motion of the nuclei for several values of the energy parameter $\varepsilon$. Broken line: classical motion of the nuclei for $\varepsilon = 3$.

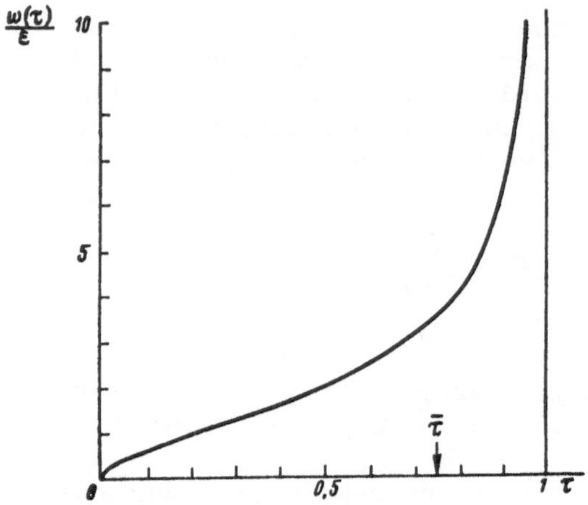


Fig. 11.4. The energy distribution of the electrons in the limiting case where the energy parameter $\varepsilon$ is small (including the quantal motion of the nuclei).

In Figure 11.3, the probability $W(\tau,\varepsilon)$ is plotted as a function of $\tau$ for several values of $\varepsilon$. A weak singularity in $W(\tau,\varepsilon)$ at $\tau = 1$ does not correspond to any physical resonance and arises as a "kinematic" effect.*

For $\tau \ll 1$, formula (11.1.6) becomes

$$W(\tau,\varepsilon) = 2\varepsilon\sqrt{\tau}\ \exp(-\pi\varepsilon/2)\ \cosh\left[\varepsilon(\pi/2 - (2/3)\ \tau^{3/2}\right].$$

The latter expression is reduced, at $\varepsilon \gg 1$, to

$$W(\tau,\varepsilon) = \varepsilon\sqrt{\tau}\ \exp\left[-(2/3)\ \varepsilon\,\tau^{3/2}\right]. \qquad (11.1.17)$$

The distribution (11.1.17) is identical with that derived in Section 8.2 for the spectrum of emitted electrons if the motion of the nuclei is treated classically. For comparison, the graph of this function for $\varepsilon = 3$ is also shown in Fig. 11.3 (a broken curve). It is clear that, for $\varepsilon$ as low as 3, equation (11.1.17) already gives quite satisfactory results.

We note that the classical treatment of the nuclear motion gives an energy spectrum of emitted electrons which is not bound at the high energy end. In real situations, however, energy conservation must be taken into account, as it is in equation (11.1.16) where $\tau < 1$. If the kinetic energy of the colliding nuclei is much greater than the electron affinity (i.e., when the classical treatment of the

---

* This singularity arises as a result of the transformation from a distribution in terms of $\theta$ (see equation (11.1.9)) to that in terms of $\tau$.

nuclei is justified), the probability that a large amount of energy is transferred from the nuclei to the electron is exponentially small and may be neglected.

If $\varepsilon$ is large, the average energy of the electrons is small in comparison with the kinetic energy of the nuclei (this has already been pointed out in Section 8.2). In this case, the average value of $\tau$ can be immediately obtained:

$$\bar{\tau} = \varepsilon \int_0^\infty \tau^{3/2} \exp(-\tfrac{2}{3}\varepsilon\,\tau^{3/2})\, d\tau$$

$$= (2/3)^{1/3}\,\Gamma(2/3)\,\varepsilon^{-2/3} = \frac{1.183}{\varepsilon^{2/3}},$$

or, returning to the original variables, $\bar{E}_{el} = 1.183\, E_c^{2/3}\, \mathcal{E}^{1/3}$.

If $\varepsilon$ is small, the probability that reflection takes place without detachment increases ($w \approx 1 - \pi\varepsilon$). In this extreme case, the distribution of electrons with respect to $\tau$ takes the form

$$W(\tau) = 2\,\varepsilon \sqrt{\tau} \,/\, \sqrt{1 - \tau}\,. \qquad (11.1.18)$$

This distribution is shown in Fig. 11.4. It is easy to check that the normalization condition is satisfied by the probability expression (11.1.18) thus:

$$\int_0^1 W(\tau)\, d\tau = 2\,\varepsilon \int_0^1 \frac{\sqrt{\tau}\, d\tau}{\sqrt{1-\tau}} = \pi\,\varepsilon\,.$$

If the elastic reflection is neglected, the average value of the energy transmission coefficient becomes

$$\bar{\tau} = \frac{1}{\pi\varepsilon} \int_0^1 \tau\, W(\tau)\, d\tau = \frac{3}{4}\,.$$

This result indicates that if the kinetic energy of the colliding atoms is near to the threshold, the dominant part of the energy in the nonelastic reaction will be transferred to the electron. The situation created in this case will be opposite to that of fast colliding atoms. This result is rather general and does not depend on the specific model used in the discussion. Finally, we shall point out that similar conclusions can be derived, in a qualitative way, from Fig. 11.4 for the energy distribution of the emitted electrons where the energy parameter $\varepsilon$ is small.

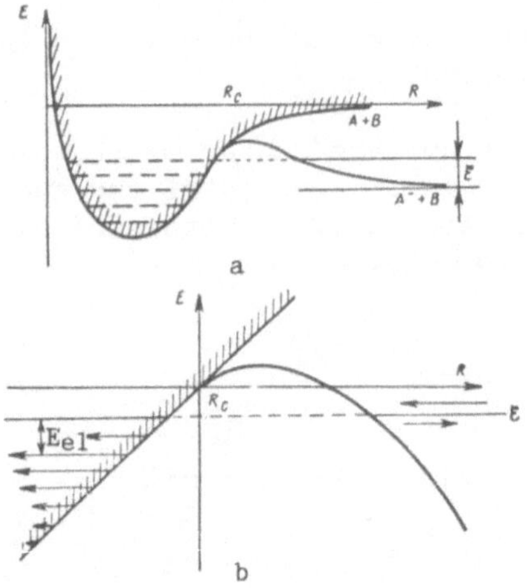


Fig. 11.5. Molecular terms for the case where the term of the molecule AB corresponds to attraction between the nuclei, and the term of the negative molecular ion $AB^-$ corresponds to repulsion between the nuclei. (a) - real position of the molecular terms; (b) - their schematization.

It is of considerable interest to investigate the general case where the continuum boundary $E_{AB}(R)$ is not horizontal in the neighborhood of the critical separation $R = R_c$. The mathematical difficulties increase significantly, which reflects the variety of physical situations arising in this case. Figure 11.5 shows an attractive molecular term AB and a repulsive molecular ion term $AB^-$. If the energy of the colliding system lies below the activation barrier, detachment of electron via a tunnel transition accompanied by the formation of the molecule will take place. If the energy lies above the barrier, the process will gradually go into that considered above or into one corresponding to the absence of the barrier.

The case where both terms AB and AB are repulsive is also possible (see Fig. 11.6). Then electron detachment followed by disassociation of the quasi-molecule will take place. As before, we have two choices. If the energy of the colliding system lies below the activation energy, the process will occur by tunneling through the barrier, whereas in the case where the energy lies above the barrier, the critical separation $R = R_c$ will be reached in the classical motion of the nuclei.

The cases discussed above exhaust the possible processes which can take place during the course of simple merging of the bound state with the boundary of the continuum spectrum of the quantal system.

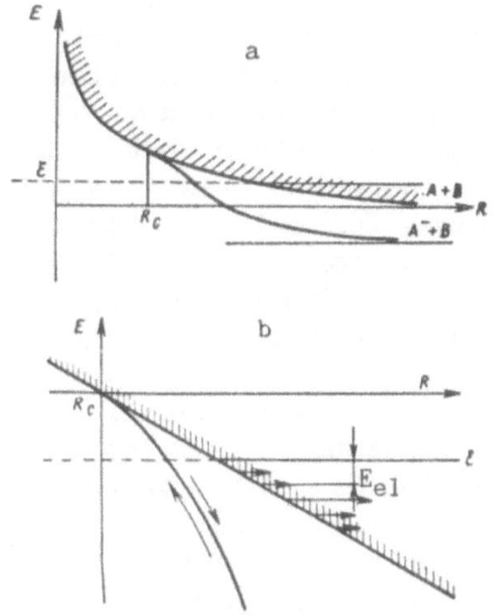


Fig. 11.6. Molecular terms for the case where both the term of the molecule AB and that of the negative molecular ion $AB^-$ correspond to repulsion between the nuclei. (a) - real position of the molecular terms; (b) - their schematization.

## 11.2 TIME-INDEPENDENT QUANTUM MECHANICAL PROBLEMS SOLVABLE BY CONTOUR INTEGRATION

Let us now consider a class of time-independent problems which can be exactly solved by contour integration and which are analogous to the time-dependent problems discussed in Chapter 9. We shall write X to denote an additional degree of freedom associated with the motion of the nuclei, and M for the corresponding reduced mass. Retaining all other assumptions of Section 9.1 about the Hamiltonian H with a separable potential, without change we obtain the following stationary problem [162, 170]:

$$\left\{-\frac{1}{2M}\frac{\partial^2}{\partial X^2} + H_o + |\phi\rangle BX \langle\phi|\right\}|\psi\rangle = \mathcal{E}|\psi\rangle , \qquad (11.2.1)$$

where $\mathcal{E}$ is the total energy of the system, $H_o$ is the Hamiltonian of the active electron which does not depend on X, and B > 0. The operator $|\phi\rangle\langle\phi|$ is the projection operator onto an electronic state $|\phi\rangle$.

We first consider the case where the Hamiltonian $H_o$ has only a discrete spectrum. Then the wavefunction $|\psi\rangle$ in (11.2.1) can be represented as $X \to \pm\infty$ in terms of plane waves (in X) multiplied by the wavefunctions of the asymptotic electronic states $|n\rangle$ defined in Section 9.1. All channels with energy $\lambda_n$ exceeding the total energy $\mathcal{E}$ will be closed. We assume that there are reaction channels corresponding to the reflected and transmitted waves. We shall denote these two waves by n $(X \to -\infty)$ and n' $(X \to +\infty)$. We shall further assume that the initial plane wave is incident from the region of negative X. Then the channel 0' is always closed: reflection will occur, with certainty, for sufficiently large $\underline{X}$ if the motion takes place along a non-horizontal term. Figure 9.1, for instance, corresponds to the case where channels 0, 1, 1', 2, and 2' are open and channels 3, 3', and 0' are closed.

Our task is to determine the probabilities of transitions between the open channels. Making use of the method of contour integration discussed in Section 9.1, the solution can be obtained in the following form:

$$|\psi\rangle = N \int_C G(E)\,|\phi\rangle\, Z(E)\, \exp\left( iB^{-1} \int^E Z(E')\, dE' + iKX \right) dE, \qquad (11.2.2)$$

where

$$K = \sqrt{2M(\mathcal{E} - E)}$$

and

$$Z(E) = \langle\phi|G(E)|\phi\rangle^{-1} \frac{dK}{dE} .$$

For the transition to the classical limit of the motion along X, if $\mathcal{E} \gg E$, we have to put

$$X = V_o t, \qquad K \approx (2M\mathcal{E})^{1/2} - EV_o^{-1}, \qquad \mathcal{E} = MV_o^2/2,$$

$$\frac{dP}{dE} \approx V_o^{-1}, \qquad \beta = BV_o .$$

Then equation (11.2.2) reduces to equation (9.1.4). The function Z(E) in the integrand of equation (11.2.2) has poles at the same points $\lambda_n$ as in the case of equation (9.1.4), and an additional branch point $E = \mathcal{E}$ which corresponds to reflection when the system moves along the zero term.

The choice of the integration contour C in equation (11.2.2) and investigation of the asymptotic properties of the wavefunction $|\psi\rangle$ are carried out using the same technique as that described for the time-dependent problem. The residues of Z(E) at $\lambda_n$ which determine the probability of the elementary processes are

$$h_n^2 \left(\frac{dK}{dE}\right)_{\lambda_n} = h_n^2 V_n^{-1}, \qquad V_n = \left[2(\mathcal{E} - \lambda_n)/M\right]^{-1/2} .$$

Introducing the notation

$$p_n = \exp\left( -2\pi h_n^2 v_n^{-1} B^{-1} \right), \qquad (11.2.3)$$

we obtain the same expressions (9.1.15) and (9.1.18) for the probabilities of transitions between open channels n and m' in the presence of nuclear motion. Instead of equation (9.1.16), we now have

$$|S_{oo'}|^2 = 0,$$

and for the probabilities of transitions with reflection, we find

$$\begin{aligned} |S_{nm}|^2 &= q_n p_{n+1} \cdots p_{n_o} p_{n_o} p_{n_o-1} \cdots p_{m+1} q_m , \\ |S_{oo}|^2 &= p_1 \cdots p_{n_o} p_{n_o} \cdots p_1 , \quad \lambda_{n_o} < E < \lambda_{n_o+1}, \\ |S_{n'm'}|^2 &= 0. \end{aligned} \qquad (11.2.4)$$

In equation (11.2.4), time-reversal invariance holds so that, for instance, $S_{nm'} = S_{m'n}$ , etc.

The following conclusions can be drawn from the present discussion:

(a) transitions "from left to right" with energy decreasing, and transitions "from right to left" with energy increasing are forbidden,

(b) there will be no reflection if the system moves "from right to left."

Equation (9.1.19) for the expansion of the S-matrix including the reflection at the point $E$ now takes the form

$$S = S_1 S_2 \cdots S_{n_o} S_{n_o} S_{n_o-1} \cdots S_1 . \qquad (11.2.5)$$

Therefore, the properties of the S-matrix established in Section 9.1 remain valid in the present case. The only modifications required are those which take into account (i) the existence of the turning point and (ii) that the velocities $V_n$ have different values at the intersection points of unperturbed terms. If $E \gg E$, all $V_n \approx V_o$ and we can neglect the product $p_1 p_2 p_3 \cdots p_N$ and, on account of $\beta = B V_o$ , we obtain the old expressions (9.1.15), (9.1.16), and (9.1.18).

An example of the application of this theory can be found in [171], where the time-independent Landau-Zener problem with one horizontal term was solved. It was that particular case where the triangular property of the S-matrix, discussed in Section 9.1, was first noticed.

It is easy to generalize the theory developed above to the case where the Hamiltonian $H_o$ has a continuous spectrum. Below we shall express our final results through adiabatic terms in a way similar to that of Section 9.2, where the adiabatic approximation was considered for the case of the time-dependent Hamiltonian.

For given total energy $\mathcal{E}$, we introduce the adiabatic momentum of the nuclei, K(X). First, consider the energy of the adiabatic term E(X) which is measured relative to the continuum boundary assumed to be horizontal. Then we shall define K(X) by

$$\mathcal{E} - \frac{1}{2M} K^2(X) = E(X) . \qquad (11.2.6)$$

The inverse function X(K) has branch points at $K = \pm (2M\mathcal{E})^{1/2}$. For instance, in the simplest case of the problem considered in Section 11.1 we have $E(X) = -\tilde{\beta}^2 X^2/2$; hence $X(K) = [2(K^2/2M - \mathcal{E}) \times \tilde{\beta}^2]^{-1/2}$. The adiabatic terms for the Hamiltonian (11.2.1) are defined in an implicit form by the relation

$$\langle\phi|(H_o - E)^{-1}|\phi\rangle = -\frac{1}{BX} . \qquad (11.2.7)$$

Making use of equation (11.2.2) for the wavefunction $|\psi\rangle$, we can then express the probability amplitude b(K) for a transition of the nuclei into a state with momentum K accompanied by the electron transition into the continuum, by

$$b(K) = \{2\,\mathrm{Im}\,X(K)\}^{1/2} \exp\left( i \int_{(C)\,K_i}^{K} X(K')\,dK' \right). \qquad (11.2.8)$$

In equation (11.2.8), the integration contour C starts at the point $K_i$ on the real positive semi-axis, which corresponds to the initial momentum of the nuclei. It is assumed that the cut in the complex K plane runs along the real axis between the two branch points $K_b$, and that the contour C lies above the cut. Positive and negative K correspond to the motion of the nuclei in the positive and negative directions along X, respectively. The momentum of the nuclei and the energy of electrons emitted into the continuum are related to each other through the conservation law for the total energy. Therefore, the energy distribution for the electrons can be directly obtained from equation (11.2.8). In particular, for the problem considered in Section 11.1, we can derive equations (11.1.14) and (11.1.15).

Equation (11.2.8) is analogous to equation (9.2.4) for the time-dependent problem. It is also not difficult to show that the probability amplitude b(K.K') for free-free transitions is

$$b(K,K') = \{2\,\mathrm{Im}\,X(K)\}^{1/2}\;\{2\,\mathrm{Im}\,X(K')\}^{1/2} \times \exp\left\{ i \int_{K'}^{K} X(P)\,dP \right\} . \qquad (11.2.9)$$

This equation is similar to equation (9.2.11) for the time-dependent case, and it can be used in the theory of energy exchange between slow atoms and electrons.

## REFERENCES


1. E. Fermi, Nuovo Cim., 11, 157 (1934).
2. E. Fermi, Ric. Sci., 7, 13 (1936).
3. G. Breit, Phys. Rev., 71, 215 (1947).
4. G. F. Chew and G. C. Wick, Phys. Rev., 85, 636 (1952).
5. K. A. Brueckner, Phys. Rev., 89, 834 (1953).
6. M. L. Goldberger and F. Seitz, Phys. Rev., 71, 294 (1947).
7. H. A. Bethe, Elementary Nuclear Theory, Wiley, New York (1947).
8. J. M. Blatt and V. F. Weisskopf, Theoretical Nuclear Physics, Wiley, New York (1952).
9. A. I. Akhiezer and I. Ya. Pomeranchuk, Some Aspects of Nuclear Theory [in Russian], Gostekhizdat, Moscow (1950).
10. D. A. Kirzhnits, Applications of Quantum Field Methods in the Theory of Many-Particle Systems [in Russian], Gosatomizdat, Moscow (1963).
11. G. V. Skornyakov and K. A. Ter-Martirosyan, Zh. Eksp. Teor. Fiz., 31, 775 (1956).
12. R. A. Minlos and L. D. Faddeev, Dokl. Akad. Nauk SSSR, 141, 1335 (1961); R. A. Minlos and L. D. Faddeev, Zh. Eksp. Teor. Fiz., 41, 1850 (1961).
13. H. A. Bethe and R. Peierls, Proc. R. Soc. (London), A148, 146 (1935).
14. L. D. Landau and E. M. Lifshits, Zh. Eksp. Teor. Fiz., 18, 750 (1948).
15. H. A. Bethe and C. Longmire, Phys. Rev., 77, 647 (1950).
16. B. H. Armstrong, Phys. Rev., 131, 1132 (1963).
17. B. M. Smirnov and O. B. Firsov, Zh. Eksp. Teor. Fiz., 47, 232 (1964).
18. Yu. N. Demkov, Zh. Eksp. Teor. Fiz., 46, 1126 (1964).
19. A. I. Baz', Ya. B. Zel'dovich, and A. M. Perelomov, Scattering, Reactions, and Decays in Nonrelativistic Quantum Mechanics [in Russian], Nauka, Moscow (1971).
20. K. Huang, Statistical Mechanics, Wiley (1963).
21. E. H. Lieb and D. C. Mattis (eds.), Mathematical Physics in One Dimension: Exactly Soluble Models of Interacting Particles, Academic Press, New York (1966).
22. R. G. Newton, Scattering Theory of Waves and Particles, McGraw-Hill (1966).
23. M. Rotenberg and J. Stein, Phys. Rev., 182, 1 (1969).

24. S. A. Adelman, Phys. Rev., A5, 508 (1972).
25. T. Ohmura and H. Ohmura, Phys. Rev., 118, 154 (1960).
26. B. M. Smirnov, Atomic Collisions and Processes in Plasma [in Russian], Atomizdat, Moscow (1968).
27. C. L. Pekeris, Phys. Rev., 112, 1649 (1958).
28. S. Geltman, Astrophys. J., 136, 935 (1962).
29. Yu. N. Demkov and G. F. Drukarev, Zh. Eksp. Teor. Fiz., 47, 918 (1964).
30. G. Lucovsky, Solid State Commun., 3, 299 (1965).
31. A. A. Berezin, Fiz. Tverd. Tela, 11, 1587 (1969).
32. A. A. Berezin and V. B. Kiri, Fiz. Tverd. Tela, 11, 2118 (1969).
33. B. M. Smirnov, Physics of Weakly Ionized Gases [in Russian], Nauka, Moscow (1971).
34. A. Temkin and J. C. Lamkin, Phys. Rev., 121, 788 (1961); A. Temkin, Phys. Rev., 126, 130 (1962).
35. N. F. Mott and H. S. W. Massey, The Theory of Atomic Collisions, Clarendon Press, Oxford (1965).
36. Y. Yamaguchi, Phys. Rev., 95, 1628 (1954).
37. A. G. Sitenko and V. F. Kharchenko, Preprint ITF-68-11, Kiev (1968); A. G. Sitenko and V. F. Kharchenko, Usp. Fiz. Nauk, 103, 469 (1971).
38. P. Beregi, B. N. Zakhar'ev, and S. A. Niyazgulov, Elem. Chast. At. Yadro, 4, 512 (1973).
39. V. N. Ostrovskii, Adiabatic Treatment of Electron Detachment in Slow Atomic Collisions and the Model of Separable Potentials, Ph.D. Thesis, Leningrad University (1972).
40. P. G. Burke and H. M. Schey, Phys. Rev., 126, 147 (1962).
41. C. Schwartz, Phys. Rev., 124, 1468 (1961).
42. G. F. Drukarev, The Theory of Electron Atom Collisions, Academic Press, New York (1965).
43. P. Swan, Proc. R. Soc. (London), A228, 10 (1955).
44. V. G. Neudachin and V. F. Smirnov, Current Problems in Optics and Nuclear Physics [in Russian], Naukova Dumka, Kiev (1974).
45. A. Herzenberg and H. S. M. Law, J. Phys. B1, 327 (1968).
46. R. K. Peterkop, Zh. Eksp. Teor. Fiz., 54, 1581 (1968).
47. V. N. Ostrovskii, Vestn. Leningrad. Univ., Ser. Fiz., No. 10, 10 (1971).
48. A. G. Sitenko, Lectures on Scattering Theory [in Russian], Vishcha Shkola, Kiev (1971).
49. R. J. Eden and P. V. Landshoff, Phys. Rev., 136, B1817 (1964).
50. Yu. N. Demkov and G. F. Drukarev, Zh. Eksp. Teor. Fiz., 49, 691 (1965).
51. G. F. Drukarev, Zh. Eksp. Teor. Fiz., 21, 59 (1951).
52. H. M. Nussenzveig, Nucl. Phys., 11, 499 (1959).
53. E. M. Ferreira and A. F. Teixeira, J. Math. Phys., 7, 1207 (1966).
54. L. D. Landau and E. M. Lifshitz, Quantum Mechanics, Pergamon Press, Oxford (1965).
55. V. N. Ostrovskii and E. A. Solov'ev, Zh. Eksp. Teor. Fiz., 62, 167 (1972).

56. A. M. Perelomov and V. S. Popov, Zh. Eksp. Teor. Fiz., 61, 1742 (1971).
57. A. B. Migdal, A. M. Perelomov, and V. S. Popov, Yad. Fiz., 14, 874 (1971); ibid., 16, 222 (1972).
58. V. M. Galitsky and V. F. Cheltsov, Nucl. Phys., 56, 86 (1964).
59. A. B. Migdal, Yad. Fiz., 16, 427 (1972).
60. J. Humblet, Mem. Soc. R. Sci. Liege, 12, 1 (1952).
61. P. Beregi, Lett. Nuovo Cim., 2, 233 (1971).
62. Yu. N. Demkov and V. N. Ostrovskii, in: Problems in Theoretical Physics. I. Quantum Mechanics, ed. by M. G. Veselov, Yu. V. Novozhilov, and P. P. Pavinskii [in Russian], Izd. Leningrad. Univ. (1974), pp. 279-293.
63. M. N. Adamov, Yu. N. Demkov, V. D. Ob'edkov, and T. K. Rebane, Teor. Eksp. Khim., 4, 147 (1968).
64. F. I. Dalidchik and G. K. Ivanov, Teor. Eksp. Khim., 8, 9 (1972).
65. S. V. Maleev, Fiz. Tverd. Tela, 7, 2990 (1965).
66. F. I. Dalidchik and G. K. Ivanov, Teor. Eksp. Khim., 7, 147 (1971).
67. L. N. Labzovskii, Opt. Spektrosk., 35, 988 (1973).
68. T. K. Rebane and R. I. Sharibdzhanov, Teor. Eksp. Khim., 10, 444 (1974).
69. Yu. N. Demkov, G. F. Drukarev, and V. V. Kuchinskii, Zh. Eksp. Teor. Fiz., 58, 944 (1970).
70. E. A. Solov'ev, Vestn. Leningrad. Univ., Ser. Fiz., No. 4, 20 (1975).
71. T. K. Rebane and R. I. Sharibdzhanov, Teor. Eksp. Khim., 10, 435 (1974).
72. T. K. Rebane and R. I. Sharibdzhanov, Teor. Eksp. Khim., 11, 291 (1975).
73. R. Subramanyan, Vestn. Leningrad. Univ., Ser. Fiz., No. 10, 10 (1970).
74. R. Subramanyan, Some Applications of Zero-Range Potentials in Quantum Mechanics, Ph.D. Thesis, Leningrad University (1968).
75. H. Nakamura, J. Phys. Soc. Jpn., 35, 848 (1973).
76. J. B. Keller, J. Math. Phys., 2, 262 (1961).
77. I. V. Komarov and D. I. Abramov, Teor. Mat. Fiz., 22, 253 (1975).
78. Yu. N. Demkov and V. S. Rudakov, Zh. Eksp. Teor. Fiz., 59, 2035 (1970).
79. V. N. Ostrovskii, Vestn. Leningrad Univ., Ser. Fiz., No. 4, 7 (1974).
80. Yu. N. Demkov, Variational Principles in the Theory of Collisions, Pergamon Press (1963).
81. J. M. Ziman, Principles of the Theory of Solids, Cambridge Univ. Press (1964).
82. K. H. Johnson, Adv. Quant. Chem., 7, 143 (1973).
83. W. John and P. Ziesche, Phys. Status Solidi B, 47, 555 (1971).
84. W. John, G. Lehman, and P. Ziesche, Phys. Status Solidi B, 53, 287 (1972).
85. W. John and P. Ziesche, Dubna Preprint JINR P4-8086 (1974).

86. V. G. Baryshevskii, V. L. Lyuboshits, and M. I. Podgoretskii, Dubna Preprint JINR P-2111 (1965).
87. I. V. Amirkhanov, V. F. Demin, B. N. Zakhar'ev, and I. I. Kuz'min, Dubna Preprint JINR P-2564 (1965).
88. V. L. Lyuboshits, Zh. Eksp. Teor. Fiz., 52, 926 (1967).
89. R. Subramanyan, Zh. Eksp. Teor. Fiz., 53, 363 (1968).
90. Yu. N. Demkov and V. N. Ostrovskii, Zh. Eksp. Teor. Fiz., 59, 1765 (1970).
91. G. F. Drukarev and I. Yu. Yurova, in: Problems in the Theory of Atomic Collisions, ed. by Yu. N. Demkov [in Russian], Izd. Leningrad. Univ. (1975), pp. 28-41.
92. R. H. Dalitz, Strange Particles and Strong Interactions, Oxford University Press (1962).
93. M. I. Petrashen' and E. D. Trifonov, Applications of Group Theory in Quantum Mechanics [in Russian], Nauka, Moscow (1967).
94. M. Hamermesh, Group Theory and Its Application to Physical Problems, Addison-Wesley (1962).
95. W. Kolos and L. Wolniewicz, J. Chem. Phys., 43, 2429 (1965).
96. F. H. Read, in: The Physics of Electronic and Atomic Collisions. Invited Lectures and Progress Reports, VIII. ICPEAC, Beograd (1974).
97. C. Ramsauer and R. Kollath, Ann. Phys., 4, 91 (1929).
98. D. E. Golden, H. W. Bandel, and J. A. Salerno, Phys. Rev., 146, 40 (1966).
99. R. L. Wilkins and H. S. Taylor, J. Chem. Phys., 47, 3532 (1967).
100. J. C. Tully and R. S. Berry, J. Chem. Phys., 51, 2056 (1969).
101. S. Hara, J. Phys. Soc. Jpn., 27, 1009 (1969).
102. L. A. Edelstein, Nature (London), 182, 932 (1958).
103. D. C. Cartwright and A. Kuppermann, Phys. Rev., 163, 86 (1967).
104. S. P. Khare and B. L. Moiseiwitch, Proc. Phys. Soc. (London), 88, 605 (1956).
105. S.J.B. Corrigan, J. Chem. Phys., 43, 4381 (1965).
106. K. Takayanagy and Y. Itikawa, Adv. At. Mol. Phys., 6, 105 (1970).
107. G. F. Drukarev, Zh. Eksp. Teor. Fiz., 67, 38 (1974).
108. N. F. Lane and S. Geltman, Phys. Rev., 184, 46 (1969).
109. H. Ehrhardt and F. Linder, Phys. Rev. Lett., 21, 419 (1968).
110. I. Yu. Yurova, Resonance Phenomena in Electron Scattering by Molecules, Ph.D. Thesis, Leningrad University (1975).
111. G. F. Drukarev and I. Yu. Yurova, Proc. IX ICPEAC, Seattle (1975), pp. 262-263.
112. J. N. Bardsley, A. Herzenberg, and F. Mandl, Proc. Phys. Soc. (London), 89, 305 (1966); ibid., 321 (1966).
113. Yu. N. Demkov and V. V. Kuchinskii, Opt. Spektrosk., 35, 804 (1973).
114. J.C.Y. Chen and J. L. Peacher, Phys. Rev., 167, 30 (1968).
115. Yu. N. Demkov and R. Subramanyan, Zh. Eksp. Teor. Fiz., 57, 698 (1969).
116. K. V. Bhagwat and R. Subramanyan, Phys. Status Solidi B, 47, 317 (1971).

117. Z. A. Kasamanyan, Zh. Eksp. Teor. Fiz., 61, 1215 (1971).
118. Yu. N. Demkov, V. N. Ostrovskii, and E. A. Solov'ev, Zh. Eksp. Teor. Fiz., 62, 501 (1974).
119. B. K. Vainstein, Diffraction of X-Rays on Long Molecules [in Russian], Izd. Akad. Nauk SSSR, Moscow (1963).
120. V. G. Baryshevskii, V. L. Lyuboshits, and M. I. Podgoretskii, Dubna Preprint JINR P-2230 (1965).
121. Encyclopedic Dictionary of Physics [in Russian], Sov. Éntsiklopediya, Moscow (1965).
122. V. A. Fock, Diffraction and Propagation of Electromagnetic Waves [in Russian], Sov. Radio, Moscow (1970).
123. G. K. Ivanov, Teor. Eksp. Khim., 10, 450 (1974).
124. V. A. Oparin, R. N. Il'in, I. T. Serenkov, and E. S. Solov'ev, Zh. Eksp. Teor. Fiz., 66, 2008 (1974).
125. V. S. Vinogradov, Fiz. Tverd. Tela, 13, 3266 (1971).
126. I. A. Eganova and M. I. Shirokov, Ann. Phys., 21, 225 (1968).
127. I. A. Eganova and M. I. Shirokov, Dubna Preprint JINR P4-5438 (1970); I. A. Eganova, Preprint No. 1, Phys. Inst. Acad. Sci., Azerb. SSR, Baku (1971).
128. Yu. N. Demkov and G. F. Drukarev, Zh. Eksp. Teor. Fiz., 49, 257 (1965).
129. Yu. A. Bychkov, Zh. Eksp. Teor. Fiz., 39, 689 (1960).
130. V. G. Skobov, Zh. Eksp. Teor. Fiz., 37, 1467 (1959).
131. A. M. Ermolaev, Zh. Eksp. Teor. Fiz., 54, 1259 (1968).
132. G. F. Drukarev and B. S. Monozon, Zh. Eksp. Teor. Fiz., 61, 956 (1971).
133. I. V. Komarov, Application of Asymptotic Methods in the Theory of Slow Collisions of Atomic Particles, Ph.D. Thesis, Leningrad University (1968); I. V. Komarov, Abstracts of VI ICPEAC, Boston, MIT (1969), pp. 1015-1017.
134. Ya. B. Zel'dovich, Fiz. Tverd. Tela, 1, 1637 (1959).
135. L. P. Presnyakov, Phys. Rev., A2, 1720 (1970).
136. I. V. Komarov, P. A. Pogorelyi, and A. S. Tibilov, Opt. Spektrosk., 27, 198 (1969).
137. T. M. Kereselidze and M. I. Chibisov, Zh. Eksp. Teor. Fiz., 68, 12 (1975).
138. F. I. Dalidchik and G. K. Ivanov, Zh. Prikl. Spektrosk., 13, 363 (1970).
139. F. I. Dalidchik and G. K. Ivanov, Opt. Spektrosk., 34, 863 (1973).
140. F. I. Dalidchik, Teor. Eksp. Khim., 10, 579 (1974).
141. G. K. Ivanov, Teor. Eksp. Khim., 10, 303 (1974).
142. R. K. Janev and Z. Maric, Phys. Lett., 46A, 313 (1974).
143. Yu. N. Demkov and V. N. Ostrovskii, Zh. Eksp. Teor. Fiz., 60, 2011 (1971).
144. V. I. Osherov, Zh. Eksp. Teor. Fiz., 49, 1157 (1965).
145. Yu. N. Demkov, Zh. Eksp. Teor. Fiz., 49, 885 (1965).
146. A. M. Perelomov, V. S. Popov, and M. V. Terent'ev, Zh. Eksp. Teor. Fiz., 50, 1393 (1966).
147. A. M. Perelomov, V. S. Popov, and M. V. Terent'ev, Zh. Eksp. Teor. Fiz., 51, 309 (1966).

148. G. B. Lopantseva and O. B. Firsov, Zh. Eksp. Teor. Fiz., 50, 975 (1966).
149. A. V. Chaplik, Zh. Eksp. Teor. Fiz., 45, 1518 (1963).
150. A. V. Chaplik, Zh. Eksp. Teor. Fiz., 47, 126 (1964).
151. E. C. Titchmarsh, Introduction to the Theory of Fourier Integrals, Oxford (1937).
152. G. A. Breit, Ann. Phys. (New York), 34, 377 (1965).
153. G. H. Herlig and Y. Nishida, Ann. Phys. (New York), 34, 400 (1965).
154. Y. Nishida, Ann. Phys. (New York), 34, 415 (1965).
155. S. K. Zhdanov and A. S. Chikhachev, Dokl. Akad. Nauk SSSR, 218, 1323 (1974).
156. A. Z. Devdariani, Zh. Tekh. Fiz., 399 (1973).
157. R. Z. Vitlina and A. V. Chaplik, Zh. Eksp. Teor. Fiz., 52, 959 (1967).
158. V. M. Borodin, Proc. of V ICPEAC, Leningrad (1967), pp. 326-327.
159. Yu. N. Demkov, Dokl. Akad. Nauk SSSR, 166, 1076 (1966).
160. Yu. N. Demkov, Vestn. Leningrad. Univ., Ser. Fiz., No. 4, 7 (1966).
161. V. I. Osherov, Fiz. Tverd. Tela, 10, 30 (1968).
162. Yu. N. Demkov and V. I. Osherov, Zh. Eksp. Teor. Fiz., 53, 1589 (1967).
163. F. I. Dalidchik, Zh. Eksp. Teor. Fiz., 66, 849 (1974).
164. V. N. Ostrovskii, Vestn. Leningrad. Univ., Ser. Fiz., No. 16, 31 (1972).
165. Yu. N. Demkov and I. V. Komarov, Zh. Eksp. Teor. Fiz., 50, 286 (1966).
166. F. B. Bronfin and A. M. Ermolaev, Vestn. Leningrad. Univ., Ser. Fiz., No. 22, 22 (1971).
167. A. Z. Devdariani and Yu. N. Demkov, Vestn. Leningrad. Univ., Ser. Fiz., No. 16, 23 (1971).
168. A. Z. Devdariani, Teor. Mat. Fiz., 11, 213 (1972).
169. A. Z. Devdariani and Yu. N. Demkov, Teor. Mat. Fiz., 21, 74 (1974).
170. V. I. Osherov, Dokl. Akad. Nauk SSSR, 168, 1291 (1966).
171. M. Ya. Ovchinnikova, Opt. Spektrosk., 17, 821 (1964).

INDEX